Motion and the English Verb

OXFORD STUDIES IN THE HISTORY OF ENGLISH

The Early English Impersonal Construction
Ruth Möhlig-Falke

Information Structure and Syntactic Change in the History of English
Edited by Anneli Meurman-Solin, María José López-Couso, and Bettelou Los

Spreading Patterns
Hendrik De Smet

Constructions and Environments
Peter Petré

Middle English Verbs of Emotion and Impersonal Constructions
Ayumi Miura

Language Between Description and Prescription
Lieselotte Anderwald

Motion and the English Verb
Judith Huber

Motion and the English Verb

A Diachronic Study

JUDITH HUBER

OXFORD
UNIVERSITY PRESS

OXFORD
UNIVERSITY PRESS

Oxford University Press is a department of the University of Oxford. It furthers the University's objective of excellence in research, scholarship, and education by publishing worldwide. Oxford is a registered trade mark of Oxford University Press in the UK and certain other countries.

Published in the United States of America by Oxford University Press
198 Madison Avenue, New York, NY 10016, United States of America.

Library of Congress Cataloging-in-Publication Data
Names: Huber, Judith, 1981– author.
Title: Motion and the English verb : a diachronic study / Judith Huber.
Description: New York, NY : Oxford University Press, [2017] |
Series: Oxford studies in the history of English
Identifiers: LCCN 2016047957 | ISBN 9780190657802 (hardback : alk. paper) |
ISBN 9780190657826 (epub) | ISBN 9780190657819 (updf)
Subjects: LCSH: English language—Verb | English language—Old English—
Verb. | English language—Middle English—Verb. | Motion—Terminology.
Classification: LCC PE1271 .H83 2017 | DDC 429/.56—dc23
LC record available at https://lccn.loc.gov/2016047957

ACKNOWLEDGEMENTS

This book is a reworked version of my doctoral dissertation finished at LMU Munich (Ludwig-Maximilians-Universität München) in the summer semester of 2013. My heartfelt gratitude goes to Ursula Lenker, my first supervisor, who sparked off my interest in the topic of motion verbs in language contact and who has continued to be an inexhaustible source of inspiration, criticism, and enthusiasm. I also thank my co-supervisor Roland Schmidt-Riese and my third reviewer Hans Sauer for many valuable remarks and their useful advice.

The book has benefitted greatly from the help and comments of numerous people. I would specifically like to thank Paul Schaffner from the *Middle English Compendium* for facilitating my corpus work with the *CME* in sending me the text files. Thanks also go to Andreas Dufter, Christine Elsweiler, Teresa Fanego, Klaus Grübl, Franziska Günther, Caroline Harthan, Christoph Hauf, Christine Haunz, Kerstin Kazzazi, Robert Mailhammer, Elke Ronneberger-Sibold, Michael Skiba, and Stefan Thim for their feedback and constructive criticism. Thanks are also due to Terttu Nevalainen for accepting the book in the series, to two great anonymous reviewers for many insightful remarks, and to Hallie Stebbins and Hannah Doyle for their help and patience in guiding the book through the publication process.

Finally, I thank my parents for their love and support, my friends, and especially my husband Bastian, whose love and cooking kept up the good spirits in the final weeks both of finishing the dissertation and the book. The book is dedicated to our daughter Elsa, who was born in between, and who keeps me *climbing up / sliding down / dancing across / stumbling over* all sorts of GROUND elements.

CONTENTS

ABBREVIATIONS

AN	Anglo-Norman
AND	*Anglo-Norman Dictionary*
BNC	*British National Corpus*
BT	Bosworth & Toller, *An Anglo-Saxon Dictionary*
BTS	Bosworth & Toller, *Supplement*
CH	Clark Hall, *A Concise Anglo-Saxon Dictionary*
CM	caused motion
CME	*Corpus of Middle English Prose and Verse*
DET	determiner
DMF	*Dictionnaire du Moyen Français (1330–1500)*
DMLBS	*Dictionary of Medieval Latin from British Sources*
DOE	*Dictionary of Old English*
DOEC	*Dictionary of Old English Corpus*
EModE	Early Modern English
F	French
FEW	*Französisches etymologisches Wörterbuch*
HC	*Helsinki Corpus*
HTE	*Historical Thesaurus of the OED*
IM	intransitive motion
IMC	intransitive motion construction
It	Italian
L	Latin
L1	first language
L2	second language
lit	literal
ME	Middle English
MED	*Middle English Dictionary*
MF	Middle French
MLG	Middle Low German
NCA	*Nouveau Corpus d'Amsterdam*

N	noun
non-lit	non-literal
NP	noun phrase
Od	direct object
OE	Old English
OED	Oxford English Dictionary
OF	Old French
ON	Old Norse
PDE	Present-Day English
Port.	Portuguese
PP	prepositional phrase
PPCEME	*Penn–Helsinki Parsed Corpus of Early Modern English*
PPCME2	*Penn–Helsinki Parsed Corpus of Middle English, second edition*
REFL	reflexive
Sp	Spanish
S-language	satellite framing language
V-language	verb framing language
TL	Tobler–Lommatzsch *Altfranzösisches Wörterbuch*
TLF	*Trésor de la langue Française*
TLL	*Thesaurus Linguae Latinae*
TOE	*Thesaurus of Old English*

Motion and the English Verb

CHAPTER 1 | Introduction

1.1 *Different ways of talking about motion*

Within the general interest of linguistics in SPACE and in how different languages represent SPACE differently (e.g. Levinson 2003; and the volumes by Bloom et al. 1996; Hickmann & Robert 2006; Levinson & Wilkins 2006; Evans & Chilton 2010; Filipović & Jaszczolt 2012; Vulchanova & van der Zee 2012), the study of how languages represent MOTION has been a particularly prolific area of research, starting with Talmy's (1975, 1985) work on lexicalization patterns. The basic idea in Talmy's early articles is that in a real world MOTION EVENT, there are several semantic components, and that languages differ in the kinds of components they habitually express in certain elements of the sentence. Here is an illustration (for details, see section 2.1).

Let the real world MOTION EVENT be William moving from Dives-sur-Mer to Pevensey across the Channel in a ship. The semantic components in this event are the FIGURE (William), the GROUND (Dives-sur-Mer, the Channel, Pevensey), the MOTION (William moves), the PATH (the fact that William moves *away from* Dives, *across* the Channel, and *to* Pevensey), and the MANNER (by ship). In one type of language, such as in French, you would typically talk about this event by saying (1a); in another type, such as English, by saying (1b).

(1) a. Guillaume traversait la Manche. (PATH in verb)
    b. William sailed across the Channel. (MANNER in verb, PATH in satellite)

In (1a), what is expressed in the verb is PATH: *traversait* 'traversed'. In (1b), by contrast, what is expressed in the verb is MANNER (*sail*), while PATH is expressed outside the verb in the preposition *across*. In the motion verb typology, the first type of language is consequently called PATH-conflating, the second MANNER-conflating. In later versions of the typology (Talmy 2000), this difference is recast in terms of verb-framing (for (1a)) and satellite-framing (1b) (see section 2.1.3). According to Talmy, each language follows one of these patterns predominantly—speakers of French, for instance, often use PATH verbs like *entrer* 'enter', *sortir* 'exit', *descendre* 'descend', and so forth, while speakers of English make frequent use of MANNER verbs such as *drive, creep, run* in combination with PATH satellites such as *through, out,* and *down.*

Research in the framework of the typology of motion event encoding has, so far, been most productive in the fields of linguistic relativity, first and second language acquisition, and contrastive linguistics. Surprisingly, as pointed out by Ameka (1996: 137), it has not been studied extensively with regard to contact semantics. Overall, research in the motion event typology also has as yet predominantly had a synchronic orientation; diachronic studies in this field are still scarce, with the articles on Romance languages by Schøsler (2008), Kopecka (2009, 2013), and Stolova (2008), and, on English, Fanego (2012) being notable exceptions.

1.2 *Aims of the present study*

The aim of this study is twofold. First of all, it offers an extensive diachronic analysis of how motion events are encoded in Old and Middle English, focusing on the following questions (chapters 3–6):

(a) Which verbs *can* be used as motion verbs in medieval English? The analysis of the motion verb inventories of Old and Middle English compiled on the basis of thesauri and dictionary definitions will show that MANNER verbs are extremely numerous already in earlier stages of the language and that a range of non-MOTION verbs such as OE *feohtan* 'to fight' can receive a contextual motion reading when combined with directional adverbials (i.e. satellites) such as in OE *sume [. . .] ūt afuhten* (lit. 'some would fight out'), meaning 'some would fight their way out'. Both characteristics meet the expectations for a satellite-framing, MANNER-conflating language.

(b) Which verbs and which structures *are typically* used to express motion in medieval English? The analysis of motion descriptions from selected texts will show that the dominant strategy is to encode PATH in satellites, i.e. outside the verb, but that the type and token frequency of MANNER verbs crucially depends on text type, an issue which has as yet not received much attention in motion typology studies.

On this basis, the second aim of the study is to investigate how the PATH verbs listed in (2) are accommodated as loan verbs in Middle English in the situation of 'intertypological' language contact with French in the late Middle Ages (chapters 7–9):

(2) a. ME *entren* (first attested *c.*1275) < OF *entrer* 'move in'
b. ME *ishen* (now obsolete) / *issuen* (first attested *c.*1350) < OF *issir* 'move out'
c. ME *descenden* (first attested *c.*1330) < OF *descendre*/L *descendere* 'move down'
d. ME *avalen* (now obsolete, first attested *c.*1330) < OF *avaler* 'move down'
e. ME *ascenden* (first attested *c.*1384) < L *ascendere*/MF *ascendre* 'move up'
f. ME *mounten* (first attested *c.*1300) < OF *monter* 'move up'
g. ME *amounten* (first attested *c.*1275) < OF *amonter* 'move up'[1]

The presence of these Romance PATH verbs as 'typologically different' motion verbs in English has often been commented on (e.g. Talmy 2000: 52–3; Brinton & Traugott 2005: 155; Beavers et al. 2010: 350; Traugott 2012: 174; Haselow 2012: 649–51), but has not been studied in any depth as yet.

The Middle English period (*c.*1150–*c.*1500) in which they are borrowed is marked by extensive bi- or even trilingualism among the literate population, involving English, (Anglo-)French, and Latin (cf., e.g., Crystal 2005: 121–39; Ingham 2012; Lutz 2013; and the discussion in section 6.1). This situation is generally seen as responsible for the fact that, etymologically, the lexicon of Present-Day English, different from that of other Germanic languages, is heavily mixed, with particularly large proportions of words borrowed from French—such as *very*, *easy*, or *carry*—having found their way into the English core vocabulary: Scheler (1977: 72), for instance, reports 38 per cent French loans in the *General Service List*, a list of roughly 4,000 highly frequent Modern English words; Durkin (2014: 38) finds almost half of the 1,000 most frequent words in the *British National Corpus* to have been borrowed from French and/or Latin. Words borrowed after 1500, by contrast, although numerous as well, tend not to make it into the core vocabulary of English (Durkin 2014: 426).

While the general picture of loanwords in English with regard to their percentages in sections of the vocabulary has been well researched, the exact integration of loans is not in all cases fully understood. I attempt to contribute to this question by analysing the early use of these PATH verbs in English as attested in autonomous Middle English texts and Middle English translations from French and Latin. We will see that initially these verbs appear to have been borrowed predominantly in specific contexts still compatible with the MANNER-conflating character of Middle English, and not in what is their core use in the donor languages, namely the description of GENERAL LITERAL MOTION EVENTS.

1.3 *Structure of the present study*

Chapter 2 presents the theoretical framework, starting with a summary of the original Talmy typology in its two versions and an overview of work in

[1] Dates of first attestation as given in the *Middle English Dictionary* (*MED*).

this typology relevant to the present questions (2.1). In section 2.2, the motion encoding typology is framed in a construction grammar approach, which is highly compatible with it, but appears more suitable for handling the diachronic data. The classification of motion verbs applied in the present study is outlined in section 2.3.

The four chapters (chapter 3–6) following on these fundamentals are devoted to motion expression in medieval English. Two major questions are pursued: (a) which verbs speakers of Old and Middle English have at their disposal when they talk about motion (section 5.1 on Old English and 6.2 on Middle English; see also Appendices A and B on the companion website to this book [www.oup.com/us/motionandtheenglishverb/]), and (b) which verbs and which structures they most frequently use for this purpose (section 5.2 on Old English and 6.3 on Middle English). Thus, we will investigate inventory and usage, i.e. what is possible and what is frequent in medieval English motion expression.

Medieval speakers are, of course, only accessible in their writings, and moreover, most of these writings belong to a rather restricted set of genres. Furthermore, the texts are often influenced by Latin and, in Middle English, also by French. The problems this entails for the present study are discussed in section 3.1. The other central problematic issue of how to establish the meaning of a given verb, particularly in historical stages of a language, is addressed in section 3.2.

Chapter 4 prepares the ground for the analyses of Old and Middle English motion expression that follow in chapters 5–6, laying out the specific aims, material, and method of the subsequent analyses. In addition to the investigations of the medieval English motion verb inventory (sections 5.1 and 6.2) and usage (sections 5.2 and 6.3), I conduct a case study of how Old English translators render Latin PATH verbs (section 5.3), whose results support the diagnosis that the French pure PATH verbs in (2) do not have any real native equivalents in medieval English. A separate section (5.4) discusses the changing realization of PATH satellites in the history of English. It shows that already in Old English, prefixes have become rare as a means of conveying PATH information in descriptions of LITERAL MOTION EVENTS; typically, PATH information is given in prepositional phrases and adverbs in medieval English (see also sections 5.2.4 and 6.3.4).

Section 6.1 gives a brief overview on the intricate language contact situation in the Middle English period, in which the English lexicon starts to change considerably due to the heavy influence of (Anglo-)French as well as Latin. The situation is even more complicated by the fact that Anglo-French, also known as Anglo-Norman, is itself strongly influenced by Middle English. I will attempt to disentangle some of these intricacies in sections 6.2.3–5, which treat the ways in which Middle English acquires new (MANNER OF) MOTION verbs. We will see that while some of the Middle English motion verbs (including *travel*) may look like straightforward loans from French, they actually have only acquired their motion meaning in England: either in Middle English itself, or in Anglo-Norman, most likely due to substratal influence of Middle English. The study thus provides a contribution to the field of contact semantics, which has as yet been neglected by motion verb studies (cf. Ameka 1996: 137).

After the investigation of the English system into which they are borrowed (chapters 3–6), the French PATH verbs and their integration in Middle English are at the centre of chapters 7–9. Chapter 7 focuses on motion expression in the donor languages Latin and medieval French, since for studying the inter-typological contact in the Middle English period, we need to know both sides involved. Section 7.1 gives a synthesis of previous studies on the changes in the typical ways of talking about motion from Latin to the Romance languages. It is complemented in section 7.2 by a case study on motion expression in the continental French *Aucassin et Nicolette*.

Chapter 8 introduces the hypotheses, methodology, and material (the *Corpus of Middle English Prose and Verse* (*CME*)) on which the analysis of the borrowed PATH verbs in chapter 9 is based. It will be shown in chapter 9 that in autonomous, genuinely Middle English texts, the PATH verbs exhibit a somewhat restricted use: They predominantly occur in specific, often metaphorical and abstract meanings, most of which are related to discourse domains that had been covered by French and Latin in the triglossic Middle English period (such as administration, law, business, religion, or science). Autonomous Middle English texts consequently use the PATH verbs less often as motion verbs in the prototypical sense (namely to denote a LITERAL MOTION EVENT, such as a human FIGURE moving with respect to a concrete GROUND). Middle English translations from French or Latin, by contrast, have a tendency to feature the verbs as they are used in the donor languages, i.e. rather freely and for all kinds of LITERAL MOTION EVENTS. Starting from the often abstract, technical, and metaphorical uses in autonomous texts, the scope of contexts in which the verb can be used has later widened for some of the verbs—particularly *enter*—, while others have fallen out of use (ME *avalen, ishen*) or have never really widened their usage contexts to LITERAL MOTION EVENTS (*amount*).

The present study thus offers not only a detailed analysis of the diachrony of motion verbs and of motion expression in medieval English, but also an in-depth investigation of the various cognitive and contact-linguistic aspects of the integration of individual French verbs, which—in the motion verb typology approach—can be considered 'typologically different'.

CHAPTER 2 | Theoretical framework

THIS CHAPTER PROVIDES AN overview of the typology of motion encoding and relevant research carried out in this area (2.1) and argues for a constructionist approach to the typology (2.2). The classification of motion verbs applied in the present study is presented in section 2.3.

2.1 *Typology of motion expression*

After introducing the semantic components distinguished in a motion event (2.1.1), this section presents the two perspectives that have been taken in the typology of motion encoding: a focus on typical verb semantics (2.1.2) and on typical structures (2.1.3). Section 2.1.4 provides an overview of characteristic features that have been identified in the literature as related to the type of motion encoding a language prefers.

2.1.1 The motion event and its components

The basis of Talmy's approach is to isolate the different meaning components of a motion event and to see where in the sentence these components are expressed in different languages. He splits up a motion event into four internal components and two external co-events (1975: 181–2; 1985: 60–1; 2000b: 25–6), namely:

Internal components of a motion event:	FIGURE, GROUND, PATH, MOTION
External co-events of a motion event:	MANNER, CAUSE

(3) a. The pencil rolled off the table.

b. The pencil blew off the table.
(Talmy 1985: 61)

The FIGURE is defined as a moving, or at least conceptually moveable object, such as the pencil in Talmy's examples (3a–b). The GROUND is an object with respect to which the FIGURE moves or is located, such as the table in (3a–b).[2] MOTION refers to the fact that the FIGURE moves; PATH to the respect in which the FIGURE moves to the GROUND (Talmy 1975: 181), i.e. it is expressed by *off* in (3a–b). PATH can be split up into INITIAL, MEDIAL, and FINAL portions (Talmy 2000a: 265–6). In other studies, these portions of PATH are also referred to as 'SOURCE', 'PATH', and 'GOAL' (e.g. Langacker 2000: 55). The motion event can happen in a certain MANNER, defined rather loosely as 'a subsidiary action or state that a Patient manifests concurrently with its main action or state' (Talmy 2000b: 152). In (3a), MANNER is, for example, expressed in *rolled*. The motion event can also happen due to some CAUSE, which is, for instance, expressed in *blew* in (3b).

Within the component of MOTION, a further distinction is drawn between 'translational motion' and 'self-contained motion': In the former—often also called 'translocational' (e.g. by Rohde 2001: 33) or 'translocative' (e.g. by Zlatev 2007: 333)—'an object's basic location shifts from one point to another in space' (Talmy 2000b: 35). In self-contained motion, by contrast, 'an object keeps its same basic, or "average", location' (ibid.). Applied to (3a), the change of location from the table to the floor, or wherever the pencil ends up, would represent translational motion. The rolling that the pencil performs, by contrast, would be seen by Talmy (2000b: 36) as the MANNER co-event, which in the case of (2a) is 'an event of [. . .] rotational self-contained motion'.

2.1.2 What's in the verb? PATH-, MANNER-, and FIGURE-conflation

The expression of the semantic components introduced in section 2.1.1 has been approached from two different perspectives in Talmy's work on conflation patterns. The earlier perspective (1975; 1985; 2000b: chapter 2) starts from the elements in the sentence and asks which semantic components of the motion event speakers of a language typically express in these elements. In this first version of the typology, Talmy focuses on what is typically conflated in the verb root, i.e. the verb without any affixes, and finds that many languages predominantly follow one of two conflation patterns: MOTION + PATH or MOTION + MANNER.[3] A third, but rarer pattern that Talmy distinguishes is that of conflating MOTION + FIGURE in the verb, constituting the most characteristic pattern in Navajo and Atsugewi. An example from Atsugewi is *-staq̓-* meaning 'for runny icky material (e.g. mud, manure, rotten tomatoes, guts, chewed gum) to move/be-located' (2000b: 58).

[2] In other studies on cognitive spatial semantics, 'trajector' and 'landmark' are used for FIGURE and GROUND respectively (e.g. Langacker (1990: 9–10), see also Zlatev's overview of basic spatial semantic concepts in the literature (2007: 326–34)).

[3] Similar observations with respect to differences between French and German had been made earlier, e.g. by Bally ([1925] 1952, [1932] 1965), Staub (1949), Tesnière ([1959] 1976: 307–10), Malblanc (1968); see also the references in Blumenthal (1987: 74–84).

Conflation of MOTION + PATH is the predominant pattern in the Romance and the Semitic languages, Japanese, Korean, Turkish, Tamil, Polynesian, etc. (Talmy 2000b: 49), as exemplified by Spanish in (4a–c).[4]

(4) a.

La	botella	**entró**	a	la	cueva	(flotando).
The	bottle	moved-in	to	the	cave	(floating)

'The bottle floated into the cave.'

b.

La	botella	**salió**	de	la	cueva	(flotando).
The	bottle	moved-out	from	the	cave	(floating)

'The bottle floated out of the cave.'

c.

La	botella	**pasó**	por	la	piedra	(flotando).
The	bottle	moved-by	past	the	rock	(floating)

'The bottle floated past the rock.'

(all from Talmy 2000b: 49–50)

Conflation of MOTION + MANNER, by contrast, is characteristic in the other Indo-European languages (except for the Romance ones), the Finno-Ugric languages, Ojibwa (an Algonquian language spoken in North America), and Warlpiri (an Australian language) (Talmy 2000b: 60). Examples of MANNER-conflation in English are (5a–b) (non-agentive and agentive respectively).

(5) a. The rock **slid/rolled/bounced** down the hill. [non-agentive]

b. I **ran/limped/jumped/stumbled/rushed** down the stairs. [agentive]

(all from Talmy 2000b: 30)

More generally, the conflation pattern in English can be said to be one of 'co-event conflation', since the verb may also express co-events relating to the motion event in ways other than MANNER (Talmy 2000b: 42–7). In (6a), for instance, *blew* expresses the CAUSE of the motion. The relation may also be one of PRECURSION, as in (6b), where the splintering happens before the glass moves onto the carpet, one of CONCOMITANCE, as in (6c), where the 'whistling event' is not related to the 'motion event' except for happening at the same time, etc.

(6) a. CAUSE:
e.g. *The napkin* ***blew*** *off the table.*

b. PRECURSION: 'the Co-event precedes the Main motion event but does not cause or assist its occurrence'
e.g. *Glass* ***splintered*** *onto the carpet.* (2000b: 42)

[4] Other Spanish PATH conflating verbs apart from the ones shown in (4a–c) are *subir* (move+up), *bajar* (move+down), *volver* (move+back), *cruzar* (move+across), *avanzar* (move+ahead/forward), etc. (Talmy 2000b: 50–51).

c. CONCOMITANCE: the co-occurring Co-event 'does not in itself pertain to the concurrent Motion'
e.g. *I **whistled** past the graveyard* (2000b: 46) (not acceptable for all speakers)

d. CONCURRENT RESULT: 'the Co-event results from [. . .] the main Motion event and would not otherwise occur'
e.g. *The rocket **splashed** into the water.* (2000b: 46–47)

e. SUBSEQUENCE (including CONSEQUENCE/PURPOSE): 'Co-event takes place directly after the main Motion event, and is enabled by, is caused by, or is the purpose of [it]'
e.g. *I'll **look** in at the stew cooking on the stove.* (2000b: 47)

The co-event conflations exemplified in (6a–e), however, are not as frequent according to Talmy; the characteristic conflation pattern for English is MANNER and MOTION in the verb (e.g. 2000b: 152).

It is important to stress that if a language is said to follow one of these principal conflation patterns (PATH + MOTION or MANNER + MOTION), this does not mean that it does so exclusively, but only that it does so typically or characteristically:

> In most cases, a language uses only one of these types for the verb in its most characteristic expression of Motion. Here, 'characteristic' means that (1) it is colloquial in style, rather than literary, stilted, and so on; (2) it is frequent in occurrence in speech, rather than only occasional; (3) it is pervasive, rather than limited—that is, a wide range of semantic notions are expressed in this type. (2000b: 27)

Thus, for instance, English, of course, has *to enter, descend,* and the other PATH-conflating verbs we will be concerned with in chapter 9. Yet (a) many of these are not 'colloquial in style' (e.g. *descend, traverse*), and (b) they do not constitute the most frequent conflation pattern in English (Talmy 2000b: 52–3).

These observations have been tested quantitatively in various studies, often on the basis of motion descriptions elicited with the help of the 'frog story', a textless picture book about a boy looking for his pet frog (Mayer 1969), whose story the test persons are asked to narrate.[5] Other studies are based on descriptions of video clips (e.g. Naigles et al. 1998; Gennari et al. 2002; Cardini 2010), on the analysis of novels (e.g. Slobin 2004b; Verkerk 2013), or on written corpus material (e.g. Cappelle 2012). Slobin (2004a: 231, 2004b: 204), for instance, reports that in English novels and frog stories, around 41 to 45 per cent of all motion events are described with a MANNER

[5] Cf. the contributions in Berman & Slobin (1994) on English, German, Spanish, Hebrew, and Turkish, and the contributions in Strömqvist & Verhoeven (2004) on Warlpiri, Tzeltal, West-Greenlandic, Basque, Swedish, Icelandic, Arrernte, Thai, and American Sign Language. Cardini (2008) uses the frog story for Italian; Berthele (2006) for German, Allemanic dialects, French, and Rhaetoromance.

verb. In Russian, which belongs to the same type as English, the proportion of MANNER verb descriptions even ranges between 56 and 79 per cent. In the predominantly PATH-conflating languages Turkish and Spanish, by contrast, MANNER verbs are only used in around 20 per cent of the motion descriptions, while PATH verbs are more frequent. Similar differences in frequency have been found, for instance, by Naigles et al. (1998: 540–1) and Gennari et al. (2002: 65) for English and Spanish, by Papafragou et al. (2002: 208–9, 2006: B80) for English and Greek, and by Cardini (2010: 1453) for English and Italian.

Note that while the individual studies confirm the general typological difference, the precise proportions of MANNER and PATH verbs found in them differ considerably from each other. This is due, first, to the slightly varying semantic classification of verbs in the individual studies (see 2.3), but probably even more so to the different types of texts and genres analysed: In Cappelle's analysis of different subcorpora of the English *BNC*, for instance, MANNER verbs account for 34 per cent in fictional texts but only for 20 per cent in biographical texts (2012: 187), with the bulk of verbs being neutral motion verbs such as *come* and *go* (2012: 186).[6] In the descriptions of video clips designed to elicit MANNER and PATH information in Cardini (2010: 1454), by contrast, English speakers use MANNER verbs in 93 and PATH verbs in 2 per cent of the clauses, while speakers of (usually predominantly PATH-conflating) Italian use MANNER verbs in 45 and PATH verbs in 46 per cent of the clauses.

2.1.3 Where is PATH encoded? Verb- and satellite-framing

The first version of the typology, described in 2.1.2, started from the verb and asked which semantic elements of a MOTION event it usually encodes in a language. This is reversed in the second version (Talmy 1991; 2000b: 221–30), which starts from semantics: Talmy isolates PATH as the 'schematic core' of a motion event and asks by which element in the sentence it is encoded, whether in the verb or outside the verb in a so-called satellite (1975: 206–22; 1985: 102–21; 2000b: 101–28). The respective languages are accordingly characterized as 'verb-framing' or 'satellite-framing' (henceforth also 'V-language' and 'S-language'). Thus, while in the first perspective focusing on the verb, English is MANNER-conflating and Atsugewi FIGURE-conflating, in this new perspective focusing on PATH, both English and Atsugewi are satellite-framing (the '*second* Talmyan typology' (Zlatev 2007: 335)).

[6] The lower proportion of MANNER verbs in Cappelle (2012) compared to Slobin (2004) may be influenced to some extent by the method used for extracting the MOTION descriptions: Cappelle searches for specific words (2012: 182–5), rather than reading the texts—the latter would hardly have been feasible with the large corpora used in his study. In the scope of Cappelle (2012), however, this is not really problematic, since the study is not primarily interested in the frequency of MANNER verbs per se, but in their different frequencies in autonomous Present-Day English texts compared to Present-Day English texts translated from French.

Different from the open-class category of verb, the satellite is a closed-class element and is defined as follows:

> It is the grammatical category of any constituent other than a noun-phrase or prepositional-phrase complement that is in a sister relation to the verb root. It relates to the verb root as a dependent to a head. The satellite, which can either be a bound affix or a free word, is thus intended to encompass all of the following grammatical forms, which traditionally have been largely treated independently from each other: English verb particles, German separable and inseparable prefixes, Latin or Russian verb prefixes, Chinese verb complements, Lahu nonhead 'versatile verbs' [. . .] Caddo incorporated nouns, and Atsugewi polysynthetic affixes around the verb root (Talmy 2000b: 102).

The Indo-European languages except for the Romance ones are thus said to typically encode the PATH of a motion event in satellites, such as in English *to run **in***, Latin ***in**currere*, or German ***hinein**laufen*. Note that Talmy's definition of satellite explicitly excludes prepositional phrases. Talmy argues that in most languages, prepositions and satellites have distinct realizations (e.g. satellites as verbal prefixes, prepositions accompanying the noun). Yet in English, both satellites and prepositions are usually placed after the verb and are often homonymous. Talmy (2000b: 106–9) gives several arguments for keeping them distinct nevertheless:

(a) There are some satellites that do not have a homonymic preposition (e.g. *away, back, together*) and vice versa (*from, of*);
(b) sometimes, a form has a different sense depending on whether it is a preposition or a satellite (e.g. satellite *over* 'rotation around a horizontal axis' as in *I fell over*; preposition *over* 'above/covering' as in *I flew over the desert*);
(c) satellites are usually stressed; prepositions are not.

Thus, while *over* and *out* would be satellites in (7a–b, d), *over* and *of* would be prepositions in (7c–d).

(7) a. It fell ***over***.
b. He ran ***out***.
c. It fell *over* the cliff.
d. He ran ***out*** *of* the room.

This distinction, however, is problematic in various ways and has been modified in many recent applications of the typology.[7] One problem is that there are good reasons for treating combinations such as *out of* (as in (7d)) as one unit, a

[7] Particularly argument (a) is problematic from a diachronic perspective: the Modern English prepositions *of* and *from*, for instance, used to have homonymous adverbs in earlier stages of the language, such as in OE *forlet se here þa burg & for **fram*** 'the army left the town and went away' (ChronA 917.46,

complex preposition, rather than as a combination of satellite and preposition (cf. Filipović 2007a: 35; Quirk et al. 1985: §§9.12, 15, 18; and also Cappelle 2001).[8] Filipović (2007a: 35) further points out that, following Talmy's definition, *across* in *He ran across* would have to be seen as a satellite, but that usually it would be clear from the context, *what* the FIGURE ran across. Therefore, she argues, prepositions and satellites are not so different after all, since both imply a GROUND—with prepositions, GROUND is explicitly expressed, while with satellites, GROUND is implicitly understood. Another important criticism is that prepositions do the same job as satellites, in that they both express PATH (see also Croft et al. 2010: 205). For these reasons, it is now a common practice to treat prepositional phrases as PATH satellites as well (e.g. Beavers et al. 2010: 337–9; Croft et al. 2010: 206; de Knop & Dirven 2008: 297; Verkerk 2013: 175).

This will also be done in the present study, where the nature of the Old English PATH expressions particularly invites this wider definition of satellite: In Old English, the PATH expression can still be placed immediately before the verb, such as *to* in (8), with the GROUND-complement (*him*) preceding it.

(8) Oft martinus geseah englas **him *to*** cuman (ÆLS (Martin) B.1.3.30, DOEC)
'Often Martin saw angels come to him.'

In cases like (8), the exact status of the PATH expression is notoriously vague (see section 5.4): Is *to* to be seen as a free particle in preverbal position or as having developed into a prefix of the verb (i.e. in both cases as satellite in Talmy's definition)? Or should it rather be analysed as an adposition? In the former case, *him* would have to be analysed as the direct object of the verb; in the latter as the complement of the adposition. The lack of any clear category boundaries here is another argument for treating directional prepositions as PATH expressions on a par with directional adverbs and prefixes.

The study of motion expressions in an increasing number of different languages has also shown that not all languages have a predominant pattern that fits into the binary distinction verb- vs. satellite-framed. A case in point are serial verb languages such as Mandarin (as in (9a)) or Thai, in which two or more verbs co-occur in the same clause, and none of them can be analysed as subordinate to the other (e.g. 'walk'-'enter'-'go(deictic)') (Zlatev & Yangklang 2004 on Thai). Croft et al. 2010 therefore suggest a more fine-grained distinction between different motion encoding patterns, with 'symmetric' strategies, in which PATH and MANNER are packaged in units of equal status, on the one

DOE s.v. *fram*). ModE *of* and *off* are originally merely unstressed and stressed variants of one and the same word. In the case of adverbs without homonymic prepositions, the fact that these have often developed from prepositional phrases that have become lexicalized may be an important reason for why they are not used as prepositions today (*away* < *on weg* 'on way'; *together* < *to gædre; back* is an aphetic form of earlier *aback* < *on bæc* (cf. *OED* s.vv.)).

[8] Beavers et al. (2010: 338), for instance, note that in an *it*-cleft, the whole PP *out of the house* has to be clefted (*It was out of the house that I went, not into the house*). This is not possible for *out* alone (**It was out that I went of the house...*). For the diachronic development of *out of*, see Elenbaas 2014.

hand, and 'asymmetrical' strategies on the other. Among the symmetric strategies, they distinguish serial verbs (9a), verbal compounding (9b), and coordination of clauses (9c):

(9) a. **Serial verbs:**

tāmen	**pǎo**	**chū**	**lái**	le. (Mandarin Chinese)
3PL	run	exit	come	PF

'They came running out.' (Croft et al. 2010: 207)

b. **Verbal compounding:**

Watashi	wa	ie	ni	**kake-**	**-konda.** (Japanese)
I	TOP	house	to	run-	-go.into:PST

'I ran into the house.' (Croft et al. 2010: 218)

c. **Coordination:**

Sanhakase	wa	hoshi	ni	**shitagat**	**-te**	besturehemu	o	**deta.** (Japanese)
three.doctor	TOP	star	to	follow	-and	Bethlehem	ACC	go.out:PST

'The wise men followed the star out of Bethlehem.' (Croft et al. 2010: 219)

Among the 'asymmetrical' strategies, they distinguish the verb-framing and the satellite-framing pattern, as well as so-called double-framing, 'in which the path or framing expression is expressed twice, once as a detached satellite and once as part of the verb' (2010: 208), as in (10). In the French example, PATH is expressed both in the verb and in a detached satellite; in the Russian example by a detached satellite (*iz*) and by a prefixed one (*vy-*).

(10) **Double-framing:**

a.

monter	***en haut*** /	**descendre**	***en bas*** (French)
go.up	above /	descend	below (Croft et al. 2010: 208)

b.

Ja	*vy-*	bežal	*iz*	doma. (Russian)
I	out-	ran	from	house.GEN

'I ran out of the house.' (Croft et al. 2010: 208)

According to Croft et al. (2010: 220), these different patterns may be ranked on a scale of morphosyntactic integration of MANNER and PATH as in (11), where the highest degree of integration is printed in boldface, and the lowest degree of integration in italics:

(11) Formal scale of morphosyntactic integration:
double framing, satellite framing < verb framing, compounding < *coordination*
(Croft et al. 2010: 220)

Coordination has a low degree of morphosyntactic integration because it makes use of two separate clauses to express MANNER and PATH (à la *She danced and crossed the street*, see the Japanese example in (9c)). Double framing (à la *She across-danced across the street*, see the Russian example in (10b)) and satellite-framing (à la *She danced across the street*) display the highest degree of integration 'in that the satellite is least like a separate clause' (2010: 222). Compounding (à la *She dance-crossed the street*, see the Japanese example (9b)) and verb-framing (à la *She crossed the street dancing*) are in between.

Croft et al. tentatively suggest that (11) can be seen as an implicational scale: 'more typical or natural [MANNER + PATH] combinations in complex events will be encoded in more highly integrated morpho-syntactic constructions' (2010: 225). They propose a parallel scale of situation types for complex motion events as in (12), where 'run out of' and 'run into' on the left would represent the pole of the most typical MANNER + PATH combinations, and where 'dance across' on the right would be the least typical.

(12) Semantic scale of conceptual situation types:
'run out of' < 'run into' < 'crawl to' < 'float into' < 'run across' < 'follow X out of' < 'dance across' (Croft et al. 2010: 223)

Thus, they suggest that, if a language uses a morphosyntactically highly integrated construction (e.g. satellite-framing, see the formal scale in (11)) for 'dance across', the implication is that it will do so as well for the more typical situation types lower on the conceptual scale (12). This would be the case for English, which uses the satellite framing pattern throughout. Japanese, by contrast, would use the intermediately integrated pattern of compounding (see (9b)) for situations up to 'run across' on the semantic scale, and would use coordination, i.e. a less integrated pattern, for the less typical situations higher on the semantic scale (see (9c)) (Croft et al. 2010: 221).

2.1.4 Concomitant characteristics of verb- and satellite-framing

Various studies have shown that whether speakers of a language predominantly use a verb- or a satellite-framing pattern to talk about motion often goes hand in hand with other characteristics of the language. These will be discussed in 2.1.4.1–4.

2.1.4.1 Boundary-crossing constraint on manner verbs in V-languages

It was pointed out quite early that PATH-conflating languages apparently have a constraint on the use of MANNER verbs, which does not allow these verbs together with telic PATH expressions (Aske 1989 for Spanish). Slobin & Hoiting (1994) call this the 'boundary-crossing constraint'. This means that a reading of (13a) as describing TRANSLOCATIONAL MOTION—from the outside of the house to the inside of the house—is not generally possible in French, due to the boundary-crossing constraint on MANNER verbs, related to the fact that French prepositions (*dans* in (13a)) alone cannot predicate TRANSLOCATIONAL MOTION (see also Tesnière ([1959] 1976: 309). The utterance would be interpreted as non-translocational 'the man ran around in the house' (13b). By contrast, (13c) is possible because it does not predicate that the FIGURE is crossing a boundary, or, in other words, because it does not predicate a GOAL in which the FIGURE ends up, which would be a change-of-state (Slobin 1997: 441).

(13) a. L'homme a couru **dans** la maison.
*'The man ran into the house.' (translocational)

b. L'homme a couru **dans** la maison.
'The man ran (around) in the house.' (non-translocational)

c. L'homme a couru **jusqu' à** la maison. (Slobin & Hoiting 1994: 496)
'The man ran up to the house.'

This constraint in V-languages holds for the majority of MANNER verbs because they are usually 'activity verbs' (Slobin 2004a: 226). The only exception Slobin (ibid.) finds are 'verbs that encode particular force dynamics—high energy motor patterns that are more like punctual acts than activities, such as equivalents of "throw oneself" and "plunge"' (see also Slobin 2006: 67; Beavers et al. 2010: 364–5; and section 7.2.2).

2.1.4.2 Multiple PATH segments in S-languages

Another difference between S- and V-languages is that speakers of S-languages tend to accumulate different PATH satellites into one clause with a single verb, whereas speakers of V-languages tend towards separate clauses for each of the different PATH segments (e.g. Slobin 1996b: 203; 1997: 442–9; 2004a: 238–9). Slobin illustrates this with a passage from *The Hobbit* in (14a) and its French (14b) and Brazilian Portuguese (14c) translation:

(14) a. He still wandered on, **out of** [S1] the little high valley, **over** [S2] its edge, and **down** [S3] the slopes beyond.

b. Il continua d'avancer au hasard, **sortit** [V1] du haut vallon, en **franchit** [V2] le bord et **descendit** [V3] la pente au-delà.
'He continued to advance aimlessly, exited from the high small valley, crossed the edge of it and descended the slope beyond.'

c. Continou avançando, **saiu** [V1] do vale alto e estreito, e **desceu** [V3] as ladeiras além.
'He continued advancing, exited from the high and narrow valley, and descended the slopes beyond.'
(all from Slobin 2005: 121)

Descriptions such as (14a) would be typical of S-languages in two respects: (1) PATH is analysed into numerous segments (*out of the high valley*, *over its edge, down the slopes beyond*), and (2) these PATH expressions are all 'stacked onto' one verb (*wandered*), i.e. they all appear in one clause.[9]

By contrast, V-languages, first, typically use a separate verb for each PATH segment, as illustrated in (14b–c) with *sortit/saiu* 'exited', *franchit* 'crossed', *descendit/desceu* 'descended'. Second, they also tend to mention fewer PATH segments overall (Slobin 1996b: 203)—the Portuguese translator of *The Hobbit* in (14c), for instance, leaves out the PATH segment *over its edge*.

What is more, V-languages tend to make more frequent use of 'bare verbs' compared to S-languages (Slobin 1996b: 200–201). To be 'bare' for a motion verb means to have no directional particles, objects, adverbs, or prepositional phrases that carry GROUND-information (e.g. bare *ils tombent* 'they fall' vs. non-bare *ils tombent par terre* 'they fall to the ground'). Slobin (1996b: 204–5) finds that this comparative 'lack' of detailed PATH segments in V-languages is 'compensated' by providing more detailed static descriptions of scenery, from which PATH can be inferred.[10] To conclude, S- and V-languages are said to differ in the amount of PATH segments they express (a) overall and (b) per clause.

2.1.4.3 Higher MANNER salience and larger MANNER verb vocabulary in S-languages

It is not only in the verb that S-languages express MANNER information more often than V-languages, but in general. Even if other means of MANNER expression, such as adverbials, are included in the comparison, 'verb-framed languages apparently are less concerned with the domain of manner of motion than are satellite-framed languages' (Slobin 2005: 123; 2004a: 250–53). S-languages therefore seem to be more 'MANNER salient' than V-languages in that MANNER information is generally given more often.

The degree of MANNER salience hinges on the 'codability' of MANNER OF MOTION (Slobin 2004a: 237). Highly codable, according to Slobin (ibid.), is what can be expressed 'by a finite rather than a non-finite verb', 'by a high-frequency rather than a low-frequency lexical item', and 'by a single verb rather than a phrase or clause'. Both satellite-framed and serial verb languages are therefore high in MANNER salience, and so are languages that regularly employ ideophones, such as *gümbür gümbür* 'splash' in Turkish or *doya-doya* 'noisily' in Japanese (Slobin 2004a: 233–5). The crucial idea about 'codability' is that in MANNER-salient languages, MANNER expression is possible at a 'low cost' (Slobin 2006: 62)—you simply put it in your main verb, for instance, a clause

---

[9] As Slobin (1996b: 214) notes, these concomitant features are not 'definitional' for V- and S-framed languages respectively. In MANNER conflating, S-framing Russian and Serbo-Croatian, for instance, 'one cannot mass all the directional phrases onto one manner verb' (Filipović 2007a: 21; 2007b: 313).

[10] Compare, for instance, the following examples from English and Spanish frog story narrations, on the scene in which a deer throws a boy into the water: *He threw him over a cliff into the pond* (Slobin 1996b: 202) and *Lo tiró. Por suerte, abajo, estaba el río. El niño cayó en el agua.* '[He] threw him. Luckily, below, was the river. The boy fell into the water' (ibid.: 204).

element that you need anyway. In V-languages, by contrast, you usually need adverbials to express MANNER information, such as in *en avion* 'by plane' in (15):

(15) Je vais aller à Munich **en avion.**
'I will go to Munich by plane.'

This, Slobin (2006: 67) argues, is (a) '"heavy" in terms of processing (production/comprehension)', and (b) entails that MANNER OF MOTION is foregrounded. Therefore, speakers of V-languages tend to leave MANNER information to be inferred and to mention it only when it needs mentioning (i.e. (15) is more probable if the FIGURE usually avoids flying, or if the distance is so short that it would not usually be covered by plane).

Furthermore, not only is the whole category of MANNER verbs more frequently used in S-languages, but verbs of this type also occur in greater variety (e.g. Slobin 1996b: 198–9): S-languages apparently tend to have a larger inventory of MANNER OF MOTION verbs as compared to V-languages. Slobin (2006: 71; see also 2005: 123) reports 'several hundred manner verbs' in the satellite-framed languages English, German, Dutch, Russian, and Hungarian, but less than hundred, and 'probably less than sixty in everyday use' in the verb-framed languages Spanish, French, Turkish, and Hebrew.

For English and Spanish, this is supported by Cifuentes-Férez (2008), who finds that English (a) has a larger inventory of MANNER OF MOTION verbs (about twice the size of the Spanish inventory, 2008: 209) and (b) tends to make more fine-grained distinctions, for instance with respect to how the FIGURE's legs are moved (2008: 221). Also Verkerk (2013), who analyses MANNER verbs in texts translated into 20 Indo-European languages, finds a correlation between the degree to which a language uses the satellite-framing pattern and its number of MANNER verbs (2013: 184, 196). The higher number of MANNER verbs in S-languages is seen by Slobin as related to a higher degree of MANNER salience:

> In a linguistic diachronic perspective, as a domain becomes more saturated in a language, speakers invent lexical items to mark distinctions that become important to them. A language with a rich manner lexicon tends to get richer over time. That is, learning and using the language engenders habitual attention to detailed analysis of a domain, leading to further lexical innovation, and increasing attention to the domain over time (Slobin 2006: 72).

This agrees with the findings of Fanego's study of the English MANNER OF MOTION lexicon in a diachronic perspective (2012). Even though the scope of the study is limited to verbs with animate agents who are in control of the motion (i.e. verbs such as *float, roll,* or *bounce* are disregarded) (2012: 27), the inventory of MANNER verbs that is discovered throughout all stages of the language is considerable: Fanego searches historical dictionaries and thesauri for equivalents of Modern English MANNER verbs and finds steadily increasing numbers of new MANNER verbs and MANNER senses—71 for Old

English, 181 new ones in Middle English, 205 new ones in Early Modern English, and 250 new ones in Late Modern English. The largest numbers of verbs belong to the semantic subdomains of MODE (e.g. OE *ahealtian* 'to limp, crawl', *climban* 'to climb', *hoppian* 'to hop, leap, dance) and VELOCITY (e.g. OE *fundian* 'to hasten', *hlēapan* 'to rush', *scēotan* 'to shoot, move rapidly, dart, run, rush'). The category of VEHICLE/CONVEYANCE (e.g. OE *rōwan* 'to row, sail', *seglan* 'to sail'), by contrast, only sees a higher number of new verbs from Early Modern English onwards, when conversion to verbs from nouns denoting vehicles becomes more common (Fanego 2012: 51), such as in *to barge* 'to journey by barge' or *to scull* 'to proceed by means of a boat propelled with a scull' (2012: 54). Sources of new MANNER verbs are word formation, borrowing, and the development of MANNER OF MOTION senses of existing verbs (e.g. ME *platten* 'to hasten, rush' is not attested as a MANNER OF MOTION verb in Old English, but only in the sense 'to give a sounding blow, to smack' (2012: 58)). This study thus lends support to Slobin's hypothesis that the semantic domain of MANNER OF MOTION in S-languages is 'constantly attracting new lexical items' (2012: 57).

2.1.4.4 Overview

To conclude this section, various studies have proposed that S-framed languages such as English or German tend to display the characteristics summarized in Table 2.1 compared to V-framed languages such as French or Spanish.

It has to be stressed, however, that these differences are not purely due to a language being verb- or satellite-framing, but also to the interacting factor of the morphosyntactic structure of a language and a specific 'rhetorical style'

TABLE 2.1. Typical concomitant characteristics of satellite-framed and verb-framed languages.

	S-LANGUAGES	V-LANGUAGES
combinability of MANNER verbs with PATH satellites	high, also for boundary-crossing events	not generally possible for boundary-crossing events
PATH segments per clause	more PATH segments per clause (i.e. satellites per verb)	one verb per PATH segment, and more bare verbs (i.e. without PATH segments)
PATH segmentation	more detailed segmentation of PATH	less detailed PATH segmentation, more scene-setting
MANNER salience	generally more information about MANNER OF MOTION	tendency to leave MANNER to be inferred
size of MANNER lexicon	larger, increasing through time	smaller

(e.g. Slobin 2006: 69) that can result from these factors and perpetuate itself (Slobin 1996b: 217, 2004a: 248).

2.2 *Constructionist approach*

In this study, the motion verb typology will be approached from a constructionist perspective (see also Narasimhan 2003, Berthele 2007, Pedersen 2009, Stefanowitsch 2013). Both approaches are highly compatible with each other—Talmy's typological classification has been characterized as 'fundamentally constructional in the sense of "construction" in current versions of construction grammar' (Croft et al. 2010: 203)—, but there is one important difference, which will be pointed out at the beginning of this section (2.2.1). The following subsections will present the intransitive motion construction as proposed by Goldberg (1995), Rohde (2001), Goldberg & Jackendoff (2004), and Stefanowitsch (2013) (2.2.2) and what this approach implies for the conceptualization of verb meaning (2.2.3). The section will end with arguing for advantages of this approach over the Talmyan one within the scope of the present study (2.2.4).

2.2.1 Differences—MOTION in the verb or in the construction?

As described in 2.1.2 above, Talmy sees the semantic component of MOTION as carried by the verb. Consequently, for a verb such as *float*, he assumes two 'usages', a basic one $float_1$ that 'refers to the buoyancy relation between an object and a medium' (2000b: 31), as in (16a), and a second one, $float_2$ that 'includes the idea of motion' (2000b: 31), as in (16b).

(16) a. The craft **floated** on a cushion of air.

b. The craft **floated** into the hangar on a cushion of air.
(both from Talmy 2000b: 31)

Although these are rather neutrally called different 'usages'—one with a locational constituent only, one with an additional directional constituent—the idea is that $float_1$ and $float_2$ are 'distinct lexical items'. This idea is explained by Talmy a few pages later with the example of transitive and intransitive *break* (2000b: 35). At the same time, however, he points out that a constructional perspective is equally possible:

> In all the preceding, where we have treated the second usage of a verb [. . .] as a lexicalization of additional components conflated into it, Aske (1989) and Goldberg (1995) treat it as the original simplex verb and treat the additional complexities of the surrounding construction as the source of the additional meanings. Perhaps the evidence adduced above can be largely reconstrued to serve as well for this constructional position. In the end, the important thing is that we correctly identify the semantic components and their interrelationships, whether these are seen

> as involving lexical conflation or constructions. [. . .] a constructionist approach should claim that no distinct lexical item for transitive *break* exists in English. Rather, it should treat the transitive causative usage of *break* as consisting of intransitive *break* in interaction with the structure of the surrounding sentence, since this would parallel its treatment of Motion-Manner verbs like *float*$_2$ (Talmy 2000b: 35).

What is sketched in this quote as an alternative to the lexicalization account corresponds, by and large, to the approach adopted here (on details concerning verb meaning, see 2.2.3): MOTION is seen as associated to 'the structure of the surrounding sentence', that is, in constructionist words, to the intransitive motion construction itself. This crucially entails that, different from Talmy, 'it is not necessary to posit an additional verb sense for each new syntactic configuration in which [a] verb appears', and, what is more, 'implausible verb senses are avoided' (Goldberg 1995: 9). This is a particularly appealing aspect of the constructionist approach in view of the data for Old and especially Middle English, where otherwise many implausible verb senses would have to be posited (see 2.2.4 and the category 'non-MOTION verbs' in Appendices A and B).

2.2.2 Constructional meaning: the intransitive motion construction

Constructionist approaches see pairings of form and meaning not only in the lexicon, but also in schematic patterns. Thus, basic sentence patterns can be seen as argument-structure constructions, which 'encode as their central senses event types that are basic to human experience' (Goldberg 1995: 39). One of those basic human experiences is that a FIGURE changes location, and this would be the central meaning of the intransitive motion construction (IMC). Argument-structure constructions, Goldberg et al. argue, 'may arise developmentally as generalizations over lexical items in particular patterns' (2004: 308, see also Goldberg 2009: 99–101). Thus, children would learn the IMC, for instance, because PDE *go* (like OE *faran* or ME *gon*, as we shall see in 5.2.4 and 6.3.4) is very frequently used with a directional, and the meaning of this frequent verb will become associated with this specific pattern (i.e. with the IMC).[11]

The intransitive motion construction has not been studied as extensively as other argument-structure constructions such as the ditransitive construction. The present account is based on the remarks in Goldberg (1995), Rohde (2001), Goldberg & Jackendoff (2004), Berthele (2007), and Stefanowitsch (2013).

The semantic side of the intransitive motion construction is 'X MOVES Y' (Goldberg 1995: 3). This means it involves two argument roles: X, i.e. the FIGURE, and Y, i.e. the PATH with respect to the GROUND. The construction also specifies the semantics of the verb slot as 'MOVE'. The schematic

[11] See also the similar account of Greek and Middle High German non-motion verbs in contextual motion readings in Paul [1886] (1920): 319–20.

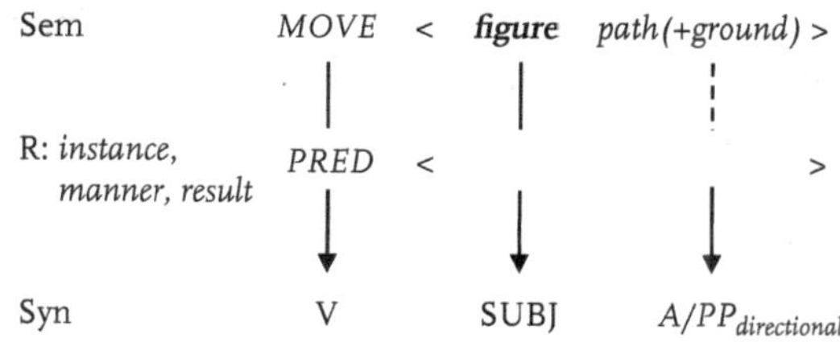

FIGURE 2.1. Intransitive Motion Construction in Present-Day English.

representation of the IMC in Figure 2.1 (adapted from Rohde 2001: 242 and Stefanowitsch 2013: 227) represents the constructional semantics in the 'Sem' line. The formal side of the construction is represented by the 'Syn' line: a verb, a subject, and a directional prepositional phrase or adverb (i.e. the satellite). Note that the order of these elements is not fixed: they will of course be in different orders in actual usage events, when combined, for instance, with topicalization constructions, interrogative constructions, etc. (e.g. Goldberg 2006: 20–2).

Instead of a directional prepositional phrase, which carries information about PATH (in the preposition) and GROUND (in the preposition complement), a directional adverb is equally possible—the difference is merely that in this case, GROUND is not expressed explicitly (as in *He went into the house* vs. *he went in*).

The construction can be instantiated by verbs which have a semantic component 'MOVE' on their own, such as PDE *move* or ME *faren* (17a–b). These verbs are in a relation of 'instance' to the semantics of the construction, 'going' being an instance of 'moving'.

(17) a. The craft **moved** into the hangar.

b. He **fared** into France (*c*.1450(?*a*.1400) *Parl.3 Ages*, MED s.v. *faren*)
'He went into France.'

Yet also other verbs which do not necessarily by themselves designate MOTION, but only MANNER, such as PDE *float* in (16b), can fill the verb slot of this construction, with the meaning 'MOVE' and the PATH(+GROUND) argument contributed by the IMC in (16b). This is indicated by the broken line in Figure 2.1, which means that the argument role of PATH(+GROUND) can be contributed by the construction and does not have to fuse with a participant role of the verb (Goldberg 1995: 50). A verb such as *float* would only have a participant role of 'floater' but not necessarily of 'path(+ground)'. The former fuses with the profiled argument role of FIGURE (*the craft* in (16b)), and the latter (*into the hangar* in (16b)) is contributed by the construction.

Argument-structure constructions also have to specify which kinds of verbs they can be associated with. This is discussed in terms of 'r-relations' by Goldberg (1995: 59–66) (see also Stefanowitsch 2008 and 2013). Next to the r-relation of INSTANCE, which is of course always possible (as in (17a–b) with PDE *move* and ME *faren*), Goldberg & Jackendoff posit the following three possible r-relations for the Present-Day English IMC (2004: 563):[12]

(18) i. MEANS: [*VERBAL SUBEVENT*]

ii. RESULT: [*VERBAL SUBEVENT*: X_1 *EMIT SOUND*]

iii. RESULT: [*VERBAL SUBEVENT*: X_1 *DISAPPEAR*]

The MEANS-relation (corresponding to the MANNER-relation in Figure 2.1) is unrestricted. As long as the verbal subevent is the means by which the motion takes place, it can be any kind of subevent, i.e. any kind of manner in which you can move—*roll, ski, float*, and so forth. According to Goldberg & Jackendoff, this is different with the RESULT-relation, in which the verbal subevent (e.g. whistling, vanishing) is the result of the motion (cf. Talmy's CONCURRENT RESULT (6d) in 2.1.2): this is only possible with a particular type of verb, namely verbs of sound emission or verbs of disappearing, as in (19):[13]

(19) a. The bullets **whistled** past the house. (2004: 540)
b. The witch **vanished** into the forest. (2004: 541)

Goldberg (1995: 65) mentions PRECONDITION as a further generally possible relation between the event type designated by the construction and the one designated by the verb (such as in the ditransitive *he baked me a cake*, where the baking is a precondition to the transfer of the cake), though not in the Present-Day English IMC. As we will see in chapters 5 and 6, this relation is possible in the IMC in older stages of the language.
What all these r-relations have in common is that they are metonymic:

> [I]f we consider certain verbs' inherent semantics to bear a *metonymic* relationship to the semantics of the construction, we may find a partial explanation. The semantics associated with the construction defines a semantic frame, and the verb must inherently designate a particular salient aspect of that frame (Goldberg 1995: 65).

As Goldberg (1995: 60) points out herself, this corresponds conceptually (though not always terminologically) to Talmy's conflation patterns, which also recognize various possible relations between 'motion event' and 'co-event' (see 2.1.2). The crucial difference lies in the fact that in Talmy's perspective, a language allows and prefers certain co-events to be conflated together with MOTION in the verb, while in the constructional perspective, a specific construction allows certain verbs in its verb slot, and contributes the MOTION

[12] For Present-Day German, see Berthele (2007: 247).
[13] The CONCOMITANCE relation (as in (5c) *I whistled past the graveyard*), in which the co-event is not a result of, but only co-occurring with the motion is judged as acceptable only by some speakers.

component itself (cf. Croft et al. (2010), who advocate an application of the typology to specific constructions rather than to whole languages).

2.2.3 Verb meaning

The preceding section has dealt with constructional meaning and its interaction with verb meaning. In this interaction, the meaning of the verb is at least as important as the meaning of the construction. This section will explain how verb meaning is conceptualized in the present study (namely, as conceptual, frame-semantic meaning). I will illustrate this important point with the following verbs:

- the MANNER verb PDE *float* in its translocational and non-translocational uses;
- the non-MOTION verb ME *labouren* in its non-MOTION and MOTION uses;
- the (at least originally) non-MOTION verb ME *travailen*, which develops into a MOTION verb;
- the predominantly CAUSED MOTION verb OE *weorpan* / ME *werpen* 'to throw' and its INTRANSITIVE MOTION uses;
- ME *hasten*, which is frequently found in both INTRANSITIVE MOTION and CAUSED MOTION uses.

The present study sees verbs as 'associated with rich frame semantic meanings' (Goldberg 1995: 29) in the sense of Fillmore's frames as 'coherent schematizations of experience' (1985: 223).[14] Verbs are thus access points to encyclopaedic knowledge—in our case, mostly knowledge about MOTION. The access to a certain frame can be seen as the decontextualized, coded meaning of a verb. This decontextualized meaning is therefore rather unspecified; it receives a specified, contextualized reading only in the actual usage event.[15] For example, the verb *run* gives access to what we know about running: this is its decontextualized meaning. We know—and frequently experience—that usually people change location when they run or, rather, that they run in order to change location (though you can also run in place). With the kind of MANNER expressed by *bounce*, a non-translocational situation will be more frequently experienced than with *run* (e.g. bouncing up and down), but a translocational situation is equally possible (e.g. a ball bouncing down a hill; see also Iwata's account of *bounce* in terms of image schema networks (2002: 78–80)). Therefore, based on experience, MANNER OF MOTION verbs are linked to a MOTION frame. This fact is taken into account, for instance, in the Fillmore-based *Framenet*,[16] where

[14] See also, e.g., Iwata: 'the verb meaning is a Fillmorean scene' (2005: 382). More generally, this cognitive view of semantics is outlined in Evans & Green (2006: 158–63), under the headings of 'Semantic structure is conceptual structure', 'Meaning representation is encyclopaedic', 'Meaning construction is conceptualisation'.

[15] See, similarly, Croft & Cruse's 'dynamic construal approach' (2004; for an overview, Cruse 2004: 261–72), or Zlatev (2003).

[16] http://framenet.icsi.berkeley.edu.

MANNER verbs (e.g. *float* or *roll*) are associated with the frame MOTION and with frame elements such as GOAL, SOURCE, or THEME (i.e. FIGURE) (see also Fried & Östman's formalization of the verb *walk* (2004: 52), whose semantic frame contains a 'walker', a 'point of departure', a 'ground', etc.).

Consequently, in the present study I will treat MANNER OF MOTION verbs as MOTION verbs as well (like, for instance, Rohde 2001: 352–4 and Berthele 2007: 247 (his category a)). At first glance, this may seem to be at odds with what has been said in section 2.2.1 (cf. the quote from Talmy 2000b: 35) on the difference between the lexicalist view of regarding MANNER + MOTION as conflated in the verb and the constructionist view of seeing MOTION as contributed by the construction. Yet, treating MANNER OF MOTION verbs as MOTION verbs crucially does not entail that the idea of conflation is introduced through the back door: While MANNER OF MOTION verbs are surely not always used to refer to MOTION in particular usage events, as in (20a), they certainly *give access to* or *evoke* a MOTION frame, as argued in the previous paragraph. Whether this MOTION reading is active in a particular context is influenced by the construction in which the verbs are used.[17] Observe again examples (16a–b), repeated here as (20a–b):

(20) a. The craft **floated** on a cushion of air. [non-motion]
b. The craft **floated** ***into the hangar*** on a cushion of air. [motion]

The 'rich frame semantic meanings' of *float* include a floating FIGURE (i.e., the participant role 'floater'), the surface on which it floats (usually water or air), and the possibility that the FIGURE moves somewhere while floating—i.e. everything we know about floating. For the two uses in (20a–b), therefore, it is not necessary to assume polysemy, because they are actually contextualized variants of one and the same meaning.

How does the association of a verb with frame-semantic meanings arise? The knowledge associated with a verb depends on the previous occurrences of the verb which the speaker has experienced, that is, it is based on usage. Frequent usages, or, in other words, the contextualized meanings in which the verb is often encountered influence the decontextualized, 'coded' meaning of a verb (in terms of Traugott & Dasher 2002: 16). Let me illustrate this with the example of ME *labouren* and *travailen*.

The decontextualized meaning of ME *labouren* can be paraphrased as 'to work, work hard, toil, take pains, suffer', i.e. it evokes a frame with a person who is working hard, taking pains, suffering, as in (21a–c). Whether the aspect of 'to toil, work hard', the one of 'to suffer', or the more general aspect of 'to

[17] This could also be put in terms of Croft & Cruse's dynamic construal approach, where a lexical item is 'associated with a body of conceptual content', i.e. with encyclopaedic knowledge, which they label 'purport' (2004: 100). A word's purport is determined by the previous uses of the word the speaker has experienced. Based on this, meaning is construed online. Thus, we could say that the 'purport' of MOTION verbs is to provide an access point to the conceptual area of MOTION.

work' is more prominent depends on the particular context (contextualized meanings).

(21) a. Thei that **labore**..may ete laufully thries in the day whan they **laubor**. (*a*.1500 *3rd Fran.Rule* (Seton) 50/28, MED s.v. *labouren*, 1a 'to perform manual or physical work, etc.')
'They who work [. . .] may lawfully eat thrice on the day in which they work.'

b. He..preyde hem to **labouren** in this nede. ((*c*.1395) Chaucer *CT.Mch.* (Manly-Rickert) E.1631, MED s.v. *labouren*, 2a 'take pains, take trouble, struggle, etc.')
'He asked them to work hard in this matter [i.e. in arranging a marriage].'

c. [They] besowght the Apostle that with his woonnte pyte he wolde succur this **laborynge** virgyne. (*c*.1425 *Found.St.Barth.* 51/3, MED s.v. *labouren*, 3a 'to endure pain, suffer')
'They beseeched the Apostle that he would help this toiling/suffering virgin with his usual pity.'

In some usage contexts, the verb *labouren* also receives a MOTION reading, such as in (22):

(22) Plato þat **laboured *þorw egipt***..not aschamed he..schuld be a pilgrime. (*c*.1450 Capgr. *Rome* (Bod 423) 1, MED s.v. *labouren*, 3b 'to journey painfully or with difficulty')
'Plato, who journeyed painfully through Egypt, not ashamed that he should be a pilgrim.'

Through the interplay of the context and the usual meaning of the verb, *labouren* in (22) is interpreted as 'to journey painfully or with difficulty'—also a contextualized meaning. In (22), this contextualized meaning hinges on the satellite *þorw egipt* 'through Egypt', i.e. on the fact that *labouren* is used in the IMC. In this example, both the verb meaning ('work hard, toil, suffer') and the constructional meaning ('FIGURE MOVES PATH(+GROUND)'), as well as the general context of pilgrimage,[18] feed into the contextualized meaning of *labouren* as 'to journey painfully': the participant role of 'labourer' is construed as the FIGURE in a motion event.

Since this particular contextualized meaning is the one the speaker/writer intends, in Traugott & Dasher's terms the interpretation of 'to journey painfully' is called an 'invited inference'—it arises 'in context "on the fly"'

[18] The sentence immediately preceding (22) is *So ded pictagoras þat trauyled as seith seynt ierome be þe prophetis memphetik* 'So did Pythagoras, who, as Saint Jerome says, travelled to the prophets of Memphis' (ed. Mills 1911: 1). It may be possible that the presence of *travailen* (a synonym of *labouren*, which is, however, much more frequent in motion uses (see below)) in the immediate context may have triggered the use of *labouren* here, for the sake of variation. In any case, the presence of *travailen* in a motion reading surely facilitates the contextualized interpretation of *labouren* as 'to travel with difficulty' here.

(2002: 17). Thus, even though the difference between the uses of *labouren* in example (22) and those in (21) is definitely greater than the difference between *float* in MOTION and non-MOTION uses (20a–b) above, it is not necessary to posit polysemy on the part of the verb for *labouren* either. More precisely, it is not necessary to posit polysemy as long as the kind of usage illustrated in (22) is not taken over by larger numbers of speakers and becoming a coded meaning for the verb, which does not seem to have been the case for *labouren*.[19]

This is different for the verb *travailen* (PDE *travel*), whose motion usage has indeed been adopted by larger numbers of speakers. Originally, ME *travailen* is a synonym of *labouren* in that it equally evokes the frame of 'to work, toil, suffer', as illustrated in (23a–c).

(23) a. Ilkone of us withouten lesyng Might win ilk wike fourty shilling; And ȝit, bot if we **travail** mare, Oft þai bete us wonder-sare. (*a*.1425(?*c*.1350) *Ywain* (Glb E.9) 3059, MED s.v. *travailen*, 1a 'work for wages')
'Each one of us, without lying, could earn forty shilling each week; and yet, unless we work more, they often beat us very heavily.'

b. Why seith holy writ þat God restide þe seueþe day, sith he **traualyd** neuere? (*a*.1500(*c*.1410) *Dives & P*.(Htrn 270) 1.263, MED s.v. *travailen*, 2a.a 'to put forth great effort, take pains')
'Why does the holy scripture say that God rested on the seventh day, given that he never took pains?'

c. Anone he **traueyled** as men done þat dyen. (*a*.1400(?*a*.1325) *Bonav.Medit.* (1) (Hrl 1701) 755, MED s.v. *travailen*, 4a.a 'to suffer pain or physical torments')
'Then he suffered like people do who are dying.'

Travailen is also found in MOTION uses, such as in (24a–c). In (24a), the MOTION reading is supported by the directional PPs *fram þe este to þe weste* as satellites, in (24b) by the combination with the satellite *far* and the context *when he come agayn*. In both examples, *travailen* is thus used in an intransitive motion construction, which gives rise to the motion reading. In (24c), by contrast, the motion reading comes about via the situational context (the elderly Earl Aubri takes leave of Guy of Warwick and Herhaud before these embark on a journey). Note, however, that in this example, different from (24a–b), *No may ich for eld trauaily* could also be interpreted as 'I cannot exert myself/work hard because of my age' instead of 'I cannot travel because of my age': the contextual motion reading is by far less stable when it is not supported by an IMC.

[19] For sense 1a of *labouren*, 'to perform manual or physical work, etc.', for instance, there are as many as 17 quotations in the *MED*, for sense 3b 'to journey painfully, with difficulty', by contrast, only four. Among the 65 attestations of the verb *labouren* in the Middle and Early Modern English parts of the *Helsinki Corpus*, only two Middle English ones allow for a motion reading.

(24) a. þe heiȝe temple..Of sonne and Mone and steorrene al-so, ***fram þe este to þe weste*** Þat **trauaillieth** [Hrl: Wendeþ] and neuere werie ne beoth ne neuere ne habbeth reste (*c.*1300 *SLeg.Kath.*(LdMisc 108) 24, MED s.v. *travailen*, 5a.a 'to make a journey, travel, go, etc.')
'The high temple [. . .] of sun, moon and also stars, that travels from the east to the west and never grows weary nor never has rest.'

b. Anaxagoras þe philosophur had **traveld *far*** for þe studie of philosophie **& when he come agayn** he fand all his gudis destroyed. (*c.*1450 *Alph. Tales* (Add 25719) 190/9, MED s.v. *travailen*, 5a.b 'to journey abroad, etc.', expanded with Banks (ed.) 1904)
'Anaxagoras the philosopher had travelled far for the study of philosophy, and when he came back he found all his goods destroyed.'

c. No may ich for eld **trauaily**: Hom ichil wende now an hye. (*c.*1330(?*c.*1300) *Guy*(1) (Auch) 5683, MED s.v. *travailen*, 5a.c 'to endure the rigors of travel, etc.')
'I cannot travel/exert myself because of [my] age: I will go home now at once.'

Compared to those of *labouren*, the MOTION uses of *travailen* seem to be far more frequent.[20] Thus, different from *labouren*, this use of *travailen* must have spread to larger numbers of the speech community, and in this, the inference 'motion' becomes generalized (cf. Traugott & Dasher 2002: 35): Any such usage event 'leaves a trace in the mental lexicon' (Geeraerts 2010: 146) of the speaker or hearer, and therefore has an effect on the coded meaning of the verb. This generalization is reflected in a crucial difference between *travailen* and *labouren*: In the attestations of *labouren* in a MOTION reading, the context always supplies very strong clues to facilitate this reading (see example (22) above). For *travailen*, by contrast, there are also attestations with very little or no contextual clues to a MOTION reading, such as in (25), where the participle *travailed* is used in adjectival function with a meaning similar to PDE *well-travelled*. This indicates that the reading 'to travel' must have become rather generalized. In (25), Guy of Warwick is on his way back from Constantinople to England, when he meets upon a sorrowful man. Guy asks him to tell him his problem, since he might be able to help him—well-travelled people having all sorts of useful knowledge:

[20] The *MED* gives as many as 41 quotations of *travailen* in the motion senses (sense 5a and 5b.a), while for *labouren*, there are only four (see footnote 19). Even if we take into account that the *MED* tends to give an increasing number of attestations for words starting with letters later in the alphabet (see Lewis et al. 2007: 19), the difference seems indicative of a greater frequency of *travailen* in this

(25) Þar auenture y myght þe say, How þy sorowe may passe away. Hyt befallyth to **trauelde** men Eyther other some gode to kenne. (*a*.1500(?*a*.1475) *Guy*(4) (Cmb Ff.2.38) 8777, MED s.v. *travailen*, sense 5b.b '*ppl.* experienced, well-traveled, worldly, etc.', expanded with Zupitza (ed.) 1875–1876)
'Perhaps I might tell you how your sorrow may pass away. It pertains to well-travelled people to know all or some good [knowledge].'

The use of *travailen* in the meaning 'to travel' then becomes a coded, decontextualized meaning of the verb, next to the one of 'to work, toil, suffer'. This means that the verb now gives access to a MOTION frame on its own; MOTION is no longer a mere inference depending on the contextual information. Note that the idea of 'difficulty, strenuousness' is also backgrounded. At this moment, both the 'work' sense and the 'travel' sense are coded meanings of *travailen*: they are polysemous. Of course it is impossible to pinpoint precisely when this happened, i.e. when *travailen* 'became a MOTION verb', as it were, because this might differ from speaker to speaker (Traugott & Dasher 2002: 35).[21] As aptly phrased by Nerlich & Clarke, '[p]olysemy always teeters on the edge of the semantic–pragmatic interface and of the synchronic–diachronic interface' (2003: 10). In the context of (26), for instance, where *to travel* is translated with Latin *itinerare* in an English–Latin wordlist—i.e. entirely devoid of any contextual clues—the 'travel' sense can safely be regarded as a (or probably the dominant) coded meaning of the verb for the compiler of this list.

(26) **To Travell**: itinerare, vbi to ga. (?*c*.1475 **Cath.Angl.*(Add 15562) 129b, MED s.v. *travailen*, sense 5a.a 'to make a journey, travel etc.')

The meaning *to travel* 'to travel, journey', as we know, has become the dominant one in English—the 'work, toil, suffer' meaning is now orthographically differentiated in PDE *to travail*, and its pronunciation also differs: /'træveɪl/ or /trə'veɪl/ as opposed to /'trævᵊl/ . Therefore, what must have started as an invited inference 'to travel with effort, with difficulty' in Middle English and Anglo-French has developed into a polysemous meaning ('to toil', 'to travel'). Later, *travel* and *travail* have evolved into different words, with distinct spelling and pronunciation.

It has been argued so far that the decontextualized meaning of a verb should be seen as a frame to which the verb gives access. In any given usage event, this receives a specified, contextualized reading, depending on the context and co-text, which includes the argument-structure construction in which the verb is used. I think that this may also hold for some of the verbs which occur both in

sense compared to *labouren*. Motion uses of *travailen* in the Middle and Early Modern English sections of the *Helsinki Corpus* are more frequent than those of *labouren* as well: in the 97 attestations of *travailen*, around a third have it as a motion verb, while in the 65 attestations of *labouren*, only two can be read as motion descriptions (see footnote 19).

[21] For this reason, in Appendix B, ME *travailen* is listed among the non-MOTION verbs.

INTRANSITIVE MOTION and CAUSED MOTION uses. Let us apply this to OE *weorpan*/ME *werpen* first. In Old English, according to Bosworth–Toller, the verb is only attested in the sense of 'to throw', as illustrated in the CAUSED MOTION uses in (27a–b), where the subject is the CAUSE (the 'thrower': *hi; se deofol*), the object is the FIGURE (the 'thrown': *hine; ænne stan*), and the adverbial PP expresses PATH + GROUND (*on ðone bat; to ðære bellan*).

(27) a. Hi **wurpon** hine on ðone bat. (Chr. 1046; Erl. 174, 17, BT s.v. *weorpan*, sense I 'to cast, throw, fling')
'They threw him into the boat.'

b. Se deofol **wearp** ænne stan to ðære bellan (Homl. Th. ii. 156, 9, BT s.v. *weorpan*, sense I 'to cast, throw, fling')
'The devil threw a stone at the bell.'

This use is still present also in Middle English, but in this period *werpen* is also found in the INTRANSITIVE MOTION construction as in (28a–b). Like in the CAUSED MOTION uses above, a satellite expresses PATH(+GROUND) (*towart ham; awaye*); the subject, however, does not express the CAUSE, but the moving FIGURE.

(28) a. Ich leote . . . a cleane mon wunien neh a cleane wummon þet ich nawhit ***towart ham*** ne **warpe** ne ne weorri, ah leote ham al iwurðen. (*c*.1225(?*c*.1200) *St.Marg.*(1) (Bod 34) 32/8, MED s.v. *werpen*, sense 1a 'to hasten, rush, etc.')
'I let a chaste man stay near to a chaste woman, so that I did not rush toward them, nor wage an attack against them, but let them all be.'

b. I rede ȝe . . . **warpes** wylily ***awaye*** (*c*.1440(?*a*.1400) *Morte Arth.(1)* (Thrn) 2746, MED s.v. *werpen*, sense 1a 'to hasten, rush, etc.')
'I advise you [. . .] hasten slyly away.'

In these uses, the frame of 'throwing' is contextualized differently. What they have in common with those in (27a–b) is that in all of them, a FIGURE moves violently to a GOAL. The only thing that is different in (28a–b) is that the FIGURE causes *itself* to move somewhere (and is expressed in the subject), rather than being caused to move by an external force (and being expressed in the object). Section 6.2.4 will treat this in more detail and discuss the factors that facilitated this type of contextualized meaning in Middle English.

Since for ME *werpen*, the INTRANSITIVE MOTION use as in (28a–b) remains rare and is only a passing phenomenon (mostly found in alliterative poetry),[22] it does not seem appropriate to assume polysemy here. The situation may be

[22] The *OED* does not give any attestation in this use later than the one from *c*.1440(?*a*.1400) in (28b) (s.v. *warp*, v.).

different, however, for verbs whose CAUSED and INTRANSITIVE MOTION uses are more or less equally widespread, such as for ME *hasten* (see 29a–b).[23]

(29) a. ***homward*** will I **hast**. (*a*.1500(*a*.1460) *Towneley Pl.*(Hnt HM 1) 28/182, MED s.v. *hasten*, sense 1c 'to go or come quickly, etc.')
'I will hasten homeward.'

b. He dooth al his labour..To **haste hem** ***fro the mete*** in subtil wise. ((*c*.1395) Chaucer *CT.Mch.*(Manly-Rickert) E.1767, MED s.v. *hasten*, sense 2a 'to drive sb., push, hasten, etc.')
'He takes great pains [. . .] to hasten them from the [wedding] meal in a clever way.'

If speakers frequently experience both usages of this verb, these will become more and more entrenched, and this may lead to the two senses being stored separately as polysemous coded meanings. On the other hand, a non-polysemous account is equally possible if the verb *hasten* is seen as giving access to a frame of 'being quick, hurrying', which can be contextualized both as 'moving somewhere quickly', as in (29a), or as 'making somebody move somewhere quickly', as in (29b). In both contextualized meanings, the argument-structure construction in which the verb is used is involved in creating the respective reading—INTRANSITIVE MOTION in (29a), with the subject *I* as FIGURE and the satellite *homward* as PATH + GROUND; CAUSED MOTION in (29b), with the subject *he* as CAUSE, the object *hem* 'them' as FIGURE, and the satellite *fro the mete* as PATH + GROUND.[24]

To conclude, it has been argued in this section that a verb's decontextualized meaning is a frame-semantic one, which is contextualized any time the verb is used. Contextualized meanings, in turn, may influence and change the decontextualized meaning, according to the usage-based approach adopted in this study. What interacts with the verb semantics in yielding these contextualized meanings is the situative context and the co-text, including the argument-structure construction in which the verb is used.[25] Therefore, translocational and

[23] Note also that none of these uses seems to be at the origin of the other: According to Gamillschegg (s.v. *hâte*), in Old French, from which the verb was borrowed, both the INTRANSITIVE MOTION and the CAUSED MOTION verb are derived from the noun *haste* 'haste'. According to the *MED* (s.v. *hasten*), the first Middle English attestations of the verb in an INTRANSITIVE MOTION context (with and without reflexive pronoun) are from *c*.1300; the first CAUSED MOTION attestation appears in a *c*.1330 manuscript of a text which was probably composed as early as *a*.1300 as well.

[24] The two accounts sketched here are related to the question whether you want to minimize or maximize polysemy, a great matter of debate in the literature, with cognitive linguistics having tended toward maximizing, i.e. to attribute many related meanings to a lexeme, instead of seeing them as contextual (see, e.g., the overview in Nerlich & Clarke (2003: 8–16), Langacker (2009), and Geeraerts (1993; 2010: 196–9) on the problematic distinction between vagueness and polysemy). This dichotomy of minimal and maximal polysemy tends to dissolve if lexical meaning is not seen as fixed, but as use potential (see Zlatev 2003).

[25] Of course it is actually the speaker/listener who draws on the co(n)text in construing the utterance meaning.

non-translocational uses of MANNER OF MOTION verbs such as *float* (see 20a–b) need not be treated as polysemous: Both are very close to the coded meaning of the verb, i.e. to the *float*-frame to which it gives access, and are specified through the co(n)text (in the case of MANNER verbs particularly through the construction). In the case of predominantly CAUSED MOTION verbs (OE *weorpan*/ME *werpen* in (27–28)) and non-MOTION verbs (ME *labouren* in (21–22)) that are used to express (INTRANSITIVE) MOTION only sporadically, the present study does not posit polysemy either: As the MOTION uses of these verbs are further removed from their coded meaning, they simply require more inferencing and probably more contextual support. The usages that speakers experience, though, are feeding back on the speakers' decontextualized meaning of a verb. Thus, if an inference becomes generalized, it can eventually also turn into a polysemous decontextualized meaning, or even into the dominant one. In this way, a verb that originally does not evoke a MOTION frame, such as ME *travailen* (or AF *travaillier*) 'to work, toil, suffer' (23–26), can become polysemous in having the coded meanings of both 'to work, toil, suffer' and 'to travel (without a necessary notion of difficulty or trouble)'—one of which may of course eventually also fall out of use.

2.2.4 Avoiding implausible verb senses

There is a specific reason why a constructionist perspective lends itself well to the present study, namely the avoidance of implausible verb senses. Sections 5.1 and 6.2 will analyse the inventory of Old and Middle English motion verbs, which mostly consists of verbs that are used in instances of the IMC. A large portion of those are, of course, verbs whose decontextualized meaning is linked to a MOTION frame itself (see sections a–c in Tables 5.1 and 6.1 and in Appendices A and B), as has been shown in section 2.2.3 to be the case for PDE *float*, ME *hasten* and, probably, for ME *werpen* 'to throw'. However, we will see that there are also quite a number of verbs whose decontextualized meaning is not associated with MOTION (as shown for ME *labouren* 'to toil' in 2.2.3), but which sometimes receive a MOTION reading in context (see section d in Tables 5.1 and 6.1 and in Appendices A and B). Two further examples for such verbs are OE *feohtan* 'to fight, combat, strive', and ME *weden* 'to be or become insane' (related with *i*-umlaut to the adjective and noun *wod* 'mad, furious'/'madness'; cf. German *wüten*); these are used in an intransitive motion construction in (30a–b).

(30) a. þa gecwædon hie þæt hie sume hie beæftan wereden & sume ***<þurh> ealle þa truman ut*** **afuhten**, gif hie mehten (Or 5 7.121.27, DOE s.v. *āfeohtan*, sense 2: *ut afeohtan* 'to fight one's way out')
'Then they said that some of them would defend them from behind, and some fight [their way] out through all the troops, if they could.'

b. Þat deor to-dede his chæfles and ***to þan kinge*** **weoðede**. (*c*.1275(?*a*.1200) Lay. *Brut* (Clg A.9) 3244, MED s.v. *weden*, sense 2 'to run wildly, charge, bolt, etc.')
'The animal opened its jaws and ran wildly to the king.'

In a lexicalization account like Talmy's we would have to postulate additional MOTION senses for these two verbs, such as those given by the *DOE* and *MED* in the respective entries (see 30a–b): 'to fight one's way out' or 'to run wildly'.[26] While it is certainly helpful for a student of Old or Middle English to have these senses listed in the dictionary, it is rather implausible that for a speaker of Old or Middle English these will have been decontextualized meanings of the verb; all the more so because the motion use of these verbs seems to have been rare and thus not very likely to have been entrenched. It is therefore more plausible to treat the meaning component MOTION as being contributed by the construction, i.e. the MOTION reading is a contextualized meaning of the two verbs, but not a coded one.

Let us recapitulate briefly how verb and construction interact, using example (30a). The IMC is associated with the general event frame of MOTION: it specifies an argument role MOVING FIGURE for its subject slot, the meaning MOVE for its verb slot, and an argument role PATH for its oblique slot. Verbs, as we have seen in 2.2.3, 'involve frame-semantic meanings; that is, their designation must include reference to a background rich with world and cultural knowledge' (Goldberg 1995: 27). In our case, the verb *feohtan* designates 'to fight, strive, combat'; the prefixed *afeohtan* 'to fight, fight against, attack' (cf. Clark Hall[27] s.v. *feohtan, afeohtan*), i.e. they have a participant role 'fighter'. Verb and construction are combined as follows: the verb's participant role 'fighter' can be construed as an instance of the argument role MOVING FIGURE (cf. 'the semantic coherence principle' in Goldberg 1995: 50; 2006: 39–40; Goldberg & Jackendoff 2004: 550). The argument role PATH is contributed by the construction alone. As for the verb slot, the verbal event 'to fight, combat' can be construed as the means by which MOTION is achieved—the FIGURE 'fighter' makes its way out through the troops by means of fighting (r-relation of MEANS, see 2.2.2).

I consider this way of accounting for the use of non-MOTION verbs in a MOTION reading to be more convincing than to postulate an extra MOTION sense for these verbs. This argument is basically the same as the well-known one in Goldberg (1995: 9–10) concerning *sneeze*: Sentences such as *He sneezed the napkin off the table, She sneezed the foam off the cappuccino* should be seen as instances of the caused motion construction in combination with the verb *sneeze*, instead of attributing the caused motion sense and the arguments *the napkin/the foam, off the table/cappuccino* to the verb *sneeze* alone.

The second example with which I want to argue for the advantage of the constructionist approach in this study are the premodals, i.e. the precursors

[26] The *DOE* here (s.v. *afeohtan*) is closer to the approach of the present study than the *MED*, since it specifies that the reading 'to fight one's way out' holds for *ut afeohtan*, i.e. for the combination with the satellite *ut* 'out'.

[27] For the purpose of giving a short description of what the verb evokes, the brief entries from Clark Hall's *Concise Anglo-Saxon Dictionary* (1916) are more adequate than the detailed ones from *DOE* or *BT*.

of PDE *must, may, shall, will*, etc. (henceforth also modal verbs). The combination of modal verbs with a directional element is generally possible in the Germanic languages (cf. Mortelmans et al. 2009: 35, 45, 65; van Riemsdijk 2002: 143) except for Modern Standard English (on remnants, see Kjellmer 2002: 344–52). In English it was available up to around the seventeenth century (cf., e.g., Traugott 1992: 194 for Old English; Fischer 1992: 263 for Middle English; Rissanen 1999: 232–3 for Early Modern English). (31a–d) are some examples from Old and Middle English.

(31) a. he **wolde** ***suð ofer sæ*** (ChroA2, 616, HC)
'He wanted [i.e. wanted to go] south across the sea.'

b. ne ***him to*** ne **dorste** sciphere on sæ. ne landfyrd (ChroE3–4, 1001, HC)
'No ship-army on sea nor land-army dared [i.e. dared to come] to them.'

c. þæt nan mann ne **mihte** ***ne inn ne ut*** (ChroE3–4, 1016, HC)
'that no one might [i.e. could get] in or out.'

d. Swa hwa swa ne bið ȝeedcenned of watere & of þam Halȝæ Gaste, ne **mæȝ** he ***into Godes rice***. (*c*.1175(?OE) *Bod.Hom*.(Bod 343) 6/2, MED s.v. *mouen* v.3)
'Whoever is not reborn through water and through the Holy Ghost, cannot [i.e. cannot come] into God's kingdom.'

To support my choice of a constructionist view here, I will first present two alternative ways of explaining this use of modal verbs with a directional element (MV + DIR), taken from studies on the modern Germanic languages—a 'lexicalist–main verb' view and an 'auxiliary and ellipsis'-view. After that, we will see how scholars of Old English have accounted for this combination, and then approach it from a construction grammar perspective, along the lines of what Berthele (2007) proposes for German dialects.

For the modern Germanic languages (apart from English, which no longer has this combination), analyses of the combination MV + DIR mostly fall into two groups: Some (such as Szumlakowski Morodo 2006) see the modal verb as a main verb with a 'full-blown subcategorization frame' (van Riemsdijk 2002: 143), i.e. they assume the pattern +DIR to be one of the valency constructions of modal verbs. This would correspond to the lexicalist approach *à la* Talmy.

The second view is to treat the modal verb as an auxiliary, which cannot in itself be complemented with a directional element. Therefore, this necessarily entails the analysis of the combination as an elliptical structure, where a full verb with MOTION meaning, such as 'come' or 'go' is to be recovered. This second analysis can be found in many grammars of German (cf. the overview in Szumlakowski Morodo 2006: 325; 339), or, also for German, in Öhlschläger (1989: 64). The problem is that there is often no specific verb in the context that could be recovered to resolve the ellipsis. Öhlschläger handles this with the idea of 'indefinite ellipsis'; van Riemsdijk argues for the presence of a 'phonetically

empty super-light motion verb GO' (2002: 143), i.e. an invisible verb which provides the subcategorization frame and takes the directional complement.
For Old English, the same question—main verb or auxiliary in MV + DIR—is raised (and left open, for that matter) by Mitchell (1985: §§1006–8):

> Are we to postulate 'ellipsis' of the infinitive of a verb of motion in examples like [. . .] *Or 86.3 þeh ic ær sæde þæt we to helle sceolden* [. . .] ['though I said earlier that we should to hell']?

Ogura's *Verbs of Motion in Medieval English* (2002: 8–10) apparently opts for the ellipsis analysis, when, under the heading of 'Verbs of Motion Unexpressed', she gives examples similar to the ones in (31) and comments:

> We can imagine which verb is to be restored: basic, colourless and common, such as *gan* and *faran*.

Mitchell is more hesitant: although he states 'I tend to agree with Visser (i, §178) that "to call this idiom elliptical, as *OED* does, is misleading"' (1985: §1007), he ends the discussion with

> [e]ach reader will have his opinion whether a particular example shows an independent usage of *magan*, **sculan*, or *willan*, or ellipsis of an infinitive. [. . .] We can leave it there. (1985: §1008).

The treatment in Lüttgens (1888: 53 for *sculan*, 86 for *willan*) is similarly hesitant, and his remarks point towards a concept similar to Öhlschläger's indefinite ellipsis.

I also 'tend to agree' with Mitchell that MV + DIR is not a matter of ellipsis, at least where there is no MOTION verb present in the context. I think that this can be very adequately argued for from a construction grammar perspective. So I would like to adopt the explanation offered for German by Berthele (2007), the only constructionist account of this structure that I am aware of (except —*avant la lettre*— Paul [1886] 1920: 320).

Berthele notes that the frog story narrations in some Southern dialects of German (he works with Swiss varieties) contain a large number of MV + DIR (2007: 231–2). He shows that an explanation of this structure via ellipsis is not satisfactory, but stresses that it is not necessary to see +DIR as part of the verbs' valency either: The argument role +DIR (i.e. PATH(+GROUND)) is simply a part of the intransitive motion construction, which is also responsible for the motion reading. Berthele points out that the construction does the same job as the invisible verb posited by van Riemsdijk (2002), while being more plausible from a cognitive-semantic perspective (2007: 240). The participant roles of the modal verbs (e.g. the 'muster' in (32) below, or the 'willer' in (31a)) can be fused with the argument role MOVING FIGURE. The interaction of the constructional meaning MOTION and the modal verb's meaning (e.g. 'obligation' in (32)) yields a reading of modally 'broken' motion, such as for instance 'obligatory motion' in (32) (similarly, 'intended motion' in (31a), 'possible motion' in (31b–d)) (see also Berthele 2007: 242).

(32) Criseyde **moste out of the town.** (*a*.1425(*c*.1385) Chaucer *TC* (Benson-Robinson) 5.5, MED s.v. *moten* v.2)
'Criseyde must [i.e. must go] out of the town.'

The meaning of the modal verbs ('ability', 'obligation', 'intention', etc.) can be construed as metonymically related to the constructional event MOTION, just as argued for the 'toil' verbs in section 2.2.3. As a more specific r-relation to license the modal verbs, Berthele posits the additional one of '*modale Rahmenbedingungen*' ('modal precondition') in the (Swiss) German IMC (2007: 247). He does not link this up with Goldberg's r-relations (see 2.2.2) explicitly, but it can of course be seen as an instance of Goldberg's PRECONDITION (with a restriction to modal verbs)—wanting, being able, or having to go somewhere as a precondition of going somewhere.

I believe that this framework is a good way of accounting for the premodals in MV + DIR in Old and Middle English as well. The intransitive motion construction contributes the meaning MOTION and specifies which r-relations are possible for the verbs in the verb slot—for Old and Middle English these include an r-relation of PRECONDITION. What is more, we will see in section 6.2.5.5 that the relation of PRECONDITION in these stages of English also licenses verbs of 'preparing, getting ready' (e.g. ME *busken, greithen, trussen*) and is thus not restricted to the premodals.

This section has argued why a constructionist approach to the motion typology is adopted in this study. The argument, roughly speaking, comes down to the fact that verbs meaning 'to fight', 'to be/become crazy', modal verbs, or verbs meaning 'to prepare' should not be seen as MOTION VERBS in a narrower sense, i.e. in the sense that they carry a semantic component MOTION or give access to a MOTION event frame. Assuming an ellipted or a zero MOTION verb in these cases does not seem convincing either. The most plausible and elegant way of accounting for these verbs in this use, therefore, is to assume an intransitive motion construction. The construction contributes (a) MOTION in the verb slot, and (b) the argument roles of MOVING FIGURE and PATH(+GROUND), which have to be fused with the verb's participant roles (or, for PATH(+GROUND), added in the case of verbs with only one participant role).

2.3 *Classifying motion verbs*

In the present study, the term 'motion verb' will be used in a wider sense as a shorthand to mean verbs that *can be used to talk about* MOTION. In this wider sense, the term 'motion verb' encompasses the categories illustrated in Figure 2.2 and described in the following.

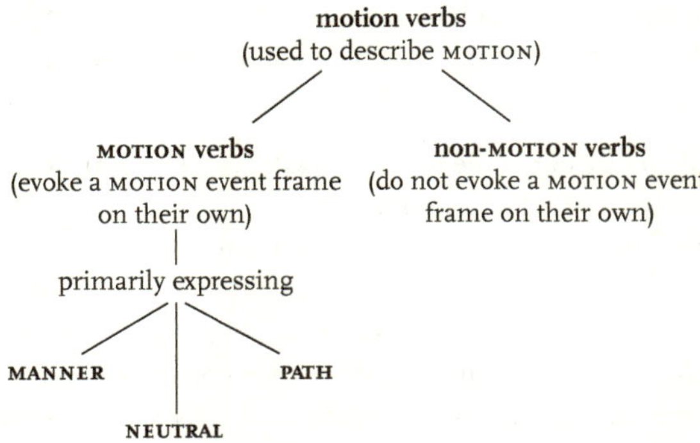

FIGURE 2.2. Idealized types of 'motion verbs'.

MOTION verbs in the narrow sense, i.e. verbs which evoke a MOTION event frame on their own (see the discussion in 2.2.3) are taken to include the following subtypes (see also, e.g., Rohde 2001: 352–4; Özçalişkan 2004: 78; Berthele 2007: 247):

(a) verbs which primarily express MANNER OF MOTION, e.g. ME *rouen* 'to row':

(33) Þe ffissher saw þe childe allane, And ***vnto him*** he **rowed** onane. (*a*.1425(?*a*.1350) *7Sages(2)* (Glb E.9) 3900, MED)
'The fisher saw the child alone, and at once rowed to it.'

(b) verbs which primarily express PATH OF MOTION, e.g. ME *entren* 'to enter':

(34) A theff..**intrede** ***in to seynt Edes seynt-warye*** (*a*.1450 *St.Editha* (Fst B.3) 2922, MED)
'A thief entered into Saint Edith's sanctuary.'

(c) verbs which do not primarily specify MANNER or PATH (like PDE *move* or OE *faran*), i.e., NEUTRAL MOTION verbs (for the term, see, e.g., Slobin 2004a: 251; Özçalişkan 2004: 78, 2005: 213; Cifuentes-Férez & Gentner 2006: 449; Fanego 2012: 37; Verkerk 2013: 175), e.g. ME *ten* 'to go, come, proceed, travel':

(35) Myche peple..at þe posterne ȝatis **Tyen** out of þe toun. (?*a*.1450(*a*.1400) *Siege Jerus.(1)* (LdMisc 656) 1158, MED)
'Many people move out of the town at the side gates.'

This third group will necessarily be larger in ancient languages compared to modern ones, because of the uncertainty in knowing exactly whether a verb such as ME *ten* has any specific MANNER or PATH associations (see also the discussion of ME *flitten* in 2.3.3)—without any native speakers available, the only thing we can be sure about is that it evokes a motion event frame.

Verbs which do not on their own evoke a MOTION event frame (non-MOTION verbs) can also receive a contextual motion reading. These include verbs which receive a motion interpretation when they are used in the IMC, i.e. in combination with a PATH(+GROUND) satellite, such as *swinken* 'to toil' with *so fer* 'so far' in (36), where the MOTION-component is contributed by the construction, which produces an interpretation of the subject (*he*) as moving FIGURE (see the discussion of *labouren* and *travailen* in 2.2.3).

(36) Iacob tolde him for-quat he **swanc** ***So fer.*** (*a*.1325(*c*.1250) *Gen.& Ex.*(Corp-C 444) 1656–7, MED s.v. *swinken*)
'Jacob told him why he had toiled [i.e. travelled] so far.'

However, non-MOTION verbs can also receive a contextual motion meaning outside an IMC. An example is the metonymical use of ME *sechen* 'to look for, try to find' in (37). The decisive criterion for including these verbs in the present study is whether according to their dictionary entries, they can be used to describe motion. For ME *sechen*, for instance, the *MED* has the sense 'to go to (a place)' (s.v. *sechen*, sense 10b).

(37) Than longen folk to goon on pilgrymages And palmeres for to **seken** straunge strondes. ((*c*.1387–95) Chaucer *CT.Prol.*(Manly-Rickert) A.13)
'Then people want to go on pilgrimages, and pilgrims to the Holy Land [want to] seek [i.e. go to] foreign countries.' (MED s.v. *sechen*, sense 10b 'go to (a place)')

In these cases, it is not any constructional context, but the general context which is responsible for the MOTION reading—in (37), the context of people going on pilgrimages and the metonymical relation of 'seeking foreign countries' to 'going to foreign countries'.

'Motion verbs' in a wider sense are thus all verbs that can be used to talk about MOTION, including verbs that are MOTION verbs in a narrower sense, that is, designating MOTION by themselves (MANNER, PATH, or NEUTRAL MOTION), but also those whose decontextualized meaning is not associated with a MOTION event frame (non-MOTION verbs). These categories are used in chapters 5 and 6 (see also Appendices A and B), which investigate how MOTION is expressed in Old and Middle English. In the following sections, we will look at the categories of MANNER verbs (2.3.1) and PATH verbs (2.3.2) in more detail, and then discuss the problem of separating the categories of MANNER-, PATH-, and NEUTRAL MOTION verb (2.3.3), which becomes even more acute when dealing with historical stages of the language.

2.3.1 MANNER

What is actually understood as MANNER information remains somewhat vague in the literature. According to Talmy's definition, 'Manner refers to a subsidiary action or state that a Patient manifests concurrently with its main action or state' (1985: 128). Slobin, rather than a definition, provides illustrative examples of what can be understood by MANNER:

'Manner' is a cover term for a number of dimensions, including motor pattern (e.g. *hop, jump, skip*), often combined with rate of motion (e.g. *walk, run, sprint*) or force dynamics (e.g. *step, tread, tramp*) or attitude (e.g. *amble, saunter, stroll*), and sometimes encoding instrument (e.g. *sled, ski, skateboard*), and so forth (Slobin 2006: 62).

It is the virtual open-endedness of possible 'subsidiary actions or states' that makes MANNER hard to subcategorize. The subcategories of MANNER that can be found in the literature are therefore rather divergent as well. (38) gives two examples.[28]

(38) **Examples for subcategories of manner in the literature**

a. Özçalişkan (2004: 81)

RAPID MOTION (e.g. *dart, plunge*), FORCED MOTION (e.g. *drag, sweep*), LEISURELY MOTION (e.g. *drift, sink*), SMOOTH MOTION (e.g. *slide, ride*), OBSTRUCTED MOTION (e.g. *stumble, leak*), FURTIVE MOTION (e.g. *creep*), MANNERS OF RUNNING (e.g. *run, flee*), MANNERS OF WALKING (e.g. *march, hurry*), MANNERS OF JUMPING (e.g. *bounce, spring*)

b. Narasimhan (2003: 135)

MODE ('motion brought about by specific types of movements of the Figural entity', e.g. *amble, hobble, run*)

ATTITUDE/DISPLAY ('expresses specific kinds of movements but additionally implies an attitude or intention on the part of the Figural entity', e.g. *sneak, swagger, toddle*)

VELOCITY ('speed (or lack thereof) as the main component', e.g. *rush*)

CONTACT ('implies (lack of) friction between the Figure and the Ground', e.g. *glide, slither*)

MEDIUM ('encodes properties of the medium through which the Figure moves', e.g. *fly, swim*)

VEHICLE (e.g. *drive, sail, canoe*)

COURSE (e.g. *hover, ramble, stray, roam*)

Due to the open-endedness of MANNER, these can never be hard and fast categories—in Özçalişkan (2004, see (38a)), for instance, *hurry* could also be seen as 'rapid motion', and it is not obvious why *ride* should be 'smooth motion'. Furthermore, since almost all researchers use their own categories (see footnote 28), there does not seem to be any classification that is agreed upon. The MANNER subcategories used in the present study are the ones that emerged as coherent groups from the Old and Middle English MANNER verbs

---

[28] Fanego (2012) builds on Narasimhan (2003), but reduces this classification to the components of *Mode, Velocity, Vehicle/Conveyance*, and *Course* only. For yet other groupings, see, e.g., Rohde (2001: 352–3), Cifuentes-Férez (2008: 140–2), Ibarretxe-Antuñano (2006); Levin (1993: 263–70), and Ikegami (1970: 108–65) for motion verbs in general; Snell-Hornby (1983: 133–70) for motion and position.

that were found in the motion verb inventories (see Tables 5.1 and 6.1 and Appendices A and B).

- speed (see Ma^{speed})
  (cf. Rohde's category SPEED (2001), Narasimhan's category VELOCITY (2003))
- slow or impeded motion (see $Ma^{slow/impeded}$)
  (cf. Özçalişkan's category OBSTRUCTED MOTION (2004))
- motion through air or liquid (see $Ma^{air/fluid}$)
  (cf. Rohde's category THROUGH AIR OR FLUID (2001), Narasimhan's category MEDIUM (2003))
- falling motion (see Ma^{fall})
  (cf. Rohde's category FALLING MOTION (2001))
- gliding motion (see Ma^{glide})
  (cf. Narasimhan's category CONTACT (2003))
- jumping motion (see Ma^{jump})
  (cf. Özçalişkan's category MANNERS OF JUMPING (2004))
- rolling motion (see Ma^{roll})
- fleeing, escaping (see Ma^{flee})
  (cf. Özçalişkan's category MANNERS OF RUNNING (2004))
- other manner of motion (see Ma^{other})

However, within the purpose of the present study, a detailed subcategorization within the domain of MANNER is not really important. After all, our focus is (a) on the manner inventory as a whole (see 2.1.4.3) and (b) on whether the pure PATH verbs borrowed in Middle English are unusual in the native system of talking about MOTION—hence, what we are looking for are native pure PATH verbs that might serve as a model for the borrowed pure PATH verbs (chapter 9). Therefore, more important than the subcategorization within the MANNER group is the distinction of the whole group of MANNER verbs from the group of PATH verbs. (On difficult cases, see 2.3.3 below.)

2.3.2 PATH

In contrast to MANNER, PATH is a less open category: There is only a limited number of ways in which a FIGURE can move with respect to a GROUND. Wälchli (2001: 300), for instance, captures what he calls kinds of 'cardinal displacement'[29] as in Table 2.2, using Latin prepositions to represent the different subtypes (F stands for FIGURE, G for GROUND): [30]

[29] He also distinguishes minor kinds of displacement, including, for instance, the MEDIUM PATHS: PRAETER (F pass by G), PRO (F pass through G), TRANS (F cross over G) (Wälchli 2001: footnote 5).

[30] Wälchli uses the deictic verbs *go* for the goal paths and *come* for the source paths. Since deixis will not be treated as PATH here, for the reasons explained below, I replaced these verbs with the more general *move* (cf. also Berthele 2004: 100).

TABLE 2.2. Types of cardinal displacement (adapted from Wälchli 2001: 300).

GOAL PATHS		SOURCE PATHS	
AD	F move to G	AB	F move from G
IN	F move into G	EX	F move out of G
SUPER	F move onto G	DE	F move down from G

In the inventory of Old and Middle English motion verbs, these subtypes of PATH will be used as well (see the subcategories of PATH 'towards', 'in', 'up', 'away', 'out', and 'down' in Tables 5.1 and 6.1 and Appendices A and B). Yet the verbs that were found required the additional categories of *back*, *forward*, *along*, *through*, and *around/in circle*.[31]

Note that deixis, or 'viewpoint',[32] will not be treated as a type of PATH, in the sense that the Old and Middle English equivalents of PDE *come* and *go* will not be treated as PATH verbs. This is in accordance with Talmy (1985: 135), who also lists deixis as a separate type of semantic component, which he calls 'direction':

> *Direction* (deictic): This refers to whether the Figure in a Motion event is moving toward or away from the speaker. [. . .] It is found incorporated in verb roots, for example English *come/go* and *bring/take* (Talmy 1985: 135).

However, in later publications, Talmy (2000b: 56–7) defines deixis as one of the components of PATH. For studies of English motion expressions, the major consequence of treating deixis as a kind of PATH or not lies in the different classification of the highly frequent verbs *come* and *go* it entails (although note that PDE *go* is not always used deictically). Again, the literature is not unanimous on this issue: PDE *come* is either treated as a PATH verb (e.g. in Özçalişkan 2004: 85; Reshöft 2011: 300, or, for both *go* and *come*, Beavers et al. 2010: 350) or not (e.g. Cardini 2008: 563; 2010: 1454; Rohde 2001: 352; Choi & Bowerman 1991: 87; Verkerk 2013: 175).

For the present study, the decision not to treat the Old and Middle English equivalents of PDE *come* (and *go*) as PATH verbs is motivated by the focus of chapter 9 on the integration of borrowed pure PATH verbs: to investigate this, we need to find out first whether there are native pure PATH verbs that may serve as models for the integration of the borrowed verbs *enter*, *descend*, *mount*, etc. *Come* and *go* are clearly different from the latter, in that they only convey the deictic component of PATH, i.e. they involve the viewpoint taken by the speaker, not PATH with respect to a GROUND element. Therefore, 'deixis should be seen

[31] Cf. also Cifuentes-Férez (2008: 139), who adds the following to Wälchli's path-types: pass/cross G, forwards, back to G/backwards, among others.

[32] 'Viewpoint' instead of 'deixis' is the term proposed by Radden (1995: 431) with respect to PDE *come* and *go* as the more comprehensive notion, including 'both the 'deictic' viewpoint taken by one of the speech act participants and the 'mental' viewpoint taken by the narrator'.

FIGURE 2.3. Viewpoints.

as a part of the *expression* of a motion event rather than part of the motion event itself' (Filipović 2007a: 17). Here is a simple example to illustrate this:

In Figure 2.3, Hatman is moving from the meadow to the forest. Hence, meadow and forest are GROUND elements, Hatman is the FIGURE, the PATH is from the meadow and to the forest. This is the event. It can be viewed from various perspectives. Describing it as 'Hatman is *coming* to the forest/from the meadow' means taking the perspective of Forestman on the right. Meadowman on the left, from his own viewpoint, may describe it as 'Hatman is *going* to the forest', but he may also shift his perspective and say 'Hatman's *coming* to the forest', for instance if he wants to alert Forestman. A narrator who is not present in the situation may describe the scene from all of these different viewpoints (cf. Radden 1995: 431). Deictic verbs, although arguably related to PATH, are therefore quite different from the kind of PATH verbs such as *enter, approach, descend* in that they involve the viewpoint taken by the speaker, rather than the respect in which a FIGURE moves to a GROUND (as in *enter the forest, approach the forest*).[33] The respect in which a FIGURE moves to a GROUND, crucially, is not dependent on viewpoint: no matter from which perspective one looks at the event in Figure 2.3, Hatman is always moving *from* the meadow and *to* the forest. Hence, deictic verbs are not treated as PATH verbs, but as NEUTRAL MOTION verbs in the present study.

Besides, the question is whether the deictic pair represented by PDE *come* and *go* can be simply assumed to have also existed in the same function in

[33] Therefore, the statement about '*enter, pass, cross, ascent, descent* [sic]' etc. in Haselow (2012: 651), that '[a]ll of these verbs implicitly express different deictic coordinates, representing a systematic distinction between movement from, toward, or around the origo' is a misrepresentation: These verbs precisely do *not* involve deixis, and do not represent motion with respect to an *origo*, but with respect to a GROUND. (If they were about origo, the verb would have to change when the origo changes, such as *come* and *go* in the example above. Yet, no matter from where you view, say, the descent of a FIGURE—whether it descends toward you or away from you—you can always use the verb *descend*, since the verb does not depend on the origo.)

earlier stages of the language.[34] Old English in particular shares many characteristics with Modern German, and Rauh's comparison of PDE *come/go* with German *kommen/gehen* comes to the conclusion that different from the Modern English pair, the deictic features of the German pair can be neutralized if there is an adverbial indicating the GOAL (for *kommen*) or the SOURCE (for *gehen*) (1981: 67). Thus, for instance, it is possible to use *kommen* 'come' in sentences such as *Ich bin noch nie nach Japan gekommen* 'I never got [lit. 'came'] to Japan' (1981: 60).

In these two sections, then, we have seen the different subtypes of MANNER and PATH as they will be used in the present study. The next section will address the problem of how to treat verbs such as PDE *climb* or *sink*, which unite MANNER and PATH information.

2.3.3 Borderline cases—MANNER and PATH verbs as prototype categories

This section will discuss the problems in separating MANNER and PATH verbs. It will arrive at the conclusion (in accordance with, e.g. Berthele 2006: 58, similarly Rohde 2001: 279–80, 348) that MANNER verbs and PATH verbs should be seen as prototype categories with some degree of overlap.

Many MOTION verbs are rather straightforward to categorize, even for historical stages of the language—ME *fusen* 'to go speedily', *slinken* 'to creep, move stealthily', *swimmen* 'to swim', for instance, are all uncontroversial MANNER verbs. Yet, there are others which unite MANNER and PATH information, such as, for instance, PDE *climb*. As discussed by Fillmore (1982: 32), this verb is prototypically used for events which unite both 'clambering' (MANNER) and 'ascending' (PATH) (such as 'a monkey clambering up a flagpole'). However, it is also used for events where either component is lacking, such as for 'a monkey clambering down a flagpole, or clambering horizontally in the rafters of a warehouse' (no PATH 'up'), or for 'a snail ascending a wall, in the way a snail usually moves' (no MANNER 'clambering').

For a verb such as *escape*, this is slightly different, since both MANNER and PATH components are usually present in situations described by *escape*. While the PATH information is 'out' or 'away', the MANNER information is that the FIGURE moves away or out against someone else's will, against some obstacle, or from undesirable circumstances, often also with danger or fear involved, as in (39).

(39) East Germany lets 7,000 take the train west: More refugees join queue to **escape** through Prague and Warsaw embassies after cave-in by Berlin (BNCweb, A1P 26)

---

[34] A related question is, of course, what the exact functional equivalent of PDE *go* is in Old English—OE *gān* is still very much restricted to movement on foot (see, e.g., Weman 1933: 175; Storjohann 2003: 164–7). The most likely equivalents for PDE *go* are probably OE *faran*, *fēran*, or *gewītan* (cf. also Ogura 2002: 25–9).

Note that the MANNER component cannot be ascribed merely to the context (such as the description of the figures as *refugees* as well as the general setting of East Germany in (39)); it remains even when the GROUND is a place commonly seen as pleasant: *He escaped from the luxury spa / from paradise* does not only mean that he left the spa, or paradise, but also implies, for instance, that he did not like it there, or that something or somebody tried to keep him there. Thus, *escape* is not a pure PATH verb (as opposed to French *sortir* 'to exit', for instance), since it is not used for situations in which only the PATH component is present, while the MANNER component is not. The reverse also holds: Even though the SOURCE–PATH need not be mentioned, such as in *He escaped into the wild*, it is still implied that he escaped from or out of somewhere.

This is similar for other verbs, such as *flee*, *fall*, *sink*, etc., which also unite both MANNER and PATH information: *Flee*, like *escape*, denotes motion 'out' or 'away', but also involves speed, fear, or danger. *Fall* is prototypically 'down' and 'due to gravity, not volitional, not propelled by the figure's own motor pattern'. *Sink* is prototypically 'down', 'in liquid', and 'rather slowly'. So should these be seen as MANNER or as PATH verbs? Some studies respond to this problem by creating an extra category of 'MANNER + PATH verbs' (e.g. Verkerk 2013: 176; Cifuentes-Férez 2007: 122; Zlatev & Yangklang 2004: 167, 178). Most studies, however, stick to the 'pure' categories and therefore (if the specific verbs are discussed at all), these MANNER + PATH verbs are classified quite variably (on the problem of differing categorizations, see also Cadierno & Ruiz 2006: 192–3). Thus, *drop* and *fall*, for instance are seen as PATH verbs by Özçalişkan (2004: 85), but as MANNER verbs by Rohde (2001: 353). Similarly, *escape* and *flee* are seen as MANNER verbs by Özçalişkan (2004: 80), whereas Rohde (2001: 354) identifies PATH as the primary information conveyed by them.

The presentation of the Old and Middle English inventory of motion verbs below (5.1 and 6.2 and Appendices A and B) classifies the 'fall' verbs (such as OE *feallan* 'to fall', *drēopan* 'to drop', or ME *dompen* 'to fall, plunge', *plumpen* 'to fall into water') and the 'flee, escape' verbs (such as OE *flēon* 'to flee', ME *escapen* 'to escape', *skeuen* 'to escape') as subcategories of MANNER verbs (Ma^{flee} and Ma^{fall}), since MANNER information is clearly part of their meaning.[35] Whichever classification is adopted though, it is crucial to bear in mind that MANNER and PATH verbs must be seen as prototype categories. These have typical members, i.e. pure MANNER verbs such as *swim*, *ride*, etc. and pure PATH verbs such as French *sortir* 'exit', *descendre* 'descend', etc. They also have marginal members, and, as we have just seen, there is some degree of overlap (see the illustration in Figure 2.4).[36]

---

[35] Comparability with studies using a different classification is ensured since the specific verbs are given in Tables 5.1 and 6.1 and Appendices A and B (subcategories Ma^{fall} and Ma^{flee}) and also in the results of the usage studies in sections 5.2 and 6.3.

[36] I therefore do not agree with the proposal in Beavers et al. that '[a] verb may lexicalize only one of manner and path' (2010: 334).

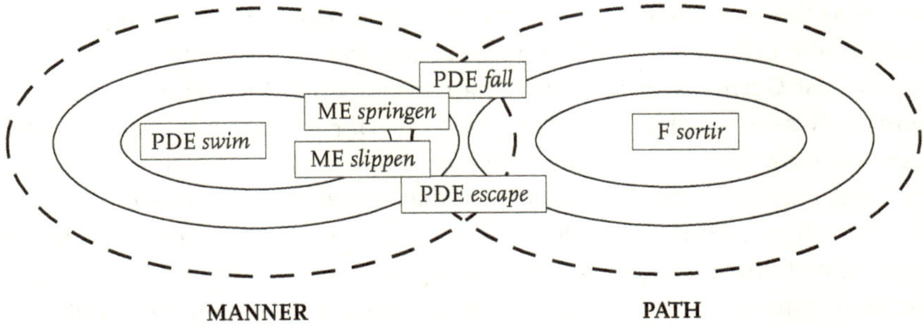

FIGURE 2.4. Overlapping prototype-categories of MANNER and PATH.

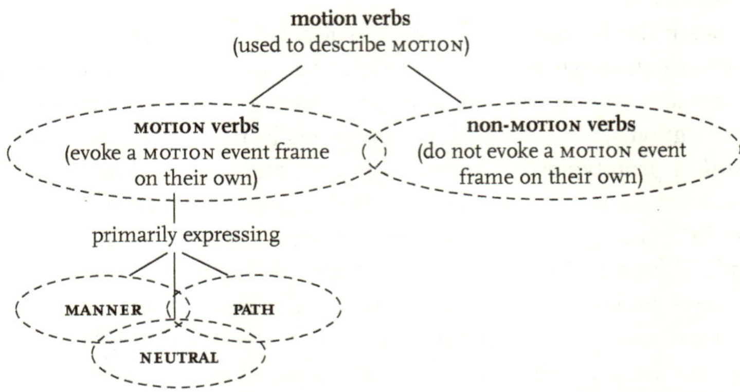

FIGURE 2.5. Types of motion verbs as overlapping prototype categories.

To introduce an additional category of 'MANNER + PATH' verbs would not have solved these problems of overlapping categories: While *fall*, *escape*, etc. would then be prototypical cases of 'MANNER + PATH' verb, the problem of overlap would resurface with verbs such as ME *springen* 'to spring up, bubble out, gush forth, flow' (Ma^{other}) or ME *slippen* 'to move quietly or stealthily; slip away, escape' (Ma^{glide}), which also have a certain 'path preference' (Verkerk 2013: 180): *springen* for upward motion, *slippen* for motion away from a source. In the classification adopted here, these are verbs primarily expressing MANNER—not as prototypical as *swim*, *ride*, but more prototypical than *fall*, *escape*, as illustrated in Figure 2.4, which can be read as a 'zoom' on the overlapping prototype categories of MANNER and PATH in Figure 2.5 below it.

Figure 2.5 is a more realistic rendering of the simplifying overview in Figure 2.2 in that it takes into account that we are dealing with overlapping prototype categories: Overlap does not only exist between the categories of verbs primarily expressing MANNER and those primarily expressing PATH, which we have just discussed. They both also overlap with the category of NEUTRAL MOTION verbs. Here are a few examples: the (often) deictic OE *cuman* 'come' (Appendix A:

NEUTRAL MOTION), for instance, would be placed between the PATH and the NEUTRAL MOTION group (see the discussion above). OE *gangan* (Appendix A: NEUTRAL MOTION) straddles MANNER and NEUTRAL MOTION, because it can mean 'to walk' as in (40a), where it is used contrastively with *rīdan* 'to ride', but it can also be used for any kind of motion, such as in (40b) for the movement of the sun, which certainly does not walk.

(40) a. se þe hors nabbe, wyrce þam hlaforde þe him fore ride oððe **gange**, oðð he ham cume (cf. *Quadr.*: *qui equitabit pro eo uel pedes eat*). (LawVIAs 5, DOE s.v. *gangan*)
'He who has no horse shall work for the lord who rides or walks for him until he [= the lord] comes home.'

b. on ærnemergen þonne seo sunne ærest up **gange** (Lch I (Herb) 111.2, DOE s.v. *gangan*)
'In the early morning, when the sun first goes up.'

Finally, there are verbs whose dictionary entries suggest that they are somewhere in between all three types of MOTION verbs, i.e. between MANNER, PATH, and NEUTRAL MOTION verb: ME *flitten* is a case in point. The intransitive motion senses[37] given by the *MED* include the ones in (41a–d):

(41) a. 'to go, come, depart' (~PATH or NEUTRAL MOTION verb)
Cadwaðlan..to scipe **flutte** mid muchelre uerde. (*c.*1275(?*a.*1200) Lay. *Brut* (Clg A.9) 30503)
'Cadwala [. . .] moved to ship with a great host.'

b. 'to flee or escape' (~MANNER verb (with MANNER and PATH information))
And feste yone fende, þat he not **flitte**. (*a.*1450 *Yk.Pl.*(Add 35290) 392/340)
'And fasten this fiend [=Satan] so that he may not escape.'

c. 'to fall (from the saddle)' (~MANNER verb (with MANNER and PATH information))
Som oute of her sadels **flette** to the grounde. (?*a.*1500(?1458) *Off alle Werkys* (Inscr) 41)
'Some fell out of her saddles to the ground.'

d. 'to be in motion' (~NEUTRAL MOTION verb)
For litelle harmeþ eny schot þat touchiþ on a thing þat meveþ & **flitteþ**. (*a.*1450(1408) **Vegetius(1)* (Dc 291) 97b)
'For any shot that touches a thing that moves and is in motion does little harm.'

(All from *MED* s.v. *flitten*)

---

[37] *Flitten* is also used as a verb of caused motion, e.g. 'to transport something', 'to expel someone', see *MED* s.v., sense 1.

What is given in brackets in these examples is the subcategory of MOTION verb which would be indicated by the respective attestation and *MED* subsense: These are not mutually compatible. For instance, the MANNER information suggested in the *MED* subsense in (41b) is 'flee, escape', which fits the attestation from the *Harrowing of Hell* in the *York Plays*, where Jesus asks Michael to bind Satan in chains so that he cannot escape (41b). In other attestations, *flitten* is used in the collocation *flen and flitten*, which equally suggests that it means 'escape' (see *MED*, s.v. *flitten*). In (41a), by contrast, the FIGURE is moving to ship with the intent of going to battle, so there is definitely no sense of 'fleeing, escaping' involved. The same is true for (41c–d). As we cannot tell which of the senses is primary, and as the only thing that the occurrences of *flitten* in (41a–d) have in common is that they all mean 'to move', *flitten* is classified in the group of 'NEUTRAL MOTION verbs'.

Note that Figure 2.5 above also indicates overlap between the group of non-MOTION verbs (i.e. verbs not evoking a MOTION event frame on their own) and the group of MOTION verbs (i.e. MANNER, PATH, and NEUTRAL MOTION verbs), an overlap that is due to our diachronic perspective. An example of a verb that would belong to this area of overlap is ME *travailen*: In Middle English, it is used to mean, on the one hand, 'to work', 'to put forth great effort, take pains', 'to put up a struggle', 'to suffer pain, endure hardships' (*MED*, senses 1a, 2a, 3, 4a, 4c), and, on the other hand, 'to travel' (*MED*, sense 5). As discussed in section 2.2.3, we cannot tell exactly when the 'travel' sense became a decontextualized meaning of the verb and stopped being contextual, i.e. conveyed through the interplay of construction, context, and verb semantics.

In sum, however, the problem of overlapping categories outlined above is not essential for the present study, particularly for our question whether the PATH verbs that are borrowed in the Middle English period (*entren, ishen, mounten, descenden*, etc.) are truly new to the native system or not. This question entails that we are looking for pure PATH verbs in the native inventory of MOTION verbs. If the language already had such pure PATH verbs, these could have served as a model in analogy to which the borrowed PATH verbs may be used, semantically and syntactically. If there are no pure PATH verbs in the native system, there would have been no such model, and the question therefore is in which way the borrowed PATH verbs are integrated (chapter 9). With regard to this question, therefore, it does not really matter in which category exactly the verbs that carry both MANNER and PATH information, such as ME *escapen, fallen*, etc., are classified. Those in the MANNER and in the NEUTRAL MOTION set are not candidates for pure PATH verbs anyway. Those in the PATH group will be analysed in detail in section 5.1.3 to determine whether they can be seen as pure PATH verbs or not.

CHAPTER 3 | Problems with historical data

THE PROBLEMS ENTAILED IN exploring the questions addressed in the present study with respect to historical stages of a language centre around two main points. The first one is the textual evidence (3.1), which is of course more limited than for present-day languages (3.1.1), and, crucially, does not include many texts from the registers and genres that would be most informative for our purpose (3.1.2): Studies that apply the motion verb typology to languages of the present day usually either use spoken material or non-translated novels (see 2.1.2). Spoken material is obviously not available from historical stages of the language, and also non-translated narrative fiction is rare, particularly for Old English.

The second issue is meaning (3.2): Section 3.2.1 will address general problems of establishing the exact meaning of a verb from an earlier stage of a language. Section 3.2.2 will discuss the more specific question of how to treat the meanings given in dictionary entries, particularly in very elaborate ones such as the *MED* or *DOE*, which tend to list each and every contextualized meaning in which a verb is attested (for contextualized and decontextualized meaning, see 2.2.3).

3.1 *The textual evidence*

In applying the motion verb typology to historical data, one obvious problem lies in the limits of our textual evidence. In a way, this is a problem of quantity (3.1.1), but, far more so, a problem of register (3.1.2).

3.1.1 Quantity

The extant record of Old English consists of a good three million words of text (according to the word count of the *Dictionary of Old English Corpus*, which contains at least one manuscript of each extant Old English text).[38] This is

[38] http://doe.utoronto.ca/pages/pub/web-corpus.html.

substantial compared to other languages of the same period. Still, if we bear in mind that many Present-Day English motion verbs are very rare,[39] and assume that this will have been similar for earlier stages of the language, we can expect that even an exhaustive list of Old English motion verbs will never be as long as those for living languages of the present day, simply because many words will be unrecorded.

For the Middle English period as a whole, the problem of quantity is less acute: Denison & Hogg (2006: 35) mention that the largest electronic collection of Middle English texts consists of as many as nineteen million words, and even this does not contain all the Middle English material that is available. But it is crucial to keep in mind that the bulk of this material is from the second half of the Middle English period (i.e. *c.*1350–*c.*1500). The early Middle English period is seriously 'underattested' (cf., e.g., Nevanlinna et al. 1993). This is particularly so for the time right after the Norman Conquest, when 'for a century or more, there was virtually no new writing in English' (Laing & Lass 2006: 419): The few texts extant from this early period are basically still Old English, because they are either copied from[40] or modeled on[41] texts from before the Conquest. Other written material from this time is in Latin or French. English only gradually gained back ground in the various domains of written discourse in the course of the Middle English period, and it is only toward the second half of the fourteenth century that it is widely established as a written language again (see also the Helsinki and Penn-Helsinki corpora of Middle English, where the number of words for 1350–1420 (ME3) is twice or more as high as the one for 1250–1350 (ME2)).

Thus, while there is a very large number of texts from the later Middle English period, for early Middle English—similar to what has been sketched for Old English above—it is likely that rarer motion verbs are not attested at all, due to the limited extant material. But we have to make use of what there is, and the inventory of 189 motion verbs from the Old English (see section 5.1 and Appendix A) and 465 from the Middle English period (see section 6.2 and Appendix B) is certainly not negligible.

3.1.2 Register and genre

What is more important than size, though, is the question of register. Recall that the motion verb typology is based on what the 'characteristic' expression of motion is in a language (see 2.1.2). Talmy crucially defines 'characteristic' not

[39] Cf. the frequencies per million words of the following PDE verbs in the 100-million word *BNCweb* (note that these frequencies also include attestations where the respective verbs are not even used as motion verbs): *canoe* (0.22), *careen* (0.1), *clomp* (0.1), *frisk* (0.37), *mosey* (0.11), *punt* (0.36), *scoot* (0.37), *stampede* (0.35). These verbs thus clearly belong to what Slobin (1997: 459) calls the 'second tier' of the MANNER verb lexicon, namely the rarer and more expressive MANNER verbs.

[40] An example is *The Old English Herbarium and Medicina de Quadrupedibus*, surviving in a manuscript from *c.*1150 (Lewis et al. 2007: 7).

[41] An example are the continuations of the *Peterborough Chronicle*.

only as 'frequent', but also as 'colloquial in style, rather than literary, stilted' (2000b: 27). Yet, as noted, for instance, by Gneuss (1991: 47), it is obviously 'next to impossible to say [. . .] what colloquial Old English must have been like' (see also, e.g., Kastovsky 1992: 351), and the same is also true for Middle English (see Blake 1992: 532–6), although it is at least possible to identify an informal style characterized, for example, by repetitions, ellipses, parataxis, and loose structures in general (Blake 1992: 534; Schlauch 1952). Overall, the problem is that the evidence, particularly for Old English, is strongly biased towards certain registers and genres (see 3.1.2.1) and that those registers and genres that the present analysis is most interested in are underrepresented. This section will discuss this problem with respect to the different analyses undertaken in the present study and will end with an illustration of the register problem using the example of the semantic changes of OE *wealcan*/ME *walken*.

3.1.2.1 Relevance for the usage studies

The focus within the motion verb typology on what is 'colloquial, rather than literary, stilted' is most relevant for the usage studies, i.e. for the token-based analyses of Old and Middle English motion expression (sections 5.2 and 6.3 below), which address the question of which kind of verbs and structures are *frequent* in talking about MOTION. For this aim, the crucial question is of course which texts to choose for the analysis—which texts can be seen as representative of 'colloquial, rather than literary, stilted' language? This is most difficult for the Old English period: Old English poetic texts are not at all useful for this purpose, because they are characterized by a specific vocabulary and style (e.g. Gneuss 1991: 47–9; Mitchell & Robinson 2006: §247; Kastovsky 1992: 351–5). Old English prose mainly falls into the following 'genres' (Mitchell & Robinson 2006: §250–1):

(a) The *Anglo-Saxon Chronicles*
(b) Translations [traditionally attributed to] Alfred and his circle (e.g. Pope Gregory's *Cura Pastoralis*, Boethius' *De Consolatione Philosophiae*, Augustine's *Soliloquia*, Bede's *Ecclesiastical History*)
(c) Homiletic writings (e.g. *Catholic Homilies*, *Lives of Saints*, *Homilies* of Wulfstan)
(d) Other religious prose (translations of portions of the Bible and the Benedictine Office)
(e) Prose fiction (*Apollonius of Tyre*, *Alexander's Letter to Aristotle*, *The Wonders of the East*)
(f) Scientific and medical writings
(g) Laws, charters, and wills

An ideal text for the token-based analysis of motion verb usage would be one that meets the following two desiderata: (i) it should not be 'literary, stilted', and (ii) its subject matter should involve a sufficient number of MOTION events.

Unfortunately, for this purpose, none of the extant Old English texts seems to be the prime choice. Neither of the desiderata hold for the genres in (f) and (g). Generally, translations are also not an ideal choice because they are likely to be influenced by the Latin original; hence the genres in (b), (d), and (e)[42] are not optimal either.[43] So it was basically through discarding the less suitable texts that the choice for Old English fell on excerpts from the *Anglo-Saxon Chronicles* (a), the *Voyages of Ohthere and Wulfstan* (from group b), and four of Ælfric's *Lives of Saints* (from group c). In the *Helsinki Corpus of Old English*, these are all classified as belonging to the 'prototypical text category' of 'non-imaginative narration'.[44]

For Middle English, the choice fell on three of Chaucer's *Canterbury Tales* (the *Miller's*, the *Reeve's*, and the *Knight's Tale*, late 14th century), whose plots involve many motion events and which, though written in verse, represent a prime example of Middle English imaginative narrative. For the sake of comparability with the *Anglo-Saxon Chronicles*, the motion descriptions from excerpts of *Gregory's Chronicle* (15th century, from *PPCME2*) were analysed, because they both belong to the same text type. A Middle English saint's life was also chosen, the *Life of Saint Edmund Rich of Abingdon* from the *South English Legendary* (late 13th century), to compare to the saints' lives analysed for Old English. The texts are discussed in more detail in 4.2.2.

3.1.2.2 Relevance for the inventory studies

The previous subsection has addressed the problem of finding texts that are not 'literary, stilted' with respect to the analyses of usage in sections 5.2 and 6.3. Yet the question of genre and style also bears some relevance on the analysis of the Old and Middle English inventory of motion verbs (sections 5.1 and 6.2 and Appendices A and B). The inventories were gathered on the basis of the *TOE*, the *DOE*, and *BT* (for Old English) and the *MED* (for Middle English). Their compilation concentrates on the question of which verbs are *possible* in talking about MOTION, rather than on the question of which verbs are *frequent* (as in the usage studies). One objection with regard to these inventories may be that they might contain a fair number of verbs that actually pertain to a

---

[42] Also the texts in category (e) are translations from Latin (*Historia Apollonii Regis Tyri*; *Epistola Alexandri*; *De Rebus in Oriente Mirabilibus*).

[43] This caveat is supported by Cappelle (2012), who shows that English texts translated from French have fewer MANNER OF MOTION verbs compared to autonomous English texts (and compared to translations from German, for that matter). We will see in 7.1 that Latin, although structurally satellite-framing, does not make extensive use of MANNER verbs, particularly in later periods (Schøsler 2008), from which most exemplars of the translated texts here date (Beda, Gregory the Great, Orosius, Augustine; the *Historia Apollonii* is from the fifth or sixth century (Ziolkowski 1996: 551); the *Epistola Alexandri* probably dates from the early fourth century, though the earliest manuscript is from the ninth century (Rypins 1924: xxx–xxxii); the *Wonders of the East* builds to a large extent on sources from the sixth century (Rypins 1924: xlv)).

[44] The category 'imaginative narration' would be more similar to the types of texts usually studied in the framework of the motion verb typology, but the texts from this category in *HC* are all translated from Latin (see Mitchell & Robinson's category '(e) prose fiction').

specific register, e.g. to a distinctive poetic vocabulary, which is not representative of the language in general. The following paragraphs will discuss this for Old and Middle English in turn.

As mentioned above, some of the vocabulary in Old English poetry is very distinct from the one found in Old English prose, and it is 'presumably not current in speech' (Godden 1992: 498; see also Gneuss 1991: 47–9; Mitchell & Robinson 2006: §247; Kastovsky 1992: 351–5). This specific vocabulary is strongly tied to the exigencies of alliterative verse, which usually requires two or three stressed alliterating syllables per line (see (42) below). This kind of poetry therefore needs a large stock of synonyms[45] for recurrent concepts (such as 'man', 'ship', 'battle', 'sea', etc.), which is related to the frequent use of metaphorical compounds and phrases, formulae, and archaic words (Mitchell 1995: §206; Kastovsky 1992: 352–4). Furthermore, Old English poetry makes heavy use of variation as a rhetorical figure (e.g. Kastovsky 1992: 352). This constitutes an additional factor involved in the necessity of a large stock of synonyms. Since several of the words that are specific to poetry in Old English are equally present in the poetry of Old Saxon, Old High German, and Old Norse, this specific alliterative vocabulary must have been common Germanic tradition (Gneuss 1992: 48).

For the purpose of the present study, though, this undeniably fundamental difference in Old English poetic and prose vocabulary is not crucial, as we are interested in verbs. Old English poetry is 'essentially nominal rather than verbal' (Godden 1992: 501): The specifically poetic words are predominantly nouns and adjectives; these are also the words that normally bear primary stress in scansion. To illustrate this nominal character of Old English poetry, here is an example from *Beowulf* in which a journey is described. The syllables bearing primary stress are printed in bold:

(42)

Ofer**eo**de	**þa**	**æþ**elinga	**bearn**	
over-went	then	princes'	son	
stēap	**stān**liðo,	**stī**ge	**near**we,	
high	stone-cliffs	paths	narrow	
enge	**ān**paðas,	**un**cūð	ge**lād**,	
narrow	single-paths,	unknown	way	
neowle	**næs**sas,	**nic**orhūsa		**fel**a.
low	headlands	watermonster-houses		many

'Then the son of princes set off across towering stone-cliffs,
narrow defiles, confined single-paths, unexplored passages,
steep headlands, many a lair of watermonsters.'
(*Beowulf*, 1408–11, from Scragg 1991: 62–3 (his translation))

[45] It can only be conjectured whether there were really semantic differences between these near-synonyms, or whether it was rather a mere question of alliteration, as suggested by Niles (1981), who is quoted in Godden (1992: 500): 'A *guðbyrne* is a byrnie [corselet] which alliterates in [g]; a *heaðobyrne* is a byrnie that alliterates in [h]'.

Although the first word is indeed a finite verb here (*oferēode* 'over-went', i.e. with prefixed PATH satellite *ofer*), it is a highly general one—OE *ēode* is the suppletive preterite of OE *gān* 'to go' (see Appendix A, neutral MOTION verbs). The verb does not seem to add anything to the imagery created by these lines. The imagery is brought about rather by the complementation of the verb: the six noun phrases that follow (*stēap stānliðo . . . nicorhūsa fela*) and contain typically poetic vocabulary such as *stānliðo, ānpaðas*, and the hapax legomenon *nicorhūs* (Scragg 1991: 65). This is typical for Old English alliterative verse, where verbs are not very prominent in the first place (see also Mitchell 1996: §460). For the present study, this means that there is no need to be overly concerned about alliterative vocabulary distorting the inventory of motion verbs in section 5.1—in the 189 Old English motion verbs collected in it (see Table 5.1 and Appendix A), there are only ten, i.e., a mere five per cent that are marked as 'restricted to poetry' by the *TOE*.[46]

The status of verbs in alliterative poetry, though, changes in Middle English, which 'develops a fine array of poetic verbs and gives them great emphasis' (Godden 1992: 501; see also Blake 1992: 512, who notes that '[t]his makes the poetry seem less compressed than its Old English counterpart'). The changed status of verbs is reflected in the lists of typically alliterative words for Middle English, which, different from similar lists for Old English, usually feature a whole range of verbs for 'go, move' (cf. Matsunami 1985).

Yet the issue we are concerned with here, namely whether the motion verbs which frequently occur in alliterative verse should therefore be considered as 'literary, stilted', is not quite clear: First of all, even in a wider definition of 'alliterative verse', such as the one by Hanna (2002),[47] the *MED* entries for many of the verbs in question also list attestations (in the motion sense) in contexts where no alliteration is involved at all. I checked the quotations given in the *MED* for motion verbs usually labelled as 'typically alliterative' (Matsunami 1985; Blake 1992: 523–4) and found that for many of them, the *MED* also has attestations in motion uses from non-alliterating contexts. Table 3.1 lists these verbs with the non-alliterative texts in which they are attested in brackets.

Secondly, particularly in view of the fact that there was no traditional set of Old English alliterative verbs that might have been transmitted into Middle English alliterative poetry,[48] we cannot rule out the possibility that these verbs were recruited into the Middle English alliterative vocabulary from 'colloquial', spoken language. Hanna (2002) and Turville-Petre (1977) seem to suggest this, stating

46 They are marked (p) for 'poetry' in Appendix A, like they are in the *TOE*: OE *lǣcan* 'to move quickly, spring, leap', *scēon* 'to go quickly, fly', *scūdan* 'to run, hurry', *snēowan* 'to proceed, go, come, hasten', *snyrian* 'to go quickly, hasten', *stincan* 'to spring, leap, move rapidly', *crūdan* 'to crowd against, press forward', *treddian* 'to tread, step, walk', *pæþþan* 'to tread (a path), to traverse', *þrægan* 'to run, proceed in a course'.

47 Hanna (2002) argues against a clear-cut distinction between alliterative and other forms of poetry, since alliteration is also found outside the traditional long-line—for instance combined with end-rhyme.

48 Note also that from the ten Old English motion verbs flagged as 'poetic' by the *TOE* only two survive into Middle English; *crūdan* (> ME *crouden* 'push, jostle, crowd, press on, hurry') and *treddian* (> ME *treden* 'go by foot, walk').

TABLE 3.1. Motion verbs typical for alliterative verse with non-alliterative attestations in the *MED* (MED, s.vv.).

airen (?*a*.1425 *Mandev.(2)*)	*sechen* 'to go' (*c*.1430(*c*.1386) Chaucer *LGW*; ?*c*.1475 *Direct. Sailing* in *Hak.Soc.79*; *a*.1500(?*a*.1400) *Firumb.(2)*)
atlen (*a*.1500(?*a*.1425) *Lambeth SSecr.*; *a*.1425 *Hayle bote*)	*shaken* 'to go' (c1330(?c1300) *Bevis*; -?-(?*a*.1500) *Hunt.Hare*)
bouen 'to go' (*a*.1425(*c*.1395) *WBible(2)*)	*shouven* (*c*.1300 *SLeg.Brendan*; (*a*.1422) Hoccl. *Hen.V.& KG*; *a*.1425(*c*.1385) Chaucer *TC*; *c*.1440 *PLAlex.*)
cairen (*a*.1500(*a*.1400) *Ipom.(1)*)	*straken* (*c*.1450(1369) Chaucer *BD*)
chesen (*c*.1330(?*a*.1300) *Tristrem*; *a*.1500(?*c*.1400) *Gowther*)	*striken* (?*c*.1200 *Orm.*; *c*.1330 *7 Sages(1)*; ?*a*.1425 *Const.Masonry(1)*)
coveren (*kever*) (*a*.1450 *Parton.(1)*)	*sweien* (*a*.1325 *SLeg.*)
dressen (*c*.1460(?*c*.1400) *Beryn*)	*tournen* ((*c*.1390) Chaucer *CT.Pard.*; *c*.1475(?*c*.1425) *Avow.Arth.*)
driven (*a*.1425(*c*.1385) Chaucer *TC*; ?*a*.1425(?*c*.1350) *NHom.(3) Pass.*)	*thringen* (*c*.1440(a1400) *Eglam.*; *c*.1450(*c*.1375) Chaucer *Anel.*)
founden (*a*.1500 *Eglam.*; *a*.1500 *Amadace*)	*waden* (*c*.1350 *NPass.*; *a*.1500 **Sidrak & B.*)
gliden 'to go' ((*c*.1390) Chaucer *CT. Th.*; *c*.1440(?*a*.1400) *Perceval*)	*winnen* (*c*.1425(*a*.1420) Lydg. *TB*; *a*.1425 *NPass.*)
halen (*a*.1425 * York *MGame*)	*winden* (*c*.1475(?*c*.1425) *Avow.Arth.*; *a*.1500 *Octav.(1)*)
helden (*c*.1330(?*c*.1300) *Guy(1)*)	
louten (?*c*.1250 *Somer is comen &*)	
meten (*c*.1300 *SLeg.And.*; *a*.1425(*a*.1400) *PConsc.*)	

[alliterative poetry's] lexical acquisitiveness, particularly its reliance upon native dialectal vocabulary and Scandinavianisms (always the result of 'tulkish', conversational, transmission into English), marks it as a colloquial amalgam removed from Latinate stylistic concerns (Hanna 2002: 502).

It appears that quite a number of the Scandinavian loan-words that entered the alliterative poetic vocabulary were taken from the spoken language. They were northern dialect words (Turville-Petre 1977: 75).

Examples for Scandinavian loans among the typically alliterative motion verbs are, for instance, *skelten* and *trinen* (see Table 6.1 and Appendix B: NEUTRAL MOTION verbs). Turville-Petre concedes, however, that

[i]t is difficult at this distance to gauge the effect of introducing dialectal elements into alliterative verse that was itself, in the main, regionally based. Some words were presumably used so regularly in alliterative verse that they became poetic words, shaking off their former associations and taking on connotations they had not earlier possessed (Turville-Petre 1977: 75).

So there seems to be no conclusive, final answer to the question whether those motion verbs that are more frequent in Middle English alliterative poetry than in other texts should be seen as 'literary, stilted' and whether they should therefore not be in the focus of the motion verb typology. Since they are taken to have come from somewhere 'outside' the traditional poetic vocabulary, perhaps from regional dialects, and since most of them are not entirely restricted to alliterative contexts (see Table 3.1), they will be included in the analysis of the Middle English motion verb inventory (section 6.2 and Appendix B). Even if these verbs were entirely poetic and not found outside alliterative poetry, they would not seriously skew the inventory of Middle English motion verbs: The present study found 465 verbs used to express motion in Middle English, and the lists of typically alliterative motion verbs consist of only around 30 to 40 words.

3.1.2.3 Illustration: *walk*

The consequence of the general problem that only a limited number of registers is available in the textual evidence, particularly for Old English, shall be illustrated by the example of *walk*. The Old English verb *wealcan* means 'to roll, toss' and is used for the rolling sea, as in (43a), or for thoughts, as in (43b) (see Table 5.1 and Appendix A: Ma^{other}):

(43) a. þa geseah he an scip ut on þære sæ swa swiðe torfigende fram þan **wealcendum** sæs yðum (LS 9 (Giles) B3.3.9)
'Then he saw a ship out on the sea tossing so heavily because of the rolling sea waves.'

b. Ða ingeðoncas ðe **wealcaþ** in ðæs monnes mode (*quando cogitationes volvuntur in mente*) (Past. 21; Swt. 155, 22, BT s.v. *wealcan*, sense IIb.2 'of other movement: metaphorical: of thought')
'The thoughts that roll in the man's mind.'

In Middle English (see Table 6.1 and Appendix B, Ma^{other}), this sense still exists (44a), but *walken* is also used for motion events in which no literal 'rolling' or 'tossing' can be involved, such as in (44b–c).

(44) a. þe grete wawes That as wyndes and wedres **walkeþ** aboute. (*c*.1400 *PPl.B* (Trin-C B.15.17) 8.40, MED s.v. *walken*, sense 1 'to roll, toss, toss about etc.')
'The great waves, which toss about like winds and weathers.'

b. Gij as pilgrim ȝede; He **welke** about wiþ glad chere Þurch mani londes fer & nere (*c*.1330(?*a*.1300) *Guy(2)* (Auch) p.416, MED s.v. *walken*, sense 2a 'to wander, travel, journey')
'Guy went as a pilgrim; he wandered about happily through many countries, near and far.'

c. Elpes arn in Inde riche . . . Hu he resteð him, ðis der, ðanne he **walkeð** wide, herkne wu it telleð her. (*a*.1300(*a*.1250) *Bestiary* (Arun 292) 505, MED s.v. *walken*, sense 3a 'to go on foot, walk, etc.')

> 'Elephants are in the realm of India [. . .] How this animal rests itself when it walks far: listen to what it says here.'

There has thus been a considerable semantic change of *walk* between Old and Middle English. Traugott & Dasher (2002) stress that semantic change is a product of language use: A speaker innovates a new use of a word by drawing on invited inferences, which can become generalized when adopted by the speech community, until they may eventually become the (or one of the) 'coded meanings' of the word (see 2.2.3). Mapped onto *walken* this would mean that a speaker uses the verb, whose coded meaning is 'to roll', to refer to 'travelling', 'going on foot'—probably first for expressive reasons. That the meaning 'to travel', 'to go on foot' is intended rather than 'to roll' will be evident from the context, i.e. an invited inference—in (44b), for instance, a reading in which Guy literally 'rolled, tossed' through many countries can be ruled out. If this use of *walken* catches on and is also adopted by other speakers, it will become a generalized invited inference, although the coded meaning of 'to roll, toss' will still be dominant. In the case of *walk*, this earlier coded meaning 'to roll, toss' must have become less frequent, and the former inference became a decontextualized meaning. Of course, such usage of a verb whose meaning actually is 'to roll' is only possible in the first place when MANNER verbs may be used to express MOTION anyway.

How does the semantic development of *walk* relate to the question of register and textual evidence? The *OED3* entry for *walk* points out that the senses 'to roam about', 'to travel', 'to go on foot' might represent an Old English colloquial usage (i.e. probably an invited inference generalized in certain registers):

> It is remarkable that to the end of the Old English period the primary sense of the verb [*walk*] (strong and weak) is 'to roll', and that from the beginning of the Middle English period it is 'to move about, travel'. The explanation of this apparently sudden change may perhaps be that the latter sense had, in fact, arisen in Old English as a colloquial usage, and as such was not deemed fitting to be used in writing until the changed literary circumstances of the Middle English period. The development of sense appears to have been from 'to roll, fluctuate, move back and forth' to 'to move to and fro, roam about, wander here and there' to 'to journey, travel, go on foot' (OED3, s.v. *walk*).

As the senses 'to travel', 'to roam about', and 'to walk' probably have not suddenly arisen out of nowhere, it is likely that *walk* had already been in use in these meanings in more colloquial Old English registers which have not come down to us. This illustrates that the registers that would perhaps be the most interesting for the present study are often not visible in the textual evidence. Yet, semantic changes like the one of *walk* might thus shed a little more light on what was colloquial usage in Old English.

To conclude, section 3.1 has discussed various problems ensuing from the limited textual evidence. The most important one is that, particularly for Old English, only certain registers are attested, which is why it is not possible to tell what colloquial Old English must have been like. Consequently, several Old English colloquial motion verbs must also be lost to us. It is only through the resurfacing of a verb like ME *walken* in the coded meaning of 'to travel, walk' that one may deduce that OE *wealcan* 'to toss, roll' was probably used colloquially for human motion on foot as well.

Furthermore, this section has also addressed the issue of words that are peculiar to alliterative, poetic registers, and which are not found in other genres. We have seen that for the purpose of gathering an inventory of Old English motion verbs, this issue is of lesser importance, because it is nouns and adjectives, far more than verbs, which distinguish the Old English poetic vocabulary from the prose vocabulary. In Middle English, by contrast, the typically alliterative vocabulary also includes verbs. These, however, apparently do not stem from a poetic tradition, but have been newly acquired into the alliterative vocabulary. Many of them are also attested in motion uses outside alliterative contexts (see Table 3.1). Even if these typically alliterative verbs do include a few exclusively poetic verbs, this is not problematic for the analysis of the inventory, which aims at exploring which verbs were *possible* to describe motion in Middle English (section 6.2). After all, what is possible also includes what is possible in poetry. Of course, the analyses of usage in 5.2 and 6.3, by contrast, will not include alliterative poetry, since their focus is on what is *frequent*.

3.2 *Meaning*

As argued in detail in 2.2.3, the present study is based on the assumption that a verb has a frame-semantic decontextualized, coded meaning. In every usage event, this coded meaning is specified in interaction with the co- and context (including the argument-structure construction in which the verb is used) as a contextualized meaning. The classification in Appendices A and B aims to categorize the verbs as MANNER, PATH, NEUTRAL MOTION, or non-MOTION verbs on the basis of their coded, decontextualized meaning.

For many verbs, even without native speaker intuition, this categorization does not pose any problem. Thus, OE *crēopan*/ME *crepen* 'to creep, crawl, etc.', or OE *rōwan*/ME *rouen* 'to row, go by water, etc.' undoubtedly belong to $Ma^{slow/impeded}$ and $Ma^{air/fluid}$ respectively. Yet, for other verbs, this is not as straightforward. Semantic categorization always runs the risk of being subjective to some degree, as we have seen in 2.3.3 in connection with the heterogeneous classifications of verbs such as *flee* or *drop* in the literature. With respect to more remote historical stages of a language, categorization is even more difficult, since we lack native speaker intuition and since the number of attestations which we can use to capture the meaning of a word is often very small. This problem comes in two guises. For one thing, it is simply often difficult or even impossible to precisely gauge the meaning of a verb (3.2.1). For another, the question is how to treat the meanings given in the dictionary entries, particularly in very elaborate ones such as those from the *MED*, the *DOE*, and *BT* (3.2.2).

3.2.1 General problems

The problem of not being able to determine exactly what a verb means is certainly most pronounced for verbs that have only one attestation, such as OE *cleacian* in (45a), which the *DOE* gives with a question mark as '?to hurry ?to proceed' (see Appendix A, Ma[speed]). A Middle English example of such a hapax legomenon is *rothelen* (45b), whose meaning '?to go uncertainly, walk blindly' (*MED*) equally comes with a question mark (see Appendix B, Ma[slow/impeded]).

(45) a. and he þanon **cleacode** swiðe earhlice to porte (LS 34 (Seven Sleepers) 446, DOE, s.v. *cleacian*)
'And thence he ?hurried very timidly to town.'

b. Uch on **roþeled** to þe rest þat he reche moȝt. (*c*.1400(?*c*.1380) *Cleanness* (Nero A.10) 890, MED s.v. *rothelen* (v.2))
'Each one ?went uncertainly to the rest that he could reach.'

But the meaning of more frequent verbs is also often hard to pin down. A good example for such a verb is OE *scrīþan*. *Scrīþan* is given in Clark Hall as 'to go, move, glide'. Bosworth–Toller has the meanings 'to go, take one's way' (I), 'to go hither and thither, go about, wander' (II), and furthermore lists the FIGURES with which the verb is used: ships, clouds, heavenly bodies, light, and time (III–V). The *OED*'s entry has '[t]o go, pass; in Old English also, to glide, creep; to wander' (still *OED1* from 1911). Cognates of *scrīþan* are German *schreiten* 'to tread, stride, step, pace' and Icelandic *skríða* 'to creep, crawl; to glide' (Cleasby-Vigfusson, s.v. *skríða*). In the context of how to translate its occurrences in *Beowulf* into Modern English, Mitchell (1995: 297–8) describes the complications in capturing the meaning of *scrīþan* as follows:

> The poet uses the verb *scrīþan* four times—of hellish monsters (l. 163), of shadows (l. 650), of Grendel, who is both a hellish monster and a *sceadugenga* 'shadow-goer' (l. 703a), and of the dragon (l. 2569). The word seems to imply smooth and graceful movement (it is used elsewhere of the sun, clouds, and stars, of a ship skimming over the sea, and of darting salmon in a pool) and an element of mystery (other poets use it of the coming of May, of the beginning and ending of the day, and of the gradual passing of human life). In *Beowulf*, there is also a suggestion of menace and danger which is echoed in other poems, where the word refers to the spread through the body of a disease which could be cancer and to flames raging unchecked.

And in the footnote to *Cōm on wanre niht scrīðan sceadugenga* 'There came going/creeping (?) the shadowgoer in the dark night', Mitchell therefore comments: 'the only translation I can suggest is 'Came . . . scrithing'!' Since these MANNER components 'smooth and graceful' or 'dangerous' cannot be established with certainty as being part of the verb's decontextualized meaning, both OE *scrīþan* as well as its reflex ME *scrithen* are classified as NEUTRAL MOTION verbs in the Appendices (on the overlap of the categories of MANNER verbs and NEUTRAL MOTION verbs, see 2.3.3). This is the general practice I adopted with respect to those verbs for which it is impossible to tell whether the MANNER or PATH components suggested in some of their dictionary definitions are merely contextual or are really part of the verb's coded meaning. Section 3.2.2 will discuss this in more detail.

3.2.2 Detailed dictionary definitions

As already shown in section 2.3.3 with the example of ME *flitten*, historical dictionaries such as the *MED* and the *DOE* or *BT* often list the whole range of contextualized meanings in which a verb is attested.[49] For the present study, it is therefore necessary to not base the classification of the Old and Middle English verbs on the single subsenses, because these will usually represent a specific kind of contextualized meaning. Rather, the entire dictionary entry must be taken into account, including the attestations it provides, with the aim of getting an idea of the decontextualized meaning of the verb.[50]

For many verbs, this was very clearcut. An example is OE *flicerian/flicorian*, as in (46), whose meaning the *DOE* defines as 'of a bird, an angel: to flutter the wings, hover, flicker' and says that its Latin equivalent in manuscripts is *volitare* 'to fly around, flutter'. All the information therefore suggests clearly that OE *flicerian/flicorian* should be classified as a MANNER verb ($\text{Ma}^{\text{air/fluid}}$ in Appendix A).

(46) witodlice an blac þrostle. **flicorode** ymbe his neb. (ÆCHom II, 11 93.4, DOE s.v. *flicerian, flicorian*)
'Truly, a black throstle fluttered around his face.'

[49] Another good example for this is OE *floterian* where, after sense 1 'to float to and fro, be tossed on the waves' and sense 2 'to flutter, move restlessly', the *DOE* gives as sense 2.b 'of St. Vincent's body' (who is floating in the sea). Such a specific entry would be unexpected in a dictionary of a present-day language. That it occurs in a period dictionary is an almost natural consequence of the fact that the number of attestations of an Old or Middle English verb is finite, as a result of which the number of its contextualized meanings is also finite. In this, the period dictionaries are therefore different from dictionaries of the present-day languages, which cannot be equally specific since they have to take into account the fact that speakers will use a word in new contexts all the time.

[50] Of course, including for the study of the Old and Middle English motion verb inventories, each verb could have been investigated with the help of corpora. But (a) this would not have been feasible for the altogether 189 Old English and 465 Middle English verbs examined here, and (b) this has been done by the compilers of the dictionaries anyway. Since the dictionary entries, particularly those from the *DOE* and *MED*, also provide large numbers of attestations, it does not seem too problematic to use them as the basis for the classification of the verbs.

For other verbs, categorization was less straightforward. The problem and the way it was handled in the present study shall be illustrated in the following subsections by the examples of OE *stīgan*, *hweorfan*, *sīþian*, and ME *hippen*.

3.2.2.1 OE *stīgan*

The subsenses of OE *stīgan* as given in *BT* are listed in (47) (there is as yet no *DOE* entry):

(47) Bosworth–Toller, s.v. *stígan*:

I. intrans. to go
(1) without implying ascent or descent
(2) implying ascent, to go from a lower to a higher level, to ascend, mount
(2 a) of getting into a vessel, etc., climbing a tree, etc.
(3) where the movement is downwards, to descend
II. trans. to ascend, mount

The crucial question is how to treat the Present-Day English verbs used to describe the meanings of the Old English verbs. I think it is important to keep in mind that, in the end, saying that *an Old English verb x 'means' the Present-Day English verbs y and z* only amounts to saying that *the Present-Day English verbs y and z can be used to translate Old English x*, or in other words, that *OE x is used in contexts where Present-Day English would have y or z*. Perhaps the point will be clearer if it is transferred to living languages: To put it bluntly, if I look up French *monter* in my dictionary and find that it means English 'to go up, climb' (and PDE *climb* undoubtedly is a verb that contains MANNER information), does that entail that French *monter* is a MANNER verb? All it entails is that English *climb* can often be used to translate French *monter*, or, in other words, that there are contexts of use for French *monter* in which *climb* would be a suitable equivalent in English. This 'equivalence fallacy' is a well-known inherent problem of bilingual dictionaries (cf., e.g., Herbst & Klotz 2003: 107–23; Svensén 2009: 253–79). Period dictionaries are of course also a kind of bilingual dictionary. Thus, if the dictionary entry for OE *stīgan* in (47) gives, among other equivalents, PDE *climb*, does that mean that *stīgan* is a MANNER verb?

Exactly such an approach is chosen in Fanego (2012, see 2.1.4.3), whose lists of Old and Middle English MANNER verbs (and consequently, also 'manner senses' (2012: 44)) are based on the Present-Day English equivalents of the verbs given in the dictionaries (2012: 40). Thus, OE *stīgan* is classified as a MANNER verb 'to climb' in Fanego: it occurs in her list of MODE verbs, i.e. verbs that express 'specific types of movements of the figural entity' (2012: 41), together with completely uncontroversial MODE verbs such as *hoppian* 'to hop, leap, dance' or *slincan* 'to creep, move stealthily'. This is, of course, in line with her approach of looking for MANNER verbs as well as MANNER senses, i.e. decontextualized MANNER meanings as well as contextualized MANNER meanings, in dictionaries.[51]

The approach I adopt here is different and influenced by the practice of constructionist approaches to attribute different verb senses (as contextualized meanings) to the interplay of verb semantics and constructional semantics (as well as context in general) (see 2.2.3). If all the intransitive motion senses that *BT* gives for OE *stīgan* are viewed together, this suggests that neither PATH nor MANNER is the primary information that is carried by the verb, but that *stīgan* should best be seen as belonging to the category of 'NEUTRAL MOTION verb'. This is supported by the attestations given in the dictionary entry. First of all, there are attestations in which the 'clambering' MANNER suggested by the translation with PDE *climb* cannot be present, e.g. when the FIGURE is rain, as in (48a), or the dead going to hell, as in (48b).

(48) a. Đa **stah** and com smylte reng (Bd. 4, 13 ; S. 582, 34)
'Then descended and came gentle rain.'

b. Đa bioð genemde deade, & ða **stigaþ** on helle (*in infernum descenderent*) (CP B9.1.3 [2245 (55.429.25), expanded with *DOEC*[52])
'They are called 'dead' and they go to hell.'

(both from BT s.v. *stīgan*, sense I.3 'to descend')

Second, let us look at PATH information: Even in the attestations that illustrate the PATH senses 'ascend' and 'descend' (senses I.(2) and I.(3) in (47)), the idea of upward or downward motion is typically supported (or perhaps even carried entirely) by PATH satellites, such as in *up on an treow* 'up on a tree' and *adune* 'down' in (49a–b).

(49) a. He **stah** ***up on an treow*** (*ascendit in arborem*) (Lk. Skt. 19.4, BT s.v. *stīgan*, sense I.2 'to ascend, mount')
'He went up on a tree.'

b. Zachee **stih** ardlice ***adune*** (Homl. Th. i. 580, 33, BT s.v. *stīgan*, sense I.3 'to descend', expanded with *DOEC*)
'Zacheus, get down quickly.' [Zacheus is in a tree.]

Also in the rarer cases where such a satellite is lacking, PATH can be inferred from the general context: In (48a), for instance, the FIGURE is *reng* 'rain', which

[51] This approach also entails that, for instance, OE *glīdan*/ME *gliden* is counted as an Old English manner verb meaning 'glide', and as three new manner senses in Middle English, 'to walk, walk smoothly or quietly', 'to rush, charge', and 'to go, travel' in Fanego (2012), while in the present study the verb would be counted only once for Old and Middle English respectively.

[52] The attestation is also from *BT*, but quoted here in the form found in the *Dictionary of Old English Corpus*. The attestations given in *BT* are sometimes clipped, leaving out words or adverbials, which is why I checked their actual form (and also their context) with the help of *DOEC*.

usually moves downwards, and in (48b), the GOAL is 'hell', which is also imagined as 'downwards'. Furthermore, even though OE *stīgan* is often used for vertical motion (either up or down, see (49a–b)), it may, according to the first sense given in *BT*, also be used for motion in any direction (see the *BT* senses in (47)). Thus it is doubtful whether even 'verticality' is really part of the coded meaning of *stīgan*. An example is (50), where 'straying from God's laws' does not necessarily imply upward or downward motion.

(50) Đu forhogdes alle **stigende** ***from rehtwisnissum ðinum*** forðon unreht geðoht heara is (*Spreuisti omnes discedentes a iustificationibus tuis quia iniusta cogitatio eorum est*) (Ps. Surt. 118, 118, BT s.v. *stīgan*, sense I. 'to go' (1) 'without implying ascent or descent'; expanded with *DOEC*)
'You despised all those going from your righteousness, because their thinking is unjust.'

One could, of course, argue for a 'GOOD IS UP, BAD IS DOWN' metaphor to be involved here, so that straying from God's laws would be seen as downward motion. However, usually, a satellite specifies whether the motion expressed by OE *stīgan* is 'upward' or 'downward'.[53] Also the fact that in most other Old English glosses of this Latin verse Latin *discedentes* is rendered with forms of *gewītan*[54] 'to go, depart' supports the interpretation of non-vertical motion in (50).

In the light of the overall information on OE *stīgan*, then, I think that it can neither be seen as a verb that in its decontextualized meaning primarily expresses MANNER of motion, nor PATH of motion, and therefore I categorized it as NEUTRAL MOTION verb (see Appendix A). The same is true for some other verbs whose Present-Day English equivalents given in the dictionaries suggest different possible PATH and MANNER meanings, as discussed more briefly in the following subsections.

3.2.2.2 OE *hweorfan*

The meanings of OE *hweorfan* are given by *BT* as 'to turn, change, go, return, depart, go about, wander, roam, hover about'. These meanings are more or less mutually exclusive in terms of PATH, ranging from 'no apparent goal/direction' (roam, hover about) to 'away' (depart) and 'back' (return). Moreover, judging from the attestations offered in *BT*, it seems that these PATH meanings are not carried by the verb but rather by the co-text, such as by the elements in italics in (51a–c).

(51) a. Næfre ic ***from*** **hweorfe** ac ic [. . .] mid wunige awo to ealdre (Exon. 14 b; Th. 30, 8; Cri. 476, expanded with *DOEC*)
'I will never go from [you], but I [. . .] will dwell with [you] forever.'

[53] Cf. two other glosses of the same verse, which translate the Latin *discedentes* as *ofdune astigende* (PsGlL (Lindelöf) C7.4, 118.118, DOEC) and *niðerstigende* (PsGlE (Harsley) C7.3, 118.118, DOEC).
[54] The *DOEC* has nine glossed versions of this psalm verse, in six of which a form of *gewītan* glosses *discedentes*: PsGlJ (Oess) C7.5; PsGlG (Rosier) C7.8; PsGlD (Roeder) C7.9; PsGlF (Kimmens) C7.19; PsGlF (Lindelöf) C 7.11; PsGlK (Sisam) C7.13. On the other two, see footnote 53.

b. **Hweorfaþ** ***eft to*** me (Blickl. Homl. 235, 16)
'Come back/again to me.'

c. Swa scriþende gesceapum **hweorfað** gleomen gumena ***geond grunda fela*** (Exon. 87 a; Th. 326, 28; Vid. 135.)
'Travelling thus, by [their] destinies, the singers of men wander throughout many lands.'

(all from BT, s.v. *hweorfan*)

Therefore, since no additional information apart from MOTION can be ascribed to the decontextualized meaning of OE *hweorfan*, it is also categorized as NEUTRAL MOTION verb, i.e. not primarily expressing MANNER or PATH.

3.2.2.3 OE *sīþian*

The last Old English verb I want to use as an illustration of the problems of working with dictionary entries is OE *sīþian*, which is given in *BT* as 'to journey, go, travel'. In motion verb typology studies, verbs meaning 'to travel' are often treated as MANNER verbs (e.g. in Narasimhan 2003 on Hindi, or Fanego 2012 on English, who has OE *sīþian* among the MANNER verbs (subcategory COURSE)). Yet here again, the fact that PDE *travel* is among the words used to render OE *sīþian* does not seem to be sufficient ground to assume that OE *sīþian* really denoted a specific MANNER of motion. In examples (52a–d), *sīþian* is used to describe the concrete motion of human figures (52a), the motion of people to hell (52b), of 'all flesh' to God (52c), and the motion of Christ to earth through Mary (52d)—there does not seem to be a specific MANNER information shared by them.

(52) a. Ða ða se Hælend **siðode**, sum man him cwæþ to: 'Ic wille **siðian** mid ðe and ðe folgian' (Homl. Skt. i. 16, 154)
'As the Savior was journeying, some man said to him: "I will journey with you and follow you."'

b. Hig into helle cuce **siðodon** (*descenderunt vivi in infernum*) (Num. 16, 33)
'They went to hell alive.'

c. For ðe sceal ælc flæsc forþ **siðian** (*ad te omnis caro veniet*) (Ps. Th. 64, 2)
'All flesh shall go forth in front of you.'

d. Þurh ðe Frea on ðas eorþan ut **siðade** (Exon. Th. 21, 4 ; Cri. 329)
'Through you [i.e. the Virgin Mary] the Lord went out on this earth.'

(all from BT s.v. *sīþian*)

Apart from the dictionary definition, which includes 'journey, travel' besides 'go', there is thus no strong indication that OE *sīþian* means any more than 'to go, move'. Therefore, like OE *stīgan* and *hweorfan*, OE *sīþian* is also categorized as NEUTRAL MOTION verb in the present study.

3.2.2.4 ME *hippen*

For ME *hippen*, the *MED* lists the following senses:

(53) 1. (a) Of persons, animals, things: to leap, spring, hop, bounce; ~ ***ayen***, spring back, rebound;—also refl.;
(b) to leap downward; of a thing: fall; ~ ***adoun***, ~ ***to ground***;
(c) to travel by hops or jumps; fig. of death: come speedily; ~***forth***, leap forth; ~ ***her and ther***, dash hither and thither; ***comen (gon) hipping***;
(d) ~ ***over***, leap across, jump over (sth.)
2. To limp, hobble.

I would argue that all of these senses share an idea of 'hopping'. The PATH component 'downward' suggested in subsense 1.b 'to leap downward, of a thing: fall' is unlikely to be part of the coded meaning of the verb, but a contextualized meaning, since it is supported by the context in all the attestations listed under this subsense. In (54), for instance, the prior mentioning of being led up the tower and the adverbial *to grounde* bring about the 'downward' reading.

(54) An carpenter..ladde him ***vpe þe tour an hei*** & made him **huppe** ***to grounde*** (*c*.1325(*c*.1300) *Glo.Chron.A* (Clg A.11))
'A carpenter led him up high in the tower and made him leap to the ground.'

The second sense, 'to limp, hobble', may at first seem unrelated, but in the attestations given under it the idea of 'hopping' is present as well, since the manner in which one moves with crutches (*hyppyng on two stavys*), or with one leg and a cane, as in (55), is also a kind of hopping.[55]

(55) So ***with his staf and his oþir legge***, he **hipped** forth as he myte tyl he cam to þe nexte town. ((1451) Capgr. *St.Gilb.* (Add 36704) 118/18)
'So with his staff and his other leg, he hopped on as he was able to, until he came to the next town.'

Since the *MED* entry has only three attestations in the sense 'limp, hobble' as in (55), and since these can be construed as a manner of 'hopping' (but not the other way around), ME *hippen* was categorized as manner verb in the subclass 'jumping manner' (Ma^{jump} in Appendix B), rather than in the subclass 'slow or impeded motion'.

Semantic classification is always subjective to some degree—it can never be ruled out that someone else's categorization would have turned out slightly different with respect to one or the other verb. The principles I followed—a

[55] In Fanego (2012), *hippen* 'to leap, hop' and *hippen* 'to limp, hobble' are treated as separate manner senses.

consideration of the dictionary entry as a whole, including the attestations given in it (*DOE* (for A–G) and *BT(S)* (for H–Z) for Old English; *MED* for Middle English)—have been exemplified in 3.2.1–4 with a few of the semantically more elusive verbs (OE *stīgan*, *hweorfan*, and *sīþian*, ME *hoppen*). For the classification of numerous other verbs, such as OE *siglan* 'to sail' and ME *hoven* 'to hover, float' (both in $\text{Ma}^{\text{air/fluid}}$), OE *slidrian* 'to slide' and ME *slippen* 'to slip' (both in Ma^{glide}), OE *hlēapan* 'to leap' and ME *sauten* 'to leap' (both in Ma^{other}), these problems were not as pronounced anyway. For further transparency, though, the appendices in which the verbs are listed according to their categories also include the dictionary definitions of the motion senses of all the verbs, an example, and, for Old English (Appendix A), also the categories in which the verbs are classified in the *TOE*.

The only viable alternative to the practice adopted here would have been to base the classification of the verbs on single subsenses (i.e. often contextualized meanings) taken from the dictionaries, as in Fanego (2012). This, however, would have necessarily entailed that a verb such as OE *stīgan* would receive multiple classification as (i) PATH verb 'up', (ii) PATH verb 'down', (iii) NEUTRAL MOTION verb, and (iv) MANNER verb, which would have gone against the grain of the aim of the present study and the approach it adopts (see particularly section 2.2.3 on verb meaning).

CHAPTER 4

Talking about MOTION in medieval English

Aims, material, and method

BEFORE WE LOOK AT the integration of the PATH verbs borrowed into Middle English (chapter 9), we want to see whether the claim that English typically expresses MANNER in the verb and PATH in satellites also holds for earlier stages of the language. After all, if pure PATH verbs had been available as a well-established native option in the language, the borrowed PATH verbs would not be semantically peculiar. To get a clear picture of the native situation, two different perspectives are necessary: (i) inventory and (ii) usage. The result of both analyses, for Old as well as Middle English, is that the inventory does not contain many pure PATH verbs (except for the ones borrowed from French and Latin in Middle English), and that, compared to other motion verbs, these are not frequently used either. These two different perspectives will be briefly explained in the present chapter, and then applied to Old English (chapter 5) and Middle English (chapter 6) respectively.

4.1 *Inventory*

First of all, we want to know which verbs are *possible* for talking about motion in Old and Middle English (sections 5.1 and 6.2 and Appendices A and B). This is basically a type approach. With the help of the *TOE* and a definition search in the *MED* respectively (on details, see 4.1.2), we will investigate the language's inventory of motion verbs and categorize these verbs into the broad semantic categories that were outlined in section 2.3:

Motion verbs:

1. MOTION verbs primarily expressing MANNER (with subcategories Ma^{speed}, $Ma^{slow/impeded}$, $Ma^{air/fluid}$, Ma^{fall}, Ma^{glide}, Ma^{jump}, Ma^{flee}, Ma^{roll}, Ma^{other})
2. MOTION verbs primarily expressing PATH

(with subcategories $Pa^{towards}$, Pa^{in}, Pa^{away}, Pa^{down}, Pa^{up}, $Pa^{through}$, $Pa^{forward}$, Pa^{circle}, Pa^{along}, Pa^{back}, Pa^{out})

3. NEUTRAL MOTION verbs (not primarily expressing PATH or MANNER)
4. non-MOTION verbs (not primarily expressing MOTION)

As discussed in section 3.1, the inventory will necessarily be larger for Middle English than for Old English, due to the larger amount of text that is available from the Middle English period. Furthermore, it should be borne in mind that both inventories cover a considerable diachronic as well as dialectal span: Texts from the Old English period mostly come from manuscripts from the late ninth to the twelfth century, and the Middle English period ranges from the twelfth to the fifteenth century. Particularly in the Middle English period, the texts come from the whole range of dialects (the Old English material is predominantly in the early and late West-Saxon dialect), since the emergent Old English standard of West-Saxon was broken off by the consequences of the Norman conquest, and a new, London-based standard only started to evolve by the end of the Middle English period.

This means that the inventories can certainly not be taken as representing the motion verb vocabulary of a particular speaker at a particular point of time: Some verbs will have fallen out of use in the course of a period, such as ME *fusen* 'to go speedily, hasten' (from OE *fȳsan*), whose latest attestation is in the Otho manuscript of Layamon's *Brut* from *c.*1300, and which might therefore not be current any longer for a Middle English speaker in the fifteenth century. Other verbs will belong to a particular dialect, such as the Anglian OE *lēoran* (see Appendix A: NEUTRAL MOTION) or the ME *trinen* (see Appendix B: NEUTRAL MOTION), a Scandinavian loan which according to the *MED* mostly occurs in texts from the North and Northwest Midlands.[56]

4.1.1 Aims of the inventory studies

On the one hand, thus, the inventories will contain more verbs than a particular speaker had in his or her motion verb vocabulary. On the other hand, they will also be incomplete in so far as they lack verbs for which we have no attestation. Nevertheless, on the basis of these inventories, we can examine which

[56] The *MED* does not usually give information about whether a word belongs exclusively to a specific dialect (*trinen* is the sole exception in my list of verbs), which is why dialect issues will not be considered in this study. I checked the verbs in the Middle English inventory against Kaiser's (1937) lists of 'Nordwörter' ('North-words') and 'Südwörter' ('South-words'), and found the verbs *dushen* 'to fall heavily, proceed swiftly', *gangen* 'to walk, go', *hobelen* 'to rock, walk lamely, hobble', *slippen* 'to slip', *sprenten* 'to leap', and *witen (away)* 'to go away', as well as the Scandinavian loans *busken* 'to prepare, hasten, hurry', *cairen* 'to go, move, travel', *raiken* 'to go, proceed, rush', *scopen* 'to leap', and *trinen* 'to go' to be North-words, and *neighlen/neighlechen* 'to approach' to be South-words. Since approximately 14 per cent of the verbs in the Middle English inventory are of Scandinavian origin (see the etymology column in Appendix B), a contact language whose influence was strongest in the Danelaw, i.e. the area in the North and East, it is conceivable that some of these are also dialectally marked.

verbs could be used to talk about MOTION. Specifically, the analyses will focus on the following issues:

(a) Similar to Fanego (2012), but with more data (see section 4.1.2), we will consider the Old and Middle English motion verb lexicon in the light of Slobin's observation that satellite-framed, manner-salient languages tend to have large and increasing inventories of MANNER OF MOTION verbs (see 2.1.4.3).
(b) We will investigate whether the verbs classified as predominantly expressing PATH can be considered to be pure PATH verbs (on PATH verbs as a prototype category, see 2.3.3). If they can, they might constitute a model in analogy to which the PATH verbs borrowed from French and Latin in the Middle English period might have been integrated.
(c) We will see to what degree non-MOTION verbs can be used as motion verbs.
(d) We will see how many of the Old and Middle English motion verbs have attestations with a PATH satellite, i.e. attestations in an intransitive motion construction (these verbs are printed in italics in Tables 5.1 and 6.1 and in the appendices). This is particularly interesting for the MANNER OF MOTION verbs, which are expected to occur with such satellites if the language is satellite-framing, or, in other words, if it has a productive intransitive motion construction. In verb-framed languages, by contrast, MANNER verbs are rather used without PATH expressions, particularly without telic or boundary-crossing ones (see 2.1.4.1–2).

4.1.2 Material and method

This section starts with some general remarks on which verbs are included in the inventory of motion verbs (4.1.2.1) and on their presentation in Tables 5.1 and 6.1 and the appendices (4.1.2.2). Following this, the methods for compiling the inventory for Old and Middle English respectively are addressed in 4.1.2.3–4.

4.1.2.1 What is included in the inventory?

The inventory is aimed to include all verbs that can be used to convey a sense of intransitive motion in active clauses whose subject is the moving FIGURE. Different from Fanego (2012, see 2.1.4.3), I did not restrict the lists of motion verbs to verbs with animate FIGURES only, and I also did not exclude 'non-agentive verbs of self-contained motion' (2012: 37), such as *roll, float*. Conceptually, such a restriction is not necessary: MOTION events with inanimate FIGURES (56a), or with non-agentive MOTION (56a–c) are also MOTION events,[57] and the MANNER or the PATH in which they happen can be observed

[57] Cf. Goldberg & Jackendoff (2004: 551): 'Consider the examples *Sue ran down the stairs* vs. *The water ran down the stairs*. The subject of *run* is a theme in both cases. In the former case it is also always an agent. But in the latter case the subject can be interpreted as either doing something (i.e. an agent) or having something happen to it (i.e. a patient). In other words, *run* with an inanimate subject is indifferent as to whether its subject is an agent or a patient. In fact this is the case with many intransitive

and expressed.[58] Besides, for some verbs that only appear with inanimate subjects, like *springan* in (56a), the lack of attestations with animate subjects might be a pure coincidence, due to the lack of a sufficient amount of data.

(56) a. Þeah ðu teo hwelcne boh ofdune to ðære eorþan, swa ðu hine alætst, swa **sprincþ** he ***up*** (Bt. 25; Fox 88, 24, BT s.v. *springan*, see App. A: Ma^{other})
'Though you pull some bough down to the earth, as soon as you let it go, it springs up.'

b. Yeres and dayes **fleet** this creature ***Thurghout the see of Grece vnto the strayte of Marrok.*** ((*c*.1390) Chaucer *CT.ML.*(Manly-Rickert) B.463, MED s.v. *fleten*, see App. B: $Ma^{air/fluid}$)
'Years and days this creature floated throughout the sea of Greece unto the strait of Morocco.'

c. Skaterd be the Scottes, Hoderd in thar hottes..Thai **tumbed** [vrr. toumbe; tumbeld] ***in Twede*** Thet woned by the se (*a*.1350 *Songs Langtoft* (Jul A.5 p.236), MED s.v. *tumben*, see App. B: Ma^{fall})
'The Scots be scattered, huddled in their huts, they tumbled into the Tweed, who lived by the sea.'

Another difference from Fanego (2012) is obviously that the present investigation does not focus exclusively on MANNER verbs, but also includes NEUTRAL MOTION verbs, PATH verbs, and non-MOTION verbs in contextual motion readings. In other words, *all* verbs that can be used to express MOTION of a FIGURE expressed in the subject of an active clause in Old and Middle English are taken into account.[59]

4.1.2.2 Presentation

The inventory of Old and Middle motion verbs in the present study differs in some way from similar lists in the literature (e.g. Fanego 2012: 41–2; or Ogura 2002: 113–17; 161–4): Ogura (2002) in particular lists all the prefixed versions of a verb as separate verbs, as is also customary in dictionaries and in the *TOE*. Thus, *forðrǣsan* (lit. 'to forth-rush') and *ūtrǣsan* (lit. 'to out-rush') are given separately (2002: 114, 117), for example. Similarly, also Fanego lists OE *stīgan*, *āstīgan*, and *in(n) stīgan* as three separate MANNER verbs (2012: 41).

---

motion verbs. Given that [the IMC] is based on the model of intransitive motion verbs, we believe the same distribution of thematic roles obtains with *GO* in the constructional subevent of [the IMC].'

[58] Also in Fanego (2012), the restriction to verbs with animate agents seems to be due to reasons of 'mak[ing] the data manageable' (2012: 36), rather than to a decision in principle.

[59] That is, all the verbs that could be found among the ones that are attested. The methods described in sections 4.1.2.3–4 may of course have missed a few motion verbs.

This usual way of presenting each of the prefixed forms of the verbs as different lexical item would not have made sense in the framework of the present study. Our focus lies on the verbs that can be used to express MOTION, i.e. on the verbal bases. In this framework, the prefixes are regarded as a separate element, namely as one of the possible realizations of the functional category satellite (see section 2.1.2). Satellites can be realized both before and after the verb in Old and Middle English, such as *ūt* in (57a–b). Section 5.4 will discuss Old English satellites in more detail.

(57) a. & þa **utræsde** ***on hine***, & hine miclum gewundode. (HC, CHROA2 755.13)
'And then [he] rushed out on him and wounded him severely.'

b. & þy ilcan geare **for** ælfred cyning mid scipum ***ut on sæ*** (HC, CHROA2 R882.2)
'And in the same year, King Alfred went out on the sea with ships.'

In the case of the above-mentioned *in(n) stīgan* and *ūtrǣsan* then, this means that the present study sees them as combinations of satellite (*in(n)*, *ūt*) and verb (*stīgan* and *rǣsan*). Consequently, they are not listed separately in the inventory of Old and Middle English motion *verbs*; they are pooled in the entries for *stīgan* and *rǣsan* respectively: All the prefixed satellites with which the individual verbs were found in the dictionaries are listed in the appendices as well. In the rare case where a verb is only attested in combination with a specific prefix, and not in unprefixed form, the verb is listed in the inventory with this prefix in brackets. An example is OE *(wiþ)tremman* 'to step back' (App. A: $\mathrm{Mo}^{\mathrm{neutral}}$)—the non-prefixed **tremman* (which must be a verb formed from the noun *trem* 'a step') is not attested.

4.1.2.3 Material and method: Old English

The motion verb inventory of Old English was gathered in various steps, which will be briefly explained in this section. For the basis of the inventory, I used the *Thesaurus of Old English* (*TOE*) (see 'Basis' below). In a second step, the verbs obtained with the *TOE* were checked against other lists of motion verbs in the literature in order to add verbs that had not been found through the thesaurus search (see 'Extension' below in this section). Thirdly, the dictionary entries and attestations of the verbs obtained in the previous step were all checked in order to (a) eliminate verbs that turned out not to be motion verbs according to the definitions of the present study (see section 2.3) and (b) classify the verbs semantically (see 'Classification and revision' below in this section). In this, the different prefixed formations of one and the same verbal base yielded by the *TOE* search (e.g. *æfter rōwan, berōwan, gerōwan, oferrōwan, oþrōwan*) were all counted instances of the same verb (e.g. *rōwan* 'to go by water, row, sail') with different satellites (see the discussion in 4.1.2.2).

Basis

The inventory of Old English motion verbs presented in 5.1 (and with more detail in Appendix A) is primarily based on the *Thesaurus of Old English* (*TOE*) (Roberts & Kay 2000, online since 2005). It includes the motion verbs given by the *TOE* under the following headings:

- the superordinate category 'to move, be in motion' (05.12), with the relevant subordinate categories[60]
- 'Flight' (11.10.04)
- 'Escape' (11.10.05)
- 'Stealthiness, a stealthy act' (11.09.01)[61]

The *TOE* itself is based on the dictionaries by Bosworth–Toller (henceforth *BT*) and Clark Hall (henceforth *CH*) (Roberts & Kay 2000: xvi). Therefore, a definition search in the dictionaries—easily possible in the online version of the *BT* dictionary—did not promise to yield any verbs other than those already collected in the *TOE*. This was confirmed with a few trial definition searches for some keywords like *travel* in the online *BT*.

Extension

The second step was to make sure that the search in the *TOE* had not overlooked any motion verbs. Therefore, the *TOE* list was cross-checked against the lists of motion verbs in Ogura (2002: 161–4) and Fanego (2012: 41–2). This yielded a few additional verbs,[62] namely *frician*, *sealtian*, and *tumbian*, all meaning 'to dance', *luncian* 'to limp, crawl', and *hīgian* 'to labour, strive, endeavour', which have been added to the present list from the one in Fanego

---

[60] These include categories such as 'to go, progress, travel (usually on land)' (05.12.01), 'to go on a journey/expedition/travel' (05.12.01.01), 'to wander' (05.12.01.02), 'to go, proceed on foot' (05.12.04), and 'to travel on water' (05.12.01.09). There are, of course, also verbs listed in these categories that do not relate to a moving subject; these are not taken into account (e.g. *hwemman* 'to shape a course for').

[61] 'Flight', 'Escape', and 'Stealthiness, a stealthy act' contain further motion verbs which are not listed under the heading of 'to move, be in motion' in the *TOE*, such as *bestelan*, *bestealcian* 'to move stealthily' and *oþberstan* 'to break out, escape'.

[62] Two verbs which were found in Ogura (2002), but not in the *TOE*, have not been added, for the following reasons:

(a) '*tofealdan* 'to come to land' (ÆGram 138.9 [*applicare*])' (Ogura 2002: 163). It was not found in *TOE* in this meaning but only as 'to join, attach, unite to'. The word is apparently only attested in Ælfric's *Grammar*, in the context of a demonstration of the two ways of forming the preterite of Latin *plicare* 'to fold'. The *Grammar* goes on to say that the preterite is formed in just the same way for the prefixed forms of Latin *plicare* and gives Old English equivalents for these prefixed forms: '*implico* ic on befealde, *replico* ic ongean fealde, *complico* ic samod fealde, *explico* ic fulfealde, ***applico* ic to fealde oððe ic gelende mid scipe**'. The question is whether this one single attestation of *tofealdan* really indicates that *tofealdan* is a motion verb. The whole context in Ælfric's *Grammar* here appears very much like an attempt to render transparent the morphological structure of the Latin prefixed verbs (see Kornexl 2001): Every Latin prefixed verb is coupled with an Old English prefixed verb; L *im-* with OE *on* 'in, on', L *re-* with OE *ongean* 'against, again', L *com-* with OE *samod* 'together', etc. Latin *applicare* means both 'to bring in contact, join' and 'to land with a ship'. Thus, the most plausible interpretation of the line in

(2012).[63] The fact that the usage study in section 5.2 only found five verbs that were not identified in the inventory, all of them non-MOTION verbs (see the discussion below Table 5.3), shows that the dictionaries and thesauri are an excellent source for finding out which verbs can be used to describe motion in Old English.

Classification and revision

In a third step, the verbs were sorted into the broad semantic categories listed in 4.1 above (see also the discussion in 2.3). This was done on the basis of their dictionary entries in the *DOE* (for verbs starting in A–G) and *BT* (for H–Z), taking into account the various subsenses and the attestations given (see 3.2 for a detailed discussion of the procedure). An example for every verb in a motion sense—with a PATH satellite, if there was such an attestation—is provided in Appendix A.

The consideration of the dictionary entries and the attestations within them also led to a few verbs being discarded again from the inventory of motion verbs. This shall briefly be addressed here with the examples of *gefōtian*, *astǣgan*, and prefixed forms of *blāwan* 'blow'. The reason why *astǣgan* is found in the *TOE* category 'to climb into' is probably that *CH* has it as 'to go up, embark'. Yet, there is apparently only one single attestation for this verb, namely from the Rushworth Gospels, where the form *astǣgdun* (preterite ind. pl.) glosses Latin *ascendentes* (*DOE* s.v. *astīgan*). In contrast to the *TOE* and *CH*, the *DOE* convincingly regards the form *astǣgdun* here as a variant weak form of the verb *astīgan*, based on the strong first preterite stem. Therefore, *astǣgan* was discarded since *astīgan* was already included in the list of motion verbs.

The case is similar for *gefōtian*: It is found in the *TOE* in the category of 'to hurry, hasten', probably because *CH* lists it with this meaning (the *DOE*

bold seems to be that L *applico* can be translated either with its morphological equivalent OE *ic to fealde* 'I to-fold, I bring in contact' or (*oððe*) with OE *ic gelende mid scipe* 'I land with ship'.

An interpretation of OE *tofealdan* as a motion verb (which I do not share), by contrast, must be based on a reading in which *oððe* 'or' is taken to suggest that OE *ic to fealde* is synonymous with OE *ic gelende mid scipe* 'I land with ship'. In this reading, *oððe* would relate the two Old English expressions to each other. It seems far more plausible, however, that *oððe* relates them both to the Latin *applico*, and that, therefore, *tofealdan* is not synonymous with *gelendan mid scipe*.

(b) '*hlidan* 'to come forth'' (Ogura 2002: 162). This verb is not included at all in the *TOE*. The attestation referred to by Ogura is from the *Paris Psalter*: *Ealle þa telgan ðe him of hlidað* (*PPs* 79.11.1b, Ogura 2002: 115) 'all the branches that grow/come forth from it [i.e. from the vine]'. Tinkler (1971: 25) includes *hlidan* in a list of 'ghost words' from the *Paris Psalter*, which sometimes have mistakenly been given a dictionary entry, and argues that it is more probably a form of OE *liðian* 'become' (*BT*), which is a 'known gloss to Latin *mitescere* 'to grow ripe, mature'' (and thus not a motion verb). It might perhaps also be a form of OE *hildan* 'to bend, incline' (which is attested in motion uses, cf. Pa^{down} in Appendix A).

[63] They had not been captured by the *TOE* search described above because they are listed under the supercategory 'Leisure' (18.02.04 'Dancing'), in 02.08.04.04 'State of being crippled', and in 11.03.04.03 'To labour for, strive after, endeavour' respectively.

does not). The reference in *CH* for *gefōtian* is to the Lindisfarne Gloss to Mark 15:44—apparently the only attestation of this verb. The Latin verse that is glossed in this attestation, though, involves the ablative absolute *accersito centurione* 'summon-PARTICIPLE-PASSIVE.ABL centurio-ABL', i.e. 'with the centurio summoned' or 'after the centurio was summoned'. The participle OE *gefōtad* is used in the gloss of the Latin participle *accersito* 'summoned'. The *DOE*'s treatment of *gefōtad* as a variant of *gefett, gefetod* 'fetched, summoned' is therefore more convincing than treating *gefōtad* as the participle of a potential verb *gefōtian* 'to hurry, hasten', which is why *gefōtian* was discarded.[64]

For a few other verbs, the *TOE* category is ambiguous as to whether they express motion or not, and this could be clarified with the help of the *DOE*, *BT*, or *DOEC*. For instance, the formations *forþblāwan* and *ūp āblāwan* are listed under the *TOE* categories 'to burst forth, break out' and 'to flow out, well up, erupt' respectively.[65] The *DOE* entry reveals that the latter, *ūp āblāwan*, is used with volcanoes: *Ehtna up ableow* (ÆLS (Agatha) 221) 'Ethna erupted' (s.v. *āblāwan*, see also *CH* s.v. *ūpāblāwan*). Therefore, no motion of the subject-FIGURE is involved. The former, *forþblāwan*, is only attested as a 'partial loan translation of *eructare*, in sense 'to breathe out (melodies)'' (*DOE* s.v. *forþblāwan*) and therefore does not involve intransitive motion either. Thus, the formations involving *blāwan* were discarded as well.

4.1.2.4 Material and method: Middle English

This section will argue why, for the Middle English inventory study, the *Middle English Dictionary* (*MED*) was chosen as the basis rather than a thesaurus (see 'Basis' below). The subsequent steps of the compilation are described under 'Extension' and 'Classification and revision'.

Basis

For the Old English inventory of motion verbs, I had resorted to the *Thesaurus of Old English*, one reason being that the *Dictionary of Old English* is still awaiting its completion. For the Middle English inventory, by contrast, I used the

[64] Two further verbs that were discarded are *gecringan* and *gefǣlsian*: *Gecringan* is given in the *TOE* under the category of 'to fall down/to the ground' (among others), but it is mostly used for 'falling in battle, dying by violence' and is therefore not regarded as a motion verb (see *DOE* s.v. *gecringan*). *Gefǣlsian* is given in the *TOE* under the category of 'to go in, enter', but, once again, the attestations cast serious doubt on this: *Gefǣlsian* usually means 'to cleanse, purify' (DOE s.v.v. *gefǣlsian, fǣlsian, gefǣlsod*). The *DOE* gives its second sense only with a question mark: '? to pass through, traverse'. This is most likely the sense that corresponds to the *TOE* category 'to go in, enter'. The only attestation of *gefǣlsian* in this sense (ChristA 317) is 'based on the assumption that *gefǣlsian* [. . .] has the senses of Latin *lustrare*, here specifically the sense 'to traverse' [Latin *lustrare* means 'to enlighten', 'to purify', and 'to traverse']; *gefǣlsian* has here otherwise been taken as 'to purify' and 'to glorify', and also emended to a form of *gefæstnian* 'to secure'' (DOE, s.v. *gefǣlsian*). Its motion use is thus based on one highly ambiguous attestation, and therefore it was discarded.
[65] *TOE* categories 05.12.05.03.04.01 and 05.12.05.03.04.02 (05. existence > 12. to move, be in motion > 05. to move and change direction, turn > 03. to travel away (from) > 04. to come/go out from).

MED's advanced search, searching the definition sections of all the verb entries for a range of Present-Day English MOTION verbs.[66]

As we will see below ('Classification and revision'), this procedure is somewhat more cumbersome and less elegant than using a thesaurus. However, for the present aim, the *MED* definition search has several advantages over the *Historical Thesaurus of the Oxford English Dictionary* (*HTE*), which would have been the thesaurus alternative. One of them is the different material used by the *MED* and the *HTE*: The *HTE* is based on (a) the second edition of the *OED* (*OED2*) and on (b) the *Thesaurus of Old English*, which again is based on several Old English dictionaries (Kay et al. 2009: ix; see also 4.1.2.3). The *OED2* in turn consists of the first edition of the *OED*, published between 1884 and 1928, and of the supplements from 1933 to 1986. Although a monumental achievement, *OED1* and *OED2* have a few weaknesses, among them a bias towards literary texts, the underrepresentation of some periods, and a 'teleological bias' towards words that survived into later stages of the language (cf. Thim 2011: 64–71 and references therein). Furthermore, regarding the historical record, many editions of Old and Middle English texts had not yet been available when the *OED* was started (Stanley 2000: 132–3; 2002: 27; see also Simpson 2002: 5–6) so that, 'inevitably, [the *OED*] missed some word senses that seemed unrepresented in English after 1150' (Roberts 2002: 189), i.e. 'many words were to be found in special senses in Middle English which were [. . .] not recorded in the *OED*' (Blake 2002: 54). Currently (since 1990), each and every *OED* entry is being updated and overhauled for the third edition, including 'new documentary evidence and modern developments in scholarship, and new entries to fill gaps in the historical record' (Simpson 2000). As this work is still in progress, however, the *HTE* is necessarily based on the older entries from *OED2* (which often means *OED1*).

The *MED*, by contrast, published between 1952 and 2001 as a comprehensive period dictionary, used a much greater wealth of Middle English material in its reading programme (Lewis et al. 2007: 3–4). Consequently, several verbs I found through the definition search in the *MED* would not have been identified as verbs used to express MOTION in Middle English by employing the *HTE*. Among them are, for instance, ME *auntren* (PDE *adventure*), *bronten* 'to move suddenly, rush' (PDE *brunt*), *jumpen* (because the first attestation in the still unrevised *OED* entry is 1511 instead of *a*.1460, as in *MED*), and *threpen* (PDE *threap*), meaning 'to hurry or rush away' when combined with *awei*.

One of the characteristics of the *MED*, the fact that its 'sense distinctions and divisions are [. . .] quite detailed and elaborate' (Lewis et al. 2007: 14), has sometimes been criticized, for example by Blake, who writes that 'there are cases where one can only think an editor is straining at shades of meaning

[66] http://quod.lib.umich.edu/m/med/structure.html. The definitions were searched for the following: 'walk'; 'run'; 'betake oneself'; 'wander'; 'rush'; 'swim'; 'ride'; 'dive'; 'march'; 'sail'; 'crawl'; 'go quickly'; 'leap'; 'climb'; 'go, move'; 'make one's way, go'; 'go, proceed'; 'journey'; 'enter'; 'descend'; 'go down'; 'ascend'; 'go up'; 'advance'; 'to go'.

which most of us find difficult to follow' (2002: 67) and argues that sometimes fewer senses would have been sufficient (see also section 3.2.2). For my purpose of finding out which Middle English verbs can be used to express MOTION, however, this practice has an enormous advantage. ME *auntren* (PDE *to adventure*) shall serve as an example.

In the *HTE*, the verb *adventure* is given under the category of 'undertake or set oneself to do sth. risky', similar to the *MED*'s definition 'to venture, take a chance, to attempt sth., undertake to do sth.' (s.v. *auntren*, sense 3.a). This meaning would well have been sufficient for the reader to understand uses such as (58), and 'to venture to go (to a place, to sb.)' (*MED* s.v. *auntren*, also sense 3.a) would not have had to be added for this purpose.

(58) Þe armed Attenieines **auntred hym till.** (*c.*1600(*c.*1350) *Alex.Maced.*(Grv 60))
'The armed Athenians ventured (to go) to him.'

Had it not been added, however, the procedure employed here could not have identified *auntren* as a verb used to express MOTION. The *MED*'s wealth of detailed sense distinctions was therefore very valuable for the present study. Another advantage over the *HTE* and the *OED* is the *MED*'s 'lavish provision of quotations' (Thim 2011: 82), which makes it easier to determine whether a given verb is attested in combination with a PATH satellite (see 6.2.6).

Extension

In the next step, the list of several hundred verbs obtained by the definition search was compared to the Middle English list of motion verbs in Fanego (2012: 58–63), from which an additional 25 verbs (20 MANNER, 1 PATH, 4 NEUTRAL MOTION) were added.[67] These had not been captured by the definition search described in the previous section because the definitions of these verbs did not feature the search terms (footnote 66).

Classification and revision

Like for the Old English inventory of motion verbs (4.1.2.3), the definitions and quotation sections in the entries of the Middle English verbs obtained through the definition searches (see footnote 66) were checked to see whether the verbs are used to express MOTION in the first place, and to

[67] MANNER: *balen* 'to dance', *bustelen* 'to thrash about', *clocken* 'to limp, hobble', *dauncen* 'to dance', *falteren* 'to stumble, stagger', *fiken* 'to move quickly or restlessly', *getten* 'to strut, swagger', *shoggen* 'to jog along', *snaperen* 'to stumble, stagger', *stampen* 'to tread heavily', *stumpen* 'to stumble, trip', *swalteren* 'to wade through water', *swarmen* 'to swarm, move about in great numbers', *thresten* 'to push one's way, jostle', *trampen* 'to tread heavily, stamp', *tramplen* 'to stamp with the feet', *waltren* 'to proceed with rolling or sidewise motion; move awkwardly or unsteadily', *wigelen* 'to move unsteadily', *wincen* 'to turn or move quickly', *wingen* 'to hasten'; PATH: *ringen* 'to encircle, surround'; NEUTRAL MOTION: *glauncen* 'to strike obliquely, glance', 'to turn quickly to one side, dodge', *shrinken* 'to flinch, wince, retreat', 'to fall, sink', *slingen* 'to throw, hurl fling', 'to hurl or fling oneself', *tracen* 'to move along a set path, follow a course, proceed', 'to dance'.

which subgroup they belong. Since, for Middle English, a dictionary definition search was used rather than a thesaurus, there were naturally many more false hits: When you search for verbs whose definition makes use of, for instance, *run*, you obviously also find many verbs such as *contrarien* 'to oppose, run counter to, conflict with', which are not used to describe MOTION and have to be discarded.

One verb, however, was eliminated even though it is presented as a motion verb in the dictionary entry: ME *outen*. The *MED* lists, as the second meaning of this verb, the following senses:

(a) to come out, go out; of air or gas; come out (of sth.), escape from; *netle in dokke* ~, a charm
(b) ~ *soure*, to come out sourly, act or speak harshly
(c) in proverbs

(MED s.v. *outen*, v., sense 2)

A look at the attestations given for these subsenses, though, shows that in all likelihood in none of them is *out* used as a verb form, but always as the adverb *out*: First of all, all but two of the seventeen attestations have a modal verb, such as *mouen* in (59a–b), *shulen* in (59c), or also *durren* 'dare' or *willen* 'will, want':[68]

(59) a. Þe mouþ of þis pott schal be ioyned..þat þere ***mowe*** noon eir **out** [L exalare] þerof. (*a*.1400 *Lanfranc* (Ashm 1396) 195/13)
'The mouth of this pot shall be closed so that no air may out [i.e. may come out] of it.'

b. He ***may*** not **out**..By cause the tour is so strong. (*a*.1425(?*a*.1400) *RRose* (Htrn 409) 4517)
'He may not out [i.e. may not get out] because the tower is so strong.'

c. He shal com hom to towne And she ***shal*** **out**. (*a*.1425(*c*.1385) Chaucer *TC* (Benson-Robinson) 4.210)
'He shall come home to town and she shall out [i.e. shall go out].'

These attestations thus simply look like instances of the combination of modal verbs with a PATH satellite, common in the Germanic languages (though no longer in Present-Day English), as we have seen in section 2.2.4.

The other two attestations in these senses are both examples of *out* in the idiom *nettle in, dock out*, meaning 'first one thing, then another' (cf. Benson & Robinson 1988: 544), i.e. referring to fickleness, as in (60).[69]

[68] The *MED* seems to point to this when, in the etymology section for *outen*, it says 'When used with *willen, mouen*, etc., a verb of motion is sometimes implied in the modal auxiliary', but still this is at odds with treating *outen* as a verb in these cases.
[69] See *MED* s.v. *dokke*, n., sense 4: '*Nettle in, dock out* orig., a charm spoken when driving out the sting of a nettle with a dock leaf [. . .];—here used with reference to inconstancy'.

(60) But kanstow playen raket, to and fro, Nettle **in**, dok **out**, now this, now that, Pandare? (*a*.1425(*c*.1385) Chaucer *TC* (Benson–Robinson) 4.461)
'But can you play rackets, to and fro, nettle in, dock out, now this, now that, Pandarus?'

Also here, *out* is very likely not a verb but an adverb, just like *in*. Thus, *outen* was eliminated from the inventory of Middle English motion verbs.[70]

With the methods described in 4.1.2 here, 189 Old English and 465 Middle English verbs used to express intransitive motion were identified. They will be presented and discussed in sections 5.1 and 6.2.

4.2 *Usage*

An inventory of motion verbs alone, however, is not enough in order to study how motion is expressed, for, as Filipović (2007a: 4) points out, lexicon is not usage, and 'the mere fact that some languages have richer lexicons of manner verbs [. . .] does not tell us anything about how or when those verbs can be used'. In addition to the inventory, which shows what is *possible*, we therefore also want to find out which verbs and structures are *typical* or *frequent* in talking about motion. The importance of distinguishing between lexicon and usage is demonstrated, for instance, in Narasimhan on (V-framed) Hindi, where she shows that this language has a 'sizable class of manner-of-motion verbs' (2003: 134), but that these are hardly ever combined with PATH satellites. The expectation for an S-framing, MANNER salient language is that MANNER verbs should be freely used together with PATH satellites, also with boundary-crossing ones (see 2.1.4.1–2), and this pattern of PATH in satellite and MANNER in verb should be the 'characteristic' one (see section 2.1.2).

4.2.1 Aims of the usage studies

The general question of how motion is most frequently 'talked about' breaks down into the following, more specific questions:

(a) How frequent are the different types of motion verbs (MANNER, PATH, NEUTRAL MOTION, non-MOTION)?
(b) Which are the typical complementation patterns, in particular: Are MANNER verbs usually combined with PATH satellites? Which are the typical complementation patterns of PATH verbs?

[70] Note that in the caused motion senses (MED s.v. *outen*, sense 1 'to drive (sth.) out, force out, etc.'), *outen* is indeed a verb, as, for instance, in *I wold have..outid all yeur chaffare without gruch.* (*c*.1460(?*c*.1400) *Beryn* (Nthld 55) 2408) 'I would have brought out all your merchandise without complaint', where *outen* also has an inflectional ending. There is one attestation in sense 1 which might perhaps be intransitive motion: *They oute wyth her bote ryght anone..To Bloys they Row* (*a*.1450 *Parton.(1)* (UC C.188) 6419), given under the subsense 1.d. ~ *with* 'launch (a boat), set out in (a boat)'.

The material on the basis of which these issues shall be studied is addressed in the following section.

4.2.2 Material

According to Talmy (2000b: 27), the typology of motion expression is based on what is 'characteristic' in a language. This entails that its focus is on neutral style, rather than on styles that are 'literary, stilted, and so on' (2000b: 27; see section 2.1.2). Studies which apply the motion verb typology to modern languages usually base their analysis on elicited spoken material or on novels. The obvious problems involved in finding a suitable text for the analysis of motion expressions, particularly for the Old English period, have been discussed in section 3.1.2.1, in which also the choice of texts has been addressed. Among the Old English texts that are extant, the *Anglo-Saxon Chronicles*, the *Saints' Lives*, and the *Voyages of Ohthere and Wulfstan* were chosen as the most suitable ones; for Middle English, *Gregory's Chronicle*, the *Life of Saint Edmund*, and three of Chaucer's *Canterbury Tales* are analysed. The texts will be briefly presented in sections 4.2.2.1–6.

4.2.2.1 *Anglo-Saxon Chronicles*

The *Anglo-Saxon Chronicles* were begun during the reign of King Alfred in the late ninth century, representing the 'earliest national chronicle written in a western vernacular language' (Bergs & Skaffari 2007: 6). There are seven extant manuscripts (and one single leaf) together referred to as the *Anglo-Saxon Chronicles*. These broadly share a 'Common Stock' of annals up to the entries for the year 890/892 and develop more independently in the following annals, as they were continued in different places in England (see, e.g., Jorgensen 2010: 4). The excerpts used for the present usage study are annals 444–964 from MS A (Chro A2 from the *Helsinki Corpus*) and 1070–87 and 1105–21 from MS E (Chro E4 from the *Helsinki Corpus*).

The most obvious reason why the *Chronicles* are chosen for the present usage study is that most of their material is very probably not translated from Latin, but is rather a genuine Old English production (e.g. Jorgensen 2010: 1) and that their subject matter contains a reasonable number of MOTION EVENTS (such as the Danish host moving around in England, or people travelling to England, Scotland, Normandy, or Rome). What may be problematic, though, is the style of the *Chronicles*, which has often been described as 'terse', 'laconic', 'semi-formulaic' or 'formulaic', of 'restricted diction and syntax', 'simple', and 'repetitive', or even 'lean and barren' (Plummer 1899: xxi), particularly so for the earlier annals, up to the end of the ninth century (cf., e.g., Clark 1995; Jorgensen 2010: 10; Stodnick 2010). Clark (1995), the most well-known work on the style of the *Anglo-Saxon Chronicles*, characterizes these annals as follows:

> Adjectives are sparse, and adverbs rare or [. . .] absent; nor is there any complexity of syntax, just a chain of simple sentences rendering a series of simple

propositions. Furthermore, with the events noted all falling within a narrow range, vocabulary and phrasing are correspondingly restricted, annal after annal using the same semi-formulaic language. (Clark 1995: 4).

Although already in these common-stock annals some passages stand out as more narrative, such as the episode of Cynewulf and Cyneheard (755) (cf., e.g., Bately 1990: 273–4; Clark 1995: 3), it is particularly the later annals, which also tend to be longer, that have been noted to become more varied in style and of greater rhetorical flourish. Swanton (1996: xvii, also referring to Clark 1995), describes the development from the ninth- to the twelfth-century annals as

> [. . .] the development of a distinctive prose genre, evolving from brief marginal annotations of plain fact to vivid thumbnail sketches of dramatic events. From the 890s onwards entries are fuller, syntactically more complex, and occasionally linked one with another. The style is increasingly personal and colloquial—accusations are 'thrown at' a traitor, whereupon words are said to 'shoot out' of him against his will [. . .] (Swanton 1996: xvii).

Clark (1995: 19), however, also points out that in spite of this general tendency towards more rhetoric, there is also some stylistic continuity:

> the narrative mode of the Chronicle sways continually between the austere objectivity of the original Easter table entries and a more emotive utterance, sometimes as of the pulpit, sometimes as of every day. This oscillation will continue to the very end; for the Final Continuation, best known for its moving account of the Anarchy, can also be classically objective [. . .] For all the occasions when rhetoric [. . .] breaks in, there remains nevertheless an underlying sense of annals as a genre of their own, requiring a special diction and a special attitude of mind (Clark 1995: 19).

Thus, we have to keep in mind that also the *Chronicles*, due to their specific style and repetitiveness might not be the most ideal text for studying how motion is expressed.

4.2.2.2 *The Voyages of Ohthere and Wulfstan*

The *Voyages of Ohthere and Wulfstan* appear in the Old English translation of Orosius' *Historia adversum paganos*, a history of the world which begins with a chapter on the geography of the world. The Old English translator adds, among other things (see Bately 1980: lxiii–lxx), a description of Northern Europe. The first part of this description of Northern Europe is presented as being the account of Ohthere, a Norwegian merchant who travelled to the court of King Alfred and who talks about his travels around the North Cape and along the coast of Denmark. The second part is the account of another traveller, Wulfstan, probably an Anglo-Saxon (Bately 1980: lxxi), who relates his travels to the Northeast of Europe.

Despite being embedded in a text translated from Latin, the *Voyages of Ohthere and Wulfstan* can be taken to be autonomous Old English material. The language of the two accounts differs, both from each other and from the translation of Orosius (see Bately 1980: lxxii; Fell 1984: 57), so that it is not likely that the translator himself composed them. Both accounts appear very much as if the words of the voyagers had been noted down directly, with only a few explanatory additions (see Lund 1984: 11–12; Fell 1984): There are many instances of *Ohthere/he sæde/cwæþ þæt* 'he said that . . .', particularly in Ohthere's account. Wulfstan's account even has a passage where it slips from the third into the first person (61):

(61) & þonne Burgenda land wæs **us** on bæcbord [. . .] & Weonodland wæs **us** ealne weg on steorbord oð Wislemuðan (OhtWu 3, 1.19.36 and 1.20.4)
'And then we had Bornholm on larboard [lit. was us on larboard] [. . .] and we had Wendland on starboard all the way until the mouth of the Vistula.'

Since a good part of the two accounts consists of such descriptions of the sailing routes, Old English verbs for 'to sail' can be expected to be overrepresented in *Ohthere and Wulfstan*.

Like the excerpts from the *Chronicles*, *Ohthere and Wulfstan* were also taken from the *Helsinki Corpus* (OhtWu2 and OhtWu3), from two different manuscripts. The first one (OhtWu2) is from the older MS L (Lauderdale or Tollemache MS) and contains only the very first bit of Ohthere's account; the second one (OhtWu3) is from MS Cotton and contains the rest as well as Wulfstan's account (on the manuscripts, see Bately 1980: xxxiii–lv).

4.2.2.3 Ælfric's *Lives of Saints*: Lucia, George, Oswald, Edmund

The third body of text from which an excerpt is analysed in the Old English usage study is Ælfric's *Lives of Saints*; here too, the excerpts (the Lives of St. Lucia, St. George, St. Oswald, and St. Edmund) are taken from the *Helsinki Corpus* (AELive). The *Lives of Saints* are suitable for our purpose, because they are rather narrative in character (Wilcox 2006: 243). What is more, they are probably not too 'literary, stilted' (see the discussion in section 3.1.2.1): Ælfric's style is famous for its brevity, clarity, and simplicity (see, e.g., Nichols 1971). He repeatedly refers to this stylistic ideal himself in his prefaces, such as in the one to the *Lives of Saints* in (62):

(62) Nec potuim*us* in ista translatione sem*per* uerbu*m* ex uerbo transferre, sed tamen sens*um* ex sensu, sicut inuenimus in s*anc*ta scriptura, diligenter curauim*us* uertere **Simplici *et* aperta locutione quatinus proficiat Audientibus**. Hoc sciendu*m* etiam q*u*od **prolixiores passiones breuiam*us* uerbis**, non adeo sensu, **ne fastidiosis ingeratur tedium si tanta prolixitas erit in propria lingua quanta est in latina**; *et* non semp*er* breuitas sermonem deturpat sed multotiens honestiorem reddit.

'Nor am I able, in this translation, to render everything word for word, but I have at any rate carefully endeavoured to give exact sense for sense, just as I find it in the holy writing, by means of such simple and obvious language as may profit them that hear it. It is to be further noted that I abridge the longer narratives of the Passions, not as regards the sense but in the language, in order that no tediousness may be inflicted on the fastidious, as might be the case if as much prolixity were used in our own language as occurs in the Latin; and we know that brevity does not always deprave speech but oftentimes makes it more charming.' (ed. and transl. Skeat 1891: 4–5)

Thus, although Ælfric's sources are Latin (see the references in Magennis 2009: 20–9), the *Lives of Saints* cannot be considered to be translations in any strict sense. This is shown, for instance, by Clark's comparison with the Latin *Life of St Edmund* by Abbo, which served as Ælfric's source for the Old English *Life of St Edmund*: Ælfric reduces the text to around a third, discarding much of the 'flowery and pompous rhetoric' and the 'elaborate figures of speech characteristic of his original' (Clark 1968: 30–2; see also McCrea 1976).[71] Nevertheless, the style of the *Lives of Saints* is 'rhetorically very accomplished' (Corona 2009: 297), using rhetorical devices such as alliteration and amplification (e.g. Nichols 1971; Corona 2009). This is useful for comparison with the terser style of the *Chronicles* (see 4.2.2.1).

4.2.2.4 *Gregory's Chronicle*

For the analysis of motion expression in Middle English, *Gregory's Chronicle* was included as a sample of history writing in order to ensure comparability with the Old English *Anglo-Saxon Chronicles*. The text is attributed to William Gregory, mayor of London in 1451/2 (on the question of authorship, see, e.g., McLaren 2002). It is one of the many 'Chronicles of London' which were written in the fifteenth century by London citizens, recording national and municipal events. Like the *Anglo-Saxon Chronicles*, it is written in the form of annals. The excerpt analysed here is taken from the *Penn–Helsinki Parsed Corpus of Middle English* (*PPCME2*, CMGREGOR, starting at the entry for 1397), where it is classified as non-imaginative narrative. The non-narrative parts of the text are omitted.

4.2.2.5 *South English Legendary: St Edmund Rich of Abingdon*

A Middle English saint's life was also chosen, the *Life of Saint Edmund Rich of Abingdon* from the *South English Legendary* (MS Laud misc. 108, late 13th cent. (Boffey 2002: 619), from the *Helsinki Corpus*) to compare to the saints'

[71] See also Minkoff (1976), who shows that there are differences in Ælfric's treatment of the Latin material depending on whether it is biblical or not. Since the *Saints' Lives* contain very little biblical material, they can be considered to reflect Ælfric's own stylistic ideal to a greater extent.

lives analysed for Old English. The *South English Legendary*—or parts of it—is attested in over 50 known manuscripts, so it 'must have been one of the most popular vernacular texts of the late 13 and 14 C' (Görlach 1974: 1). It is written in verse, and its genre has been described by Samson (1986: 192) as 'hagiography rubbing shoulders with romance'.

4.2.2.6 Chaucer's *Canterbury Tales*: *Knight's Tale, Miller's Tale, Reeve's Tale*

Lastly, the *Miller's*, the *Reeve's* and the *Knight's Tale* from Chaucer's *Canterbury Tales* (end of 14th cent.) were examined. They are all written in verse, the former two belonging to the genre of fabliau (cf., e.g., Cooper 1996: 95; 109), the latter being a mixture of chivalric romance and epic (cf., e.g., ibid: 63). The *Canterbury Tales* are famous for their 'fitting of style to genre' (ibid.: 24), and consequently, the *Knight's Tale* as a 'noble storie' is rather elaborate in style, while the two fabliaux as 'cherles tales' are marked by a 'low style' (ibid.: 105), which could be characterized as colloquial in its simplicity and informality. Their syntax, for instance, is rather simple, with little subordination; their vocabulary is 'homely in register' (Cooper 1996: 105); and there is a lot of direct speech with 'its most unsophisticated idioms' (ibid.: 104).[72]

4.2.2.7 Overview

Some potentially problematic issues will have to be borne in mind in the analysis of the material: the repetitive, formulaic nature of much of the *Chronicles*, the thematic slant towards sailing of *Ohthere and Wulfstan*, and the Latin sources of the *Saints' Lives*. An overview of the excerpts serving as the basis for the usage study is provided in Table 4.1 together with the number of motion descriptions extracted from them.

The Old English texts are all taken from the *Helsinki Corpus* (*HC*), and all the motion descriptions they contain have been extracted. For the Middle English period, the *Knight's*, the *Miller's*, and the *Reeve's Tale* from Chaucer's *Canterbury Tales* are taken from the edition of Benson and Robinson (1988), the *Life of Saint Edmund* from *HC*, and again, all their motion descriptions have been gathered. *Gregory's Chronicle* is taken from *PPCME2*, and only the first 200 motion descriptions have been taken into account.[73] Thus altogether, a good 550 motion descriptions could be analysed for Old and Middle English respectively. Note, however, that the number of motion descriptions in the individual texts is rather unequal—the excerpts from the *Anglo-Saxon Chronicles* together account for 80 per cent of the motion descriptions in the Old English samples, for instance. In the analyses in sections 5.2 and 6.3, the texts will therefore always be reported individually.

[72] For details, see Cooper (1996: 88, 104–107, 115–117) and Benson (1986: 64–88); on Chaucer's style in general, also Burnley (1983: 165–200).
[73] The excerpt from *Gregory's Chronicle* in *PPCME2* is longer than the one chosen for *HC*, which would only have contained around 50 motion descriptions.

TABLE 4.1. Material for the Old and Middle English usage study.

	TEXT	PERIOD	MOTION DESCRIPTIONS
OE	*Parker Chronicle* annals 444–946 (HC: COCHROA2)	850–950 (OE2)	284
	Peterborough Chronicle annals 1070–87 and 1105–21 (HC: COCHROE4)	1050–1150 (OE4)	161
	The Voyages of Ohthere and Wulfstan	850–950 (OE2)	30
	MS L (HC: COOHTWU2)	950–1050 (OE3)	
	MS G* (HC: COOHTWU3)		
	Ælfric's *Lives of Saints* Lucia, George, Oswald, Edmund (HC: COAELIVE)	950–1050 (OE3)	82
Total OE			557
ME	*Gregory's Chronicle* annals 1397–1469 (PPCME2: CMGREGOR)	1420–1500 (ME4)	200
	The Life of Saint Edmund Rich of Abingdon (HC: CMSELEG)	1250–1350 (ME2)	86
	Chaucer's *Knight's Tale* (ed. Benson & Robinson 1988)	1350–1420 (ME3)	150
	Chaucer's *Miller's Tale* (ed. Benson & Robinson 1988)	1350–1420 (ME3)	52
	Chaucer's *Reeve's Tale* (ed. Benson & Robinson 1988)	1350–1420 (ME3)	76
Total ME			564

* The information in the *Helsinki Corpus* refers to the Cotton MS as MS G, while Sweet (1883) and Bately (1980) refer to it as MS C.

Also in the usage study, like in the inventory, it would not be surprising if the results for Middle English and for Old English differed to a certain extent, due to the different kind of evidence that is available. With the texts extant from Middle English (e.g. the *Miller's* and the *Reeve's Tale*), we can come closer to what was probably 'everyday language' than is possible for Old English. It would have been amiss to discard these more informal texts only for the reason that there is no equivalent from the Old English period with which they can be compared. A common basis of comparison is ensured through the inclusion of the text type 'chronicle' and 'saint's life' in both the Old and Middle English texts that are analysed (the *Anglo-Saxon Chronicles* and Ælfric's *Lives of Saints* for Old English, and *Gregory's Chronicle* and the *Life of Saint Edmund* for Middle English).

4.2.3 Method

In order to gather the motion descriptions, I read the texts and collected all the sentences in which an intransitive MOTION EVENT is described, no matter whether the FIGURE is animate and agentive, as in (63), or not, such as the fire and the spears in (64) (see 4.1.2.1).

(63) a. sona þa inn **eode** ***an þæs cyninges þegna*** (OE, Edmund, R87)
'Then soon one of the king's thegns went in.'

b. and hoom ***he*** **rit** anon (ME, Knight, 1026)
'And he rode home rightaway.'

(64) a. æfter ðisum gebede **bærst** ut of heofonum ***swyðe færlic fyr*** (OE, George, R141)
'After this prayer, a very sudden fire burst out of the heavens.'

b. Up **spryngen** ***speres*** twenty foot on highte (ME, Knight, 2607)
'Spears spring up, twenty foot high.'

Although most motion descriptions in the texts feature an element expressing PATH (such as *inn*, *hoom*, *ut of heofonum*, and *up* in (63) and (64)), this is not a defining criterion: MOTION descriptions without PATH expression, like (65a–b), are of course taken into account just as well.

(65) a. Seldon he wolde **ridan**, ac **siðode** on his fotum (OE, Oswald, R80)
'Rarely he wanted to ride, but travelled on foot.'

b. Ther seen men who kan juste and who kan **ryde** (ME, Knight, 2604)
'There [i.e. at the tournament] people see who can joust and who can ride.'

Since in the analysis we will also be interested in how frequently the verbs combine with PATH satellites or GROUND elements, the syntactic pattern in which the verbs are used is also recorded. Thus, for instance, in (65b) ME *riden* is tagged as 'V' (bare verb, no complementation), and in (63b) as 'V+Adv_*home*' (with adverbial path satellite *home*).

Only the main motion verb in a clause is taken into account.[74] This means that in examples such as (66a–b), the participles OE *ridende* and *gangende* and ME *russhyyng* are not counted as verbs, since they function as adverbials. In these examples, the main motion verb, and thus the only one that is counted as 'verb', is *ferde* 'went' in (66a) and *cam* 'came' in (66b).[75]

[74] Examples with coordinate motion verbs, such as *No neer atthenes wolde he* ***go ne ride*** (ME Knight, 968) 'He did not want to go or ride any nearer to Athens' were counted twice—once for *go*, and once for *ride*.

[75] In these cases, the component expressed in the participle can always be construed as part of the same motion event as the one expressed in the main verb, e.g. 'went riding and walking over the Thames' in (68a). A motion participle was only counted as a separate clause in one single case in which the participle cannot plausibly be seen as an adverbial expressing a component of the same motion event as that in the main verb: *Sum wegfarende man ferde wið þone feld, þa wearð his hors gesicclod, and sona þær* ***feol, wealwigende*** *geond ða eorðan wodum gelicost* (OE, Oswald R204) 'Some travelling man went towards this field. Then his horse became sick, and soon it fell there, wallowing over the earth very much like someone mad.' Even if the comma between *feol* and *wealwigende*, which was

(66) a. Eac on þis ylcan geare wæs swa mycel ebba [. . .] swa þæt man **ferde** ***ridende*** & ***gangende*** ofer Tæmese (OE, CHROE, 1114.9)
'Also in this same year, there was a great ebb [. . .] so that people went riding and walking across the Thames.'

b. The grete Emetreus [. . .] **Cam** ***ridynge*** lyk the God of armes, mars (ME, Knight 2159)
'The great Emetreus came riding like the God of arms, Mars.'

Like in the study of the inventory (see 4.1.2), caused motion descriptions are not considered. However, motion descriptions which have a reflexive object pronoun that is coreferential with the FIGURE in the subject, as in (67a–b), are included:

(67) a. Syððon **geden** ***heom*** to scipe. (OE, CHROE4, 1070)
'Afterwards they went REFL to ship.'

b. And to the hors he **goth** ***hym*** faire and wel (ME, Reeve, 4062)
'And to the horse he goes REFL fair and well.'

In Old as well as in Middle English, this 'pleonastic reflexive' (Mitchell 1985: §272) often occurs with verbs 'implying motion or rest', such as OE *gān* / ME *gon* 'to go' in (67a–b), which otherwise normally do not take an object, and which also do not have a second participant role (see also the discussion in 6.2.4.1). The pronoun is therefore not an argument of the verb. Its function has been described as conveying 'the notion that the person denoted by the subject was particularly affected by the result of the action or event' (Visser 1963–73: §328). In the case of verbs which can be used both for intransitive and caused motion, such as OE *wendan* and ME *hasten* (see the discussion of examples (29a–b) in 2.2.3), therefore, we cannot tell whether the reflexive pronoun is pleonastic or an argument of the verb, as in (68a–b).

(68) a. þa **wende** he ***hine*** west (OE, CHROA2 R 894.38)
'Then he turned himself/REFL west.'

b. Now to the temple of dyane the chaste/As shortly as I kan, I wol ***me*** **haste** (ME, Knight, 2052)
'Now to the temple of Diana the chaste as quickly as I can I will hasten myself/REFL.'

(68a–b) could thus either be read as, 'he hastened/turned himself to it' (caused motion with AGENT and FIGURE coreferential), or as 'he hastened/turned to

---

probably inserted by the modern editor of the manuscript, is disregarded, the 'wallowing' expressed in the participle *wealwigende* can hardly refer to the MANNER in which the horse 'fell' (*feol*): *geond ða eorðan* 'over the earth' cannot be the satellite of *feol* 'fall', but must belong to *wealwigende*. (It is hard to imagine a FIGURE falling over the earth or around on the earth). Thus, the example is interpreted as describing two separate 'motion events': First the horse falls, then it wallows over the ground. In this case, therefore, both *feol* and *wealwigende* were counted as verbs.

it' (intransitive motion with pleonastic reflexive) (cf. Mitchell 1985: §273). The motion descriptions with reflexive pronouns are all included in the analysis, since they meet the criterion that the moving FIGURE is expressed in the subject of an active clause.

Finally, lexicalized uses of motion verbs in non-motion meanings are not taken into account as motion descriptions. This mostly concerns *forþfēran* (literally 'go away') in the metaphorical meaning of 'to die' such as in (69a–b), which is ubiquitous in the *Anglo-Saxon Chronicles*,[76] but also similar uses of other verbs in the meaning of 'to die', such as *gewītan* in (70) (see also the discussion of *gewītan* in 5.1.3.1).

(69) a. & þy ilcan geare Wulfhere **forþferde** (ChroA2, R.675)
'And in the same year, Wulfhere died.'

b. þrim gearum ær he **forðferde** (AELive (Edmund), R.1)
'Three years before he died.'

(70) heo wearð þa gewundod, þæt hire wand se innoð ut, ac heo ne **gewat** swa þeah, ac þurhwunode on gebedum swa lange swa heo wolde (AELive (Lucia) R.127)
'Then she was wounded [so severely] that her guts came out, but nevertheless she did not die, but continued in prayer as long as she wanted.'

OE *gefaran*, *gegān*, and *gerīdan* also have lexicalized uses in which they occur with an accusative object in the meaning of 'to take, capture, occupy, conquer'.[77]

(71) a. & þa feng Eadred æþeling his broþor to rice, & **gerad** ***eal Norþhymbra land*** him to gewealde. (ChroA, 946.2)
'Then his brother prince Eadred succeeded to the kingdom and took all Northumberland into his dominion.'

b. þa for he þonan to Snotingaham & **gefor** ***þa burg***, & het hie gebetan & gesettan (ChroA, 922.11)
'Then [King Edward] went thence to Nottingham and occupied the fortress and ordered it to be repaired and manned.'

c. On ðam ilcan steode þe God him geuðe þæt he moste ***Engleland*** **gegan** (ChroE, 1086.72)
'On the same place in which God had granted him [= William the Conqueror] that he might take England.'

---

[76] *Forþfēran* meaning 'die' is attested more than 150 times in all of the excerpts from the *Chronicles* included in the *Helsinki Corpus*, almost always in its prefixed form *forþfērde*—the separated form is only found one single time, in the annal for 734: *& ferdon forþ Tatwine & Bieda* 'and Tatwine and Bede died'. Stodnick (2010: 101) links *forþfēran* 'die' in the chronicles to its Latin equivalent *obiit*, which occurs frequently in calendars and Easter tables. Another lexicalized motion verb meaning 'die' in the *Chronicles* is *gefaran*, occurring 18 times in A2 and once in E3–4, as well as the hybrid form *geforþferde* in A2, 571 (which apparently only was *gefōr*, until a different scribe added *-þferde*, cf. Bately 1990: 277 and Stodnick 2010: footnote 30).

[77] See *DOE* s.v. *gefaran*, sense II.A.2 'to capture, occupy (a people / a city)'; s.v. *gegān*, sense B.I. 'to get / obtain / secure (something)', B.I.2 'to obtain / acquire (something *acc.*) by peaceful action',

We cannot be sure whether in contexts like (71a–c), the verbs still conveyed an idea of MOTION, or whether they had been lexicalized as 'to take, capture, occupy, conquer' to an extent where a MOTION meaning is no longer present (it is interesting that ME *entren* has similar uses; see the discussion in 9.1.4). I decided to rather be less inclusive and exclude attestations of these verbs in the 'take, capture, occupy, conquer' readings, such as (71a–c), from the quantitative study of Old English motion description. However, those attestations of the verbs in which they definitely describe motion, such as in (72), are of course included in the analysis.

(72) he **gerad** to Ecgbryhtes stane be eastan Sealwyda (ChroA, 878)
'He rode to Ecgbryht's stone to the east of Sealwyda.'

After these preliminary remarks, chapters 5 and 6 will apply the two perspectives outlined in chapter 4 here—inventory and usage—to Old and Middle English respectively.

B.I.3 'to acquire (something *acc.*) by force'; *BTS* s.v. *gerīdan*, sense III. 'to obtain by riding, ride and get, take possession of'.

CHAPTER 5 | Talking about MOTION in Old English

THE PRESENT CHAPTER STARTS with a study of the Old English motion verb inventory (5.1), followed by a study of usage (5.2), conducted in order to find how motion is typically expressed. The studies of inventory and usage will be supplemented by an exemplary analysis of how Old English translators render sentences with Latin PATH verbs (5.3). Section 5.4 will discuss the different realizations of PATH satellites in Old English.

5.1 *Motion verb inventory*

5.1.1 General results

With the method described in section 4.1.2.3, 189 Old English verbs used to express motion could be found. They are shown in Table 5.1, sorted into the broad semantic categories outlined in section 2.3. For the specific meanings and more details on the single verbs, see Appendix A on the companion website to this book [www.oup.com/us/motionandtheenglishverb/], which follows the same basic structure. In Table 5.1, verbs for which no attestation with a PATH satellite could be found (see 4.1.1) are printed in roman, while the others are italicized: Note that most of the verbs are attested with PATH satellites. Verbs which were only found with a specific prefix are given with the respective prefix in brackets (e.g. *(wiþer)hyppan*).

With 107 members, all in all, MANNER verbs (Ma) are the largest category among the Old English motion verbs, accounting for 57 per cent (see also Figure 5.1). The second largest (25 per cent) are NEUTRAL MOTION verbs ($Mo^{neutral}$, i.e. not primarily expressing MANNER or PATH) with 48 members, and the third largest (10 per cent) are PATH verbs (Pa) with 19 members. A smaller group of 15 is formed by verbs which do not primarily express MOTION.

In the following subsections, the different questions to which the inventory was subjected (see 4.1.1) will be discussed on the basis of these results. Section 5.1.2 will address the question of MANNER salience. Section 5.1.3 will take a closer look at the Old English PATH verbs to see how many of them can be considered

TABLE 5.1. Old English motion verb inventory.*

a) Verbs primarily expressing MANNER OF MOTION (Ma) (57% (107))

1. **Speed (Ma^{speed})** (30): *ærnan, blǣstan, cleacian,* drīfan, *ef(e)stan /of(e)stan, fundian, fȳsan,* hradian, *irnan/iernan/rinnan/yrnan, lǣcan, ōnettan,* pleg(i)an, *racian, rǣsan, recan, scēon, scēotan,* scotian, *scūfan,* scūdan, *scyndan, snēowan, snyrian, st(i)ellan/ styllan, stincan, swengan, tengan,* þeran, þocerian, *þringan*
2. **Slow or impeded motion ($\text{Ma}^{\text{slow/impeded}}$)** (8): *crēopan, healtian, huncettan, luncian, slincan, smūgan, snīcan,* tealtrian
3. **Implying movement through air or fluid ($\text{Ma}^{\text{air/fluid}}$)** (13): *dūfan, fiþercian, fiþerian, flēogan, flēotan, flicerian/flicorian, flogettan,* floterian, flotian, *līþan, rōwan, siglan/ segl(i)an, swimman*
4. **Falling (Ma^{fall})** (9): *drēopan,* drēopian, *drēosan, feallan,* feallettan, *gefetan, hrēosan, lūtan, sincan*
5. **Gliding (Ma^{glide})** (9): *flōwan, glīdan, gelīsian, hæppan, sicerian, seohhian/sēon, slīdan,* slidrian, *slūpan*
6. **Jumping (Ma^{jump})** (6): *hlēapan,* hlēapettan, hoppetan, *hoppian, (wiþer)hyppan, springan*
7. **Rolling (Ma^{roll})** (7): *hwearfian,* hwearftlian, *hwierfan/hwyrfan,* turnian, *tyrnan,* þrāwan, *wealwian*
8. **Fleeing, escaping (Ma^{flee})** (3): gedīgan, *flēon, sceacan/scacan*
9. **Other manner (Ma^{other})** (22): *climban, crūdan,* fēþan, *frician, lācan, rīdan,* s(e)altian, spurnan, *stæppan, (be)stealcian, stalian, (be)stelan, swēgan, swōgan, tredan,* treddan, *treddian,* treppan, tumbian, wealcan, *weallan, windan*

b) Verbs primarily expressing PATH OF MOTION (Pa) (10% (19))

1. **Towards ($\text{Pa}^{\text{towards}}$)** (4): grētan, *lendan,* genēahian, *nēahlǣcan*
2. **In (Pa^{in})** (2): innian, scipian
3. **Away (Pa^{away})** (5): *feorrian, feorsian/fyrsian,* (for)lǣtan, wīcan, *(ge)wītan*
4. **Down (Pa^{down})** (5): gryndan, *hnīgan, hyldan, līhtan, sīgan*
5. **Up (Pa^{up})** (2): *(a)rǣman, rīsan*
6. **Through ($\text{Pa}^{\text{through}}$)** (1): pæþþan

c) NEUTRAL MOTION verbs (Mo) (25% (48)): *būgan, cyrran/cirran/cierran, cuman, dragan, dwelian, dwelsian, dwolian, elþēodgian, faran, fēolan, fēran, fercian, ferian, folgian, fylgan, gān, gangan, gengan, gēotan, hweorfan, lēoran, reccan, scrīþan, settan, sīþian,* spyrian, *stīgan, strīcan, strīdan, styrian, swīcan, swician, swīfan, tēon, (wiþ)tremman, þrægan, wadan, wandrian,* wærlan, *wǣðan, weallian, wegan, wendan,* wōrian, wracian, *wracnian,* wrēcan, *wrigian*

d) Verbs not primarily expressing motion (non-MOTION) (8% (15)): *berstan, brecan,* (ge)dōn, *facian, feohtan, healdan, hīgian,* metan, nēosian, *niman, (ā)redian, sēcan, slēan, weorþan, winnan*

* For meanings, attestations, and prefixes, see Appendix A.

to be pure PATH verbs, since we also need to find out whether the French- and Latin-borrowed PATH verbs that enter the language in the Middle English period (chapter 9) meet with native equivalents or whether they can truly be seen as misfits in the system into which they come. Section 5.1.4 will discuss the

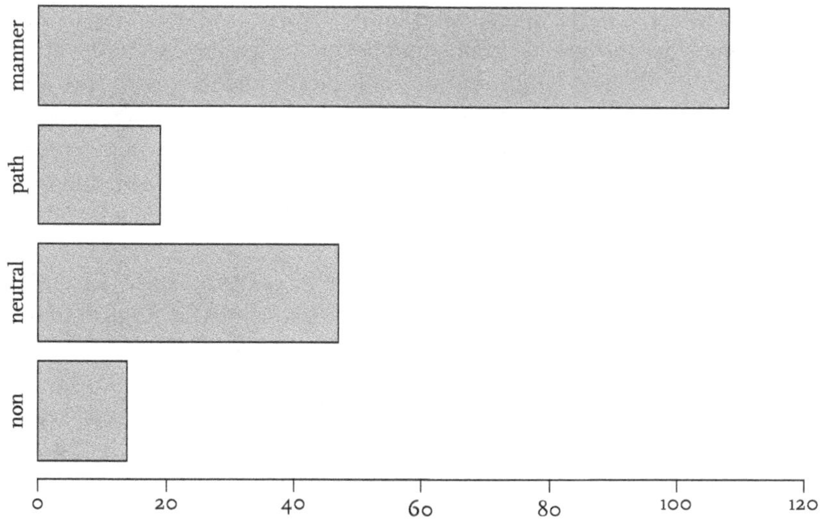

FIGURE 5.1. Types of verbs in the Old English motion verb inventory.

non-MOTION verbs that were found to be used in motion meanings, and section 5.1.5 will present the results of the inventory study with respect to the combinability of the verbs, particularly the MANNER verbs, with PATH satellites.

5.1.2 Verbs primarily expressing MANNER OF MOTION

The fact that MANNER OF MOTION verbs form the largest group among Old English motion verbs (with 107 members) already suggests that Old English was probably indeed a MANNER-salient language, just like Present-Day English (on MANNER-salient languages having large MANNER verb inventories, see 2.1.4.3). This finding corroborates that of Fanego (2012) (see 2.1.4.3), who also found numerous MANNER verbs for Old English. The fact that MANNER verbs are comparatively more numerous[78] in the inventory of the present study is mainly due to my inclusion of verbs with inanimate AGENTS (such as water in *sicerian* 'to ooze' or ships in *flotian* 'to float') and verbs for which the FIGURE is not necessarily in control of the MOTION (such as *hæppan* 'to move accidentally, to slip'), which are outside the scope of Fanego (2012) (see 4.1.2.1).

[78] Fanego (2012) lists 71 Old English MANNER verbs, or rather 64 different verb roots, if different prefixed formations of the same verb root were considered—as they are here—to be instances of the same verb. The present study with its wider scope has found 107 different Old English MANNER verbs. For a comparison with Fanego (2012), however, it has to be taken into account that, sometimes, the classifications adopted in the two studies differ (for details, see 3.2.2 and 4.1.2.2): For instance, many of the verbs that appear in the MANNER verb category of 'Course' in Fanego (2012) (mostly in the meaning 'to travel, to wander') are not categorized as MANNER verbs but as NEUTRAL MOTION verbs in the present study. This concerns the verbs *faran*, *scrīþan*, *sīþian*, *swīcan*, *swician*, *wandrian*, *weallian*, *wendan*, and *wōrian*. If these were taken into account, the difference in 'size' of the manner inventory presented in the two studies would be even larger.

Of course, some of the verbs found in the present inventory may be rarely attested, such as *cleacian*, *fiþercian*, and *fiþerian*, which have only one attestation each (cf. *DOE*, s.vv.), and some may only be attested in glosses (e.g. *fealletan*, glossing L *se concīdere*, cf. *DOE*, s.v.) or in poetry (e.g. *crūdan*, cf. *DOE*, s.v.). Yet, since the Old English textual evidence is limited anyway (see 3.1), it seemed advisable not to exclude these verbs on the basis of their limited number of attestations (note that many of the Present-Day English MANNER OF MOTION verbs are also rather rare: see footnote 39).

Not only the comparatively large number of MANNER verbs in the Old English motion inventory can be taken as indicative of Old English being a MANNER-salient language, but also the fact that many of the MANNER verbs seem to be derived from other MANNER verbs. This could cautiously be interpreted as reflecting a general tendency of introducing more fine-grained distinctions in the domain of MANNER OF MOTION. Thus, for instance, the suffix *-ettan* is involved in the following pairs:

(73) a. *hoppian* 'to hop' (Ma^{jump}) — *hoppetan* 'to jump about' (Ma^{jump})
b. *hlēapan* 'to leap' (Ma^{jump}) — *hlēapettan* 'to jump about, skip' (Ma^{jump})
c. *flēogan* 'to fly' (Ma$^{\text{air/fluid}}$) — *flogettan* 'to fly about, flutter' (Ma$^{\text{air/fluid}}$)
d. *feallan* 'to fall' (Ma^{fall}) — *feallettan* 'to fall' (Ma^{fall})

The meaning of OE *-ettan* is described as 'frequentative or intensifying' (Kastovsky 1992: 391; see also Koziol 1972: §634a: 'intensive-iterative')—thus, *hlēapettan*, for instance, denotes 'repeated jumping'. OE *huncettan* (Ma$^{\text{slow/impeded}}$) may be yet another such derivation, related to an unattested **hincian* 'to halt, limp' (Holthausen 1934, s.v. *huncettan*).

The suffixes *-lian* and *-rian* have also been described as frequentative (Koziol 1972: §634–5); they are involved in the following pairs:

(74) a. *slīdan* 'to slide' (Ma^{glide}) — *slidrian* 'to slither' (Ma^{glide})
b. *hwearfian* 'to turn' (Ma^{roll}) — *hwearflian* 'to turn' (Ma^{roll})
c. *flotian* 'to float' (Ma$^{\text{air/fluid}}$) — *floterian* 'to float to and fro' (Ma$^{\text{air/fluid}}$)

Particularly with regard to the latter two suffixes, *-lian* and *-rian*, Koziol (1972: §§26–42) points out that they are not necessarily only used for word formation, i.e. not only attached to an existing base: Due to their perceived iconic[79] quality, they also appear in coinages, where the element to which they are attached does not exist otherwise (similar to PDE *-le* (<OE *-lian*) in *bust-le, hust-le*). Koziol sees OE *flicorian* as such a *Lautbild* (1972: §42; see also Marchand 1969: 273, 323). Perhaps OE *þocerian* 'to run to and fro' (Ma^{speed}), for which no base is attested either, could also be put in this category.

---

79 This is not straightforward onomatopoeia, but a less immediate type of iconicity, a case of 'relational diagram' in Peircean terms (cf., e.g., Hiraga 1994).

The existence of these MANNER verbs which are formed by suffixation from other MANNER verbs can be taken to support Fanego's diagnosis (2012) that MANNER OF MOTION is a semantic domain that has attracted new lexemes in earlier stages of the English language as well. In other words, if speakers had not felt the need to communicate more fine-grained meaning distinctions in the domain of MANNER, or, more generally, had not wished to be more expressive in this domain, there would have been no reason to derive additional forms such as *slidrian*, instead of simply using *slīdan*. In addition to the numerous MANNER verbs in the motion verb inventory, this is another indication that Old English was a MANNER-salient language.

5.1.3 Verbs primarily expressing PATH OF MOTION

In comparison to MANNER verbs (107) and NEUTRAL MOTION verbs (48), PATH verbs are far less numerous, with only 19 members. It is important to stress, however, that this does not in itself imply that PATH verbs are marginal. After all, the number of possible PATHS is quite restricted and does not really lend itself to such fine-grained differentiation as the domain of MANNER, which virtually does not set any limits to semantic elaboration (e.g. Özçalişkan 2004: 85; Tesnière 1959: 309).

Yet it is highly noticeable in the PATH verbs found for Old English that, although they arguably predominantly encode PATH information, hardly any of them are pure PATH verbs in the sense that PDE *enter*, *descend*, or *exit* are. This is most obvious, of course, for *scipian* 'to take ship', i.e. 'go on board', which refers to 'motion into', but only with respect to very specific GROUNDS (ships). Most verbs that were classified as primarily expressing PATH carry some additional information and cannot really be seen as 'Path verbs proper' (Zlatev & Yangklang's term (2004: 167)), as will be argued in the following sections. This aspect needs in-depth investigation, since the question of whether there are pure PATH verbs in the Old English motion verb inventory is crucial to be able to determine whether the pure PATH verbs that are borrowed from French and Latin (*entrer*, *descendre*, etc.) in the Middle English period are indeed a novelty in the English motion verb inventory. The following subsections will discuss this question by analysing the single verbs in the different subcategories of PATH in alphabetical order, starting with the largest categories, those of PATH^{away} (5.1.3.1) and PATH^{down} (5.1.3.2) with five members each, and moving on to the smaller ones of PATH^{towards} with four members (5.1.3.3), PATH^{in} and PATH^{up} with two members (5.1.3.4–5), and PATH^{through} with one member (5.1.3.6).

5.1.3.1 Verbs primarily expressing PATH^{away}

OE *feorrian*, *fyrsian*/*feorsian*

Feorrian 'to go far away, depart' appears to carry the additional information of 'far away'—in (75), for instance, it is used in the collocation *feorriað hi & fleoð* 'they go far away and flee' to translate Latin *longe fugiunt* 'they flee far'. Thus, it renders the information in the Latin adverb *longe* 'a long way, far':

(75) gif hi hwylcne man on ðam landum ongitað <oððe geseoþ> oððe him hwylc folligende bið, þonne **feorriað** hi & ***fleoð*** (V feor þæt hi fleoð; cf. Orient. mirab. 12.3 *si quis persequatur,* ***longe*** *fugiunt*). (Marv 12.3, DOE s.v. *feorrian,* sense 1 'to go far away, depart')
'If they [the giants] perceive or see some person in this land, or if someone is following them, then they go far away and flee.'

Apart from this additional component of 'far', *feorrian* would be a good candidate for a pure PATH verb. However, according to the *DOE*, *feorrian* is mostly found in glosses (12 out of 15 attestations), most often as a gloss of the Latin verb *elongare*. It is conspicuous that OE *feorrian* is very similar to L *elongare* in its morphological makeup, since L *longus* 'long, far' and OE *feorr* 'far', from which both verbs are derived, can be considered equivalents. Hence, this looks very much like one of the loan formations that Kornexl (2001) describes as typical of glosses, which, as she convincingly argues, often display a 'concern for morphological correspondence' and can be seen as 'manifestations of the putative classroom practice to tell "what every syllable signifies"' (Kornexl 2001: 207–8). The same is true for OE *fyrsian/feorsian*, which, in this meaning, also glosses L *elongare* and *emigrare* (*DOE*, s.v. *fyrsian*).

OE *(for)lǣtan*

OE *(for)lǣtan* 'to leave' is very often used for abandoning other people, not only places, as seen with *fæder & modor, þæt child, hine* in (76a–c). Particularly in (76b–c), the focus does not seem to lie so much on the fact that a FIGURE went away, but rather on the fact that someone was left behind:

(76) a. for ðan **forlæt** se man ***fæder & modor***, & geðeot hine to his wife (cf. Gen: *relinquet homo patrem suum et matrem*). (Gen 2.24, DOE s.v. *for-lǣtan,* sense 12.a 'to go away from permanently, cease to reside with, etc.')
'Because of this the man leaves father and mother and joins himself to his wife.'

b. þa cyrdon heo ham mid þam & ***þæt child*** **forlæten** æt þam temple mid þam oðre femnen. (LS 18.1 (NatMaryAss 10N) 332, DOE s.v. *for-lǣtan,* sense 6.a 'to leave behind (a person *acc.*)')
'With this, they went home, and left the child at the temple with the other women.'

c. he wearð þa gebunden . . . and to ðam lande gelæd, and **forleton *hine*** þær to anum treowe huxlice gebundenne (cf. Idt 6:9 *et sic vinctum de restibus dimiserunt eum*) (ÆHomM 15 155, DOE s.v. *for-lǣtan,* sense 11.a 'to leave, go away from (a place), leave the presence of, leave behind (a person, place, or thing *acc.*)')

'Then he was fettered and led to the land, and there they left him, disgracefully bound to a tree.'

The meanings of *forlǣtan* include 'to leave behind', 'to leave undone, neglect, leave uncared for', 'to abandon, forsake, leave permanently, sever ties with' (*DOE* s.v. *forlǣtan*, senses 6–8, 11a, 12–15), i.e. the thing or person that is left usually seems to be an entity affected in some way, in the semantic role of PATIENT, rather than a mere locative/directional SOURCE/GOAL, such as the deserted wall and town in (77a) or a monastery in (77b). This PATIENT role is clearly visible also in (77c) with *þa forlætenan mynstru* (the participle being used adjectivally here) and *Godes lof ðe ær wæs forlæten*.

(77) a. hig þa **forlættan** þone wall & heora burh, & flugan onwæg (Bede 1 9.46.20, DOE s.v. *for-lǣtan*, sense 13.a 'to abandon, leave unprotected, desert')
'Then they deserted the wall and their town and fled away.'

b. ðes ilces geares com se abbot Heanri of Angeli æfter Æsterne to Burch & seide þet he hæfde **forlæten** þone mynstre mid ealle. (ChronE 1130.12, DOE s.v. *for-lǣtan*, sense 13.c 'to leave, quit, cease to occupy')
'In the same year came the abbot Henry of Angeli after Easter to Peterborough and said that he had quit this monastery [i.e. Angeli] altogether.'

c. ac ic wille nu þurh Godes wissunge þa **forlætenan** mynstru on minum anwealde gehwær mid munecum gesettan & eac mid mynecenum & Godes lof geedniwian ðe ær wæs **forlæten**. (Ch 779 8, DOE s.v. *for-lǣtan*, sense 8 'to leave uncared for, neglect; as past participle: abandoned')
'But now, through God's instruction, I want to fill the abandoned monasteries everywhere in my territory with monks, and also with nuns, and renew God's praise, which was neglected before.'

This ties in with the fact that *(for)lǣtan* is only attested with the thing or person that is left in the accusative object, never in a prepositional phrase—accusative objects prototypically do not encode direction but rather an affected entity.

OE *wīcan*

OE *wīcan* 'to yield, give way' is apparently only attested once (78a) and seems to imply a force acting upon the FIGURE (cf. also the ModG cognate *weichen* 'to yield, give way') and the FIGURE becoming weak: The FIGURE in (78a) is the 'ramparts' of the two halves of the Red Sea, which have been separated by God to allow the Israelites to cross it, and which now 'yield' to the gravitational force again and crush down on the Egyptian army.

(78) a. [. . .] **wicon** weallfæsten, wægas burston, multon meretorras (Cod. Th. 208, 14; Exod. 483, BT s.v. *wícan*, expanded with *DOEC*)
'The ramparts gave way, the waters burst, the towering walls of the [Red] Sea melted.'

b. þæt sio ecg **gewac** brun on bane, bat unswiðor þonne his ðiodcyning þearfe hæfde (Beo 0711 (2575), DOEC)
'[Beowulf struck the monster with the sword] that the edge [of the sword] gave way / became weak, burnished on the bone, and bit less strongly than its lord required.'

c. Ne gemealt him se modsefa, ne his <mæges> laf **gewac** æt wige (Beo. Th. 5251; B. 2629, BT s.v. *gewícan*, expanded with *DOEC*)
'His [i.e. Wiglaf's] courage did not melt, nor did his kinsman's legacy [i.e. his sword] fail at war.'

The prefixed form of the verb, *gewīcan* 'to give way, fail' occurs twice in *Beowulf* (78b–c), in the episode of the fight against the dragon. In the first attestation (78b), it is used to describe Beowulf's sword failing him, becoming weak when he strikes the dragon with it. His retainer Wiglaf then comes to his aid, and his sword does not fail—*ne gewac* in (78c). Neither attestation (78b–c) should necessarily be interpreted as describing a motion event. The status of *(ge)wīcan* as a verb that purely expresses PATH OF MOTION is thus rather unlikely.[80]

OE *(ge)wītan*

The dictionary entries for OE *(ge)wītan* usually do not only give the meaning 'to depart', but also the more general one 'to go': CH (s.v. ±*wītan*[1]), for instance, gives its meaning as 'to go, depart, go out, [. . .] pass away, die (often *forðgew.*)'. The definition in *BT* (s.v. *gewítan*) starts with 'to turn one's eyes in any direction with the intention of taking that direction, to set out towards'. Ogura (2000) gives it as 'to go, depart', and Weman (1933) treats it in the category of verbs meaning 'to set out, go'. All of this suggests that *gewītan* does not only denote 'motion away from somewhere' but also motion in general.

This, however, is of course next to impossible to test via the attestations, since all translocational motion is motion away from somewhere—different from upward, downward, or inward motion. An argument in support of the PATH flexibility of *gewītan* could be that the verb is often used together with prepositions and adverbs expressing PATH^{away}, such as *fram* 'from', *on weg/aweg*

[80] This is supported by the fact that its Middle English reflex, *wiken*, is given by the *MED* with the meanings 'to fail, to flag; also *refl.* give way', and the periphrasis *ben weken* as 'to be yielded up', i.e. not with motion meanings. The Middle English attestations do not invite motion readings either, cf. *Ofte when y syke, wiþ care y am þourhsoht; when y wake, y* ***wyke***, *of serewe is al mi þoht.* (c.1325 *I syke* (Hrl 2253) 53, MED s.v. *wiken*, v. 2) 'Often when I sigh, I am pervaded with grief; when I wake, I languish, all my thought is about sorrow.' The verb is therefore no longer part of the Middle English inventory of motion verbs (cf. Appendix B).

'away', as also shown by Ogura (2000: 71). Some examples from the *DOEC* are given in (79a–c).

(79) a. and se engel ***him* gewat *fram*** (ÆCHom II, 28, 221.20, DOEC)
'And the angel went from them.'

b. oþ þæt þæt scamlease wif geswænced ***aweg* gewat *fram þam ehþyrle his cytan***. (GDPref and 3 (C), 16.212.11, DOEC)
'Until the shameless woman went away, vexed, from the window of his cottage.'

c. and se fugol sona ***aweg* gewat** (ÆCHom II, 11, 93.46, DOEC)
'And the bird immediately went away.'

At least in the case of adverbs (*aweg* in 79b–c), which, different from prepositions (79a), are not needed to introduce a GROUND-NP, this would result in redundancy—'to depart away'; similar in the noun *onweggewiteness* 'away-departure' in (80b) below, or in the prefixed form *forðgewītan* pointed out by CH (lit. 'to forth-depart'). But then again, redundancy is ubiquitous in language, so this is probably a futile argument. However, a cursory look over the attestations of *gewāt* (the singular preterite of *gewītan*) in the prose part of the *DOEC* suggests that the verb is hardly ever used without a directional element (prefix, adverb, or prepositional phrase) when it refers to concrete motion. The 'bare' use of *gewāt* is only frequent when it refers to 'dying', such as in (80a) (see also Ogura 2000: 72). It is not unlikely that this might have played a role in the lexicographers' choice to render it with *depart*—a common Present-Day English euphemism for 'die'—in the dictionaries. In its concrete motion uses, *gewāt* is almost always combined with directional elements, such as in (80b–c).

(80) a. ac traianus **gewat** on þæm ilcan geare. (Æ24 B1.4.24, 111, DOEC)
'But Traian died in the same year.'

b. Ða wæs æfter noht monegum gearum æfter his **onweggewitenesse** of Breotone, þætte Wine wæs adrifen from þæm ilcan <cyninge> of his biscopseðle. Þa **gewat** he ***to Wulfhere Mercna cyninge***; ond mid feo gebohte æt him þæt biscopseðl æt Lundenceastre. (Bede 3 B9.6.5, 5.170.11, DOEC)
'Then it happened, not many years after his departure from Britain, that Wine was driven from his bishop's see by this same king. Then he went to Wulfhere, king of the Mercians, and bought the bishop's see at London from him with money.'

c. Ne con ic noht singan; & ic forþon of þeossum gebeorscipe ut eode, & ***hider* gewat**, forþon ic naht singan ne cuðe. (Bede 4 B9.6.6, 25.342.29, DOEC)
'I cannot sing, and therefore I went out of this feast and went hither, because I could not sing.'

Particularly in (80b–c), it does not seem too convincing to see *gewītan* as primarily expressing PATH^{away}, because in both, the departure is already described in the preceding verb—*wæs adrifen of his biscopseðle* 'was driven from his bishop's see' and *of þeossum gebeorscipe ut eode* 'went out of this feast'—, while the satellite that accompanies *gewat* in both cases (*to Wulfhere Mercna cyninge* and *hider*) denotes the GOAL of the motion.

Another argument is the fact that *gewītan* often collocates with the infinitive of another motion verb, as in (81).

(81) Heo feðera onsceoc, **gewat fleogan** eft mid lacum hire (Gen A,B A1.1, DOEC)
'She shook [her] feathers, went flying back with her offerings.'

Ogura discusses this in more detail and notes that, in this auxiliary-like use, when combined with MOTION verbs, 'the meaning of *gewitan* seems to be weakened and absorbed into the meaning of the verb in the infinitive' (2000: 73). Moreover, it is not only *gewītan* that combines with an infinitive in Old English but, as Visser (1969: §1317) notes under 'Go + infinitive', this combination is possible for a whole range of motion verbs, including OE *faran, fēran, gān, gecyrran, (ge)irnan*, etc. This group of motion verbs, none of which are PATH verbs, might indicate that *gewītan* also means 'to go' rather than 'to depart' in this combination. With respect to *go* + infinitive (the only such combination that remains in the language), Visser (1969: §1318) notes: 'In the course of the subsequent periods the relation between *go* and the infinitive tends to undergo an alteration, in such a way that *go* often loses some of its meaning of motion in space, and dwindles down to a mere sign of aspect.' On this matter, see also Wulff (2006: 120), who describes the constructional semantics of Present-Day English *go* + infinitive as 'initiation of action/event'. I think it is not improbable that Old English *gewītan* + infinitive might have had similar semantics of 'initation of action/event', particularly because we also have the form *uton*, originally the first person plural present subjunctive of *wītan*, which combines with infinitives as an 'adhortative' (cf., e.g., van Bergen 2012), roughly equivalent in function to Present-Day English *let's*. Thus, *gewītan* may have more of an inchoative quality rather than really mean 'to go away', although the two are of course related metonymically.

To conclude, none of the Old English PATH^{away} verbs can be considered to be a pure PATH verb applicable for all kinds of motion events away from a GROUND. OE *feorrian, fyrsian/feorsian, wīcan*, and *forlǣtan* carry additional information besides PATH^{away}: The verbs formed from OE *feorr-* convey the additional meaning of 'far' (and are furthermore predominantly found in glosses of Latin verbs such as *elongare*, and therefore quite likely to constitute artificial formations). OE *wīcan* implies a force acting upon the FIGURE or a FIGURE becoming weak; with OE *forlǣtan*, the GROUND is often 'affected' (closer to the semantic role of PATIENT). Thus, apparently none of them can be used for all kinds of 'motion away'. While these verbs carry additional meaning, the reason why *gewītan* should not be considered a pure PATH verb is different: The attestations, particularly those like (80b–c), suggest that the meaning PATH^{away} is not very prominent in this verb and that *gewītan* has more in common with the general, NEUTRAL MOTION verbs.

5.1.3.2 Verbs primarily expressing PATH^{down}

OE *gryndan*

OE *gryndan* 'to descend' is a weak verb formed from a noun denoting the GOAL-GROUND, in this case *grund* 'ground' (**grund-jan > gryndan*). The non-prefixed *gryndan* is only attested once, as a gloss for L *descendens* (*gryndende*), and the *DOE* gives one other attestation for *āgryndan*, with respect to the setting sun, and with the note 'in poetry'. Therefore, *gryndan* is not a very strong candidate for a good, common PATH verb, if only for lack of attestations.

OE *hnīgan*

Judging from the attestations of *hnīgan* in *BT*, this verb is mostly used for non-translational motion, i.e. where a FIGURE does not change its position but only its posture, as also suggested by the first three words of its definition in *BT*: 'to bend, bow down, incline'. *BT* actually only gives one single example (82) in which the verb is used for translational motion (from *Christ and Satan*), and here it is the need for alliteration (with *he* and *helle*) which might play a role.

(82) Ða he ***to helle*** **hnigan** sceolde (Sat. 375)
'When he must sink to hell.'

For the prefixed form *underhnīgan* 'to descend beneath, go lower than', *BT* has two attestations of translocational motion (also both in alliterative verse), but in a majority of attestations the verb is used metaphorically in the meaning of 'to submit to'.

OE *hyldan*

The non-prefixed OE *hyldan* 'to bend, incline' is also used for changes of posture only, such as in (83a–b):

(83) a. Hie to gebede **hyldon** (Andr. Kmbl. 2057; An. 1029, BT s.v. *hyldan*)
'They bent to pray.'

b. ðeos <wyrt> . . . hafaþ smæle leaf on fingres lencge & todælede & nyþer wið þa eorþan **ahyldende** (Lch I (Herb) 150.0, DOE s.v. *ahyldan*, sense B.1 'physical, of downward direction: to bend down, etc.')
'The plant has small leaves of a finger's length, and divided, and bending down towards the earth.'

In some of its prefixed forms, however, the verb is used to mean 'to turn away, depart' (see *DOE* s.v. *ahyldan, behyldan*), as in (84a–b), where an interpretation with 'down' would not make much sense. In these attestations, the idea of 'down' carried by *hyldan* apparently gives way to the one of 'motion away', or even of motion alone, since 'away' is also expressed in the prepositional phrases with *fram/from*.[81]

(84) a. *ne declines in ira seruo tuo* ne **ahird** [sic] þu on irre ***fram þeowe þine*** (PsGlH 26.9, DOE s.v. *āhyldan*, sense B.2.1.ii 'to turn away from someone')
'Do not turn from your servant in ire.'

b. weræs blod **behildæþ** ***from me*** *viri sanguinum declinate a me* (ABC forms of *onhyldan*, DFGIJK forms of *ahyldan*). (PsGlE 138.19, DOE s.v. *behyldan*[1] 'to turn away, depart')
'Men of blood, depart from me!'

Thus, *hyldan* is not a good candidate for a PATH verb either, since, non-prefixed, it describes non-translocational motion ('to bend'), and since in its prefixed forms, the idea of 'down' is apparently not very prominent.

OE *līhtan*

The first meaning of *līhtan* in *BT* is 'to make light or easy, to alleviate, relieve, assuage', but in view of the attestations listed under the second sense, 'to relieve of a burden, to light, alight', as well as in the second sense of the prefixed form *ālīhtan* (*DOE*, s.v. *ālīhtan*[1]),[82] it seems to be a reasonably good PATH verb, and not exclusively used for descending from horses (85a), but also from chariots (85b) and trees (85c).

(85) a. & se kyng **alihte** ***dune of his hors*** (ChronE 1123.9)
'And the king alighted down from his horse.'

b. [. . .] þa se þegen **alihte** ***of his cræte*** & eode togeanes þe (cf. IV Rg 5:26 [. . .] *quando reversus est homo de curru suo in occursum tui*). (ÆCHom I, 27 408.243)
'When the man alighted from his cart and went towards you.'

c. Zacheus þa swyftlice ***of þam treowe*** **alihte** (ÆCHom I, 38 510.86)
'Then Zacheus alighted quickly from the tree.'

(all from *DOE*, s.v. *ālīhtan*[1], sense 2. 'to alight, get down (from something)')

---

[81] This use is attested in glosses of L *declinare*, which means 'to bend, incline', but also 'to turn away, depart'. It could thus merely be due to the glossator seeing OE *ahyldan/behyldan* as equivalent to L *declinare*, since both share the meaning of 'to bend, decline', even though in these particular contexts (84a–b) the Latin verb would mean 'to turn away, depart'.

[82] However, *ālīhtan* is not very frequent either: According to the *DOE*, there are no more than five attestations in the sense 'to alight, get down', and six others where it means 'to lighten, remove (a burden), relieve (the heart)'.

It is noticeable though, that *(ā)līhtan* is not found in descriptions of longer downward motions, such as from a mountain, but only for very brief descents that can be effected in one single leap or step.[83] This restriction is probably connected to the apparent metonymic origin of its motion meaning in 'for a rider to relieve the horse of his weight', a short descent, which has become generalized to other SOURCE-GROUNDS but not to longer descents. Hence, *(a)līhtan* is restricted to the description of only a subset of downward motions.

The prefixed *gelīhtan*, in contrast, is even found in examples where PATH^{down} does not seem to be involved, as in (86) (cf. *BT*'s sense III.3 'come, approach', from which (86) is taken). For explanatory reasons, I chose to gloss (86) in a very literal way; the translation in Attenborough (1922: 156) renders it as 'comes to his aid on the high road', and the one in Liebermann (1903:174) as 'oder [ihm beim Transport] auf der Strasse beispringt'.

(86) Se þe þeof wrecan wille & æhlip gewyrce oþþe on stræte ***to*** **geliht** (LI. Th. i. 230, II, BT)
'He who wants to avenge a thief and makes an attack or comes to on the road.'

Another example of *gelīhtan* without a notion of 'down' is, according to *BT*, its occurrence as a manuscript alternative for *nēahlǣcan* 'approach' in *Sona þæs þe he* ***nealehte*** *to þæm herige, þa sceat he mid þy spere/Sona þæs þe he* ***gelyhte*** *to þam hearge, þa sceat he mid his spere* 'As soon as he came near to the pagan sanctuary, he shot with the/his spear' (B9.6.4, 10.138.7) in a manuscript of Bede's *Ecclesiastical History*. The FIGURE, the priest Cefi, however, is actually said to ride to the sanctuary on a horse, which would make the reading 'alighted at the sanctuary and shot the spear' possible. A similar reading, in which the FIGURE would jump towards the thief from a horse, is imaginable for (86), although no horse is mentioned in the context.

OE *sīgan*

Similarly, also *sīgan* 'to pass from a higher to a lower position, to sink, descend, decline, fall down' is sometimes found in contexts where no downward motion seems to be involved, cf. *BT*'s sense II. 'to move towards a point (cf. to make a descent upon a place)'. Such an example without downward motion is (87). Here, even the notion 'downstream' is unlikely, since in the preceding lines, Godwin is reported to have come via Dover and the Isle of Sheppey, i.e. he is actually moving up the Thames.

[83] The only description of a longer downward motion recorded by the *OED* is from early Middle English, and the example is actually the same for both verbs ('this/a man descended from Jerusalem to Jericho'): *Þes Mon..a* ***lihte*** *from ierusalem in to ierico* (*a*.1225 (OE) *MS Lamb.* in R. Morris *Old Engl. Homilies* (1868) 1st Ser.79 OED3 s.v. *alight*, v.[1]) and *A mon* ***lihte*** [L. *descendebat*] *from ierusalem into ierico.* (*c*.1175 *Lamb. Hom.* 79, OED2 s.v. *light*, v.). As is clear from the reference to L *descendebat* in the quote, this seems to be an attempt at a close translation of a biblical sentence that contains *descendere*. The discussion in 5.3.2.3 shows that the Vulgate often has PATH verbs for journeys to and from Jerusalem, which are usually not rendered with a PATH verb in the Old English translation.

(87) Godwine **sah** him æfre towerd Lundenes mid his liþe ðæt he com to Suþgeweorce (Chr. 1052, BT s.v. *sīgan*, sense II 'to move towards a point') 'Godwin kept moving towards London with his force until he came to Southwark.'

The prefixed form *āsīgan*, again, is mostly attested in non-translocational uses, i.e. meaning 'to sink to the ground, collapse', as in (88).

(88) sloh hine ða an hiora mid anre æxe yre on þæt heafod, þæt mid þam dynte he nyþer **asah** (ChronC 1012.12, DOE s.v. *āsīgan*, sense 2.a 'to sink to the ground') 'Then one of them hit him with an axe-iron on the head, so that he sank down with the blow.'

In sum, OE has no pure PATH^{down} verb: OE *(ā)līhtan* does not apply to all sorts of downward motion events, but only to short descents effected in one leap. The others (*hnīgan, hyldan, asīgan*) predominantly denote change of posture, i.e. non-translocational motion, are used in contexts where PATH^{down} is not involved (*ahyldan, sīgan*), or are rarely attested (and only in glosses or poetry: *gryndan*).

5.1.3.3 Verbs primarily expressing PATH^{towards}

OE *grētan*

In the PATH^{towards} verbs, *grētan* 'to approach, come near, visit' is perhaps the most problematic candidate. Though 'approach, come near, visit' is the first sense listed in *DOE*,[84] *grētan* has many other meanings, among them 'to assail, afflict' (see 89a), 'to address, speak to' (see 89b), and, as in Present-Day English, 'to greet' (see 89c), which, judging from the large number of attestations provided in the *DOE* entry (in senses 9.a–9.g s.v. *grētan*[1]), is already quite widespread in Old English.

(89) a. gif man mannan an oþres flette manswara hateþ oððe hine mid bismærwordum scandlice **grete**, scilling agelde þam þe þæt flet age. (LawHl 11, DOE s.v. *grētan*[1], sense 6e 'to assail verbally; to abuse, revile') 'If somebody in another person's house calls someone a perjurer, or assails him disgracefully with insults, he shall pay a shilling to the one who owns the house.'

[84] The reason why the apparently rare sense of 'approach, come near, visit' is given as the first one in the *DOE* is probably that it may be the historically oldest one. On the etymology of *greet*, the *OED2* writes: 'The primary sense is uncertain; the senses of early occurrence in continental Germanic are 'to approach', 'to call upon', 'to provoke or compel to action', 'to attack', 'to irritate, annoy', 'to address, salute'. In modern German and Dutch as in English the sense 'salute' has become the prominent one, such other senses as survive being now apprehended as transferred from this. [. . .] The ultimate etymology is equally uncertain with the radical meaning; many scholars refer the word to Old Aryan **ghrōd-*: *ghrēd-* 'to resound' (see *greet* v.[2]), on which supposition the primary sense should be 'to call on'; another view is that the Germanic root **grôt-* is an extension of the root which appears in Greek as *χρψ* with the sense 'to approach closely, touch', etc.' (*OED*, s.v. *greet* v.[1]).

b. [. . .] næbbe ic nænigne intingan þe to geseonne ne ðe to **gretanne** (Mart 5 (C) My 20, A.4, DOE s.v. *grētan*[1], sense 8 'to speak to, address (someone *acc.*))
'I don't have any cause to see you nor to speak to you.'

c. þa com eft to him se foresprecena his freond & mid bliðe ondwleotan hine halette & **grette** (cf. BEDA. Hist.eccl. 2.12, 180 *laetoque uultu salutans eum*) (Bede 2 9.130.2, DOE s.v. *grētan*[1], sense 9.a 'to greet (someone *acc.*) etc.')
'Then again his abovementioned friend came to him and with joyful face saluted and greeted him.'

Despite the verb being rather frequent overall (roughly 350 occurrences for *grētan* and approximately 50 for *gegrētan*, see *DOE*), the sense 'to approach, come near, visit' does not seem to be very widely attested. According to the *OED*, it is also a sense 'which did not survive beyond Old English'. In the sense 'to approach, come near, visit', *DOE* actually gives only one single attestation for *grētan*, (90), from a riddle where, moreover, a pun is involved on another use of *grētan*, namely 'to have sexual intercourse with'.

(90) bið stiþ ond heard, stede hafað godne; þonne se esne his agen hrægl ofer cneo hefeð, wile þæt cuþe hol mid his hangellan heafde **gretan** þæt he <efenlang> ær oft gefylde (with pun on sense 1.c below: 'to have sexual intercourse with'; polite solution is 'key'). (Rid 44 3, DOE s.v. *grētan*[1], sense 1 'to approach, come near to, visit')
'It is stiff and hard, it has a good place; when the man pulls up his own dress over the knee, and, with the head of his hanging thing, wants to approach the well-known hole of the same length, that he often filled before.'

For the prefixed *gegrētan*, this is similar. Most of the attestations provided in the *DOE* entry relate to the sense 'greet, salute', which seems to be the most prominent one also for *gegrētan*. Actually, even most of the attestations listed under sense 1 'to approach, visit, etc.' involve coming with gifts, as in the italicized parts of (91a–c).

(91) a. & þæt hy Godes cyrican æghwar georne griðian & friþian & ***mid leohte & lacum*** hi gelome **gegretan**, & hy sylfe þær georne to Christe gebiddan. (LawVIAtr 42.3)
'That they willingly protect and defend God's church everywhere and often visit it [i.e., the church] with light and offerings, and willingly pray to Christ there themselves.'

b. [. . .] and sece ***mid his ælmessan*** cirican gelome, and halige stowa ***mid his leohte*** **gegrete**. (Conf 4, 29.372)
'And shall often seek the church with his alms, and visit the holy places with his light.'

c. [. . .] & halige stowe **gegrete *mid leohte & mid oðre ælmyssan*.** (WCan 1.2 108)
'And visit the holy place with light and with other alms.'

(all from *DOE* s.v. *gegrētan*, sense 1 'to approach, visit (something / some place / someone *acc.*, with offerings, alms, gift, etc.)')

Thus, even the attestations which illustrate the sense 'approach, visit' seem very close to the sense 'to salute, make a sign of veneration to' (*DOE*, s.v. *gegrētan*, sense 5.c) or 'to show respect to, venerate' (*DOE*, s.v. *grētan*, sense 9.3). Overall, *grētan* does not seem to be a likely candidate for a pure PATH verb.

OE *lendan*

OE *lendan* 'to arrive, come to land' does not only express PATH either, but also carries GROUND information ('to come to land' < **land-jan*, built on the same derivational pattern as *gryndan* in 5.1.3.2). Furthermore, in the form *gelendan*, the notion of PATH^{towards} sometimes seems to recede in favour of the general idea of MOTION; this is reflected in *BT*'s sense II 'to go, proceed' and in examples such as (92), where no notion of 'arrival' can be implied.

(92) Ðæt scip wile hwilum stigan ongean ðone stream, ac hit ne mæg, buton ða rowend hit teon, ac hit sceal fleotan mid ðy streame: ne mæg hit no stille gestondan, buton hit ankor gehæbbe, oððe mon mid roðrum ongean tio; Elles hit **gelent** mid ðy streame. (CP B9.1.3, 2378-9, BT s.v. *gelendan*, expanded with *DOEC*)
'The ship sometimes wants to go against the stream, but it cannot, unless the rowers pull it, but it shall float with the stream. It cannot stand still unless it has an anchor or people pull against with oars; otherwise it goes with the current.'

OE *nēahlǣcan*, *genēahian*

OE *nēahlǣcan* and *genēahian* 'to approach' contain the adverb/preposition *nēah* 'near, nigh' in their morphological make-up.[85] OE is *nēahlǣcan* is found in combination both with prepositional phrases (93a) and with direct objects (93b).

(93) a. And he ana belaf and **nealæhte *to þære stowe*** (Hml. S. 30, 105, BTS s.v. *nēahlǣcan*)
'And he alone remained and approached to the place.'

b. Ða he ***hyre* nealæhte**, þa stod hyre big iong man fæger mid gyldenum hræglum gegyred. (Mart 5 (Kotzor) 811 (Au 2, B.8), DOEC)

[85] The gloss words *feorrian* and *fyrsian/feorsian* discussed in 5.1.3.1 might therefore also have been modelled as antonyms to the existing *nēahlǣcan* and *genēahian*, since the adverbs on which they are based respectively—*nēah* 'near' and *feorr* 'far'—are antonyms as well.

'When he approached her, a young man stood by her, clothed handsomely in golden garments.'

In the examples with direct object, the notion of PATH^{toward} can only be conveyed by the verb because there is no other PATH expression (such as a preposition or adverb). OE *nēahlǣcan* and *genēahian* can thus be considered pure PATH verbs, the only ones in the PATH^{toward} group.

5.1.3.4 Verbs primarily expressing PATH^{in}

OE *innian*

The meanings that *BT* gives for OE *innian* are 'to get within, put in, bring in, put up, lodge'. On its derivation, the *OED* remarks that 'Old English *innian, geinnian,* appear to attach themselves in part directly to the adv. *inn,* partly to be more immediately associated with the derivative, *inn* n.', i.e. to the noun meaning 'a dwelling, house, chamber, lodging', as in (94a–c). In the latter interpretation, GROUND information would be involved as well, and the derivational pattern would be like the one in *scipian* 'to take ship', *lendan* 'to come to land' (< **land-jan*) and *gryndan* (< **grund-jan*): a weak verb formed from the noun denoting the GOAL-GROUND. These uses where 'give/find lodging' is involved (i.e. the noun *inn*) are also the only ones that survive into Middle English (cf. *MED*, s.v. *innen* v.2).

(94) a. Þa hi þider comon, þa woldon hi **innian** hi þær heom sylfan gelicode. (ChronE B17.9, 1072, DOEC)
'When they came there [to Dover], they wanted to lodge themselves where it pleased themselves.'

b. [. . .] and **innian** þa gode þearfan and þa nacodan scrydan and fedan (HomS 42 (Baz-Cr) B3.2.42, 79, DOEC)
'[. . .] and shelter the good paupers and clothe and feed the naked.'

c. Þæs cynges cniht ða se ðe hæfde **geinnod** þone godspellere matheum æt his huse (ÆCHom II, 37, 276.115, DOEC)
'The king's servant who had sheltered the evangelist Matthew in his house.'

Most of the attestations of OE *innian* found in the *DOEC* are instances of CAUSED MOTION, i.e. 'to bring in, shelter, house (somebody)', as in (94b–c). Note that (94a), too, has a reflexive pronoun *hi* ('lodge themselves'). Actually, the only clearly intransitive motion attestation that I could find is (95a), the one given in *BT* from the Old English translation of Boethius' *De Consolatione Philosophiae*. Furthermore, other manuscripts of the same text have a different wording of this line: (95b), for instance, has the combination of adverb *innan* 'inside' and *bið* 'is', instead of the verb form *innaþ*.

(95) a. He werodaþ syððan he **innaþ** (*interius recepta dulcescant*) (Bt. 22, 1; Fox 76, 31, BT s.v. *innian*)
'It [i.e. the 'medicine' of the teachings of wisdom] grows sweet when it enters.'

b. He is swiðe biter on muðe, & he þe tirð on ða þrotan þonne ðu his ærast fandast; ac he weredað siððan he **innan bið** & swiðe liðe on ðæm innoð (Bo 22.51.2, DOEC)
'It [i.e. the 'medicine' of the teachings of wisdom] is very bitter in the mouth, and it pulls on the throat when you first try it, but it grows sweet when it is inside, and very gentle in the stomach.'

Thus, *innian* is not a convincing pure PATH verb either (a) because of its relation to the noun *inn* 'lodging' and (b) because it is rarely used for intransitive, but mostly for CAUSED MOTION.

OE *scipian*

Scipian 'to take ship' is not a pure PATH verb, since it contains very specific GROUND information (GOAL = *ship*), see (96). Note that *on Wiht* is not a directional, but a locative adverbial here.

(96) And se eorl Rodbeard [. . .] on Wiht **scipode** & into Normandig for (ChronE (Irvine), 1091.41, DOEC)
'And the Earl Robert [. . .] took ship in the Isle of Wight and went into Normandy.'

5.1.3.5 Verbs primarily expressing PATH^{up}

OE *ārǣman*

OE *ārǣman* 'to rise' is only attested three times: with the FIGURE of the rising sun (97a), with Abraham rising up in (97b), and in the metaphorical caused motion sense of 'improve oneself' in (97c). The intransitive example in (97b) seems to describe non-translocational motion (i.e. Abraham rises with the sword to kill Isaac).

(97) a. dæges þriddan up ofer deop wæter ***ord* aræmde**. (GenA 2873)
'On the third day, the daybreak arose over the deep water.'

b. up **aræmde** <Abraham þa>; se eorl wolde slean eaferan sinne unweaxenne, <ecgum> reodan magan mid mece (Ex 411)
'Then up rose Abraham; the man wanted to kill his young son, redden the kinsman with the blade, with the sword.'

c. ealra þissera apostola geendunge ic hæbbe awriten [. . .] ða ge mihton rædan & eow **aræman** on þam, gif ge holde wæron eowrum agenum sawlum. (ÆLet 4 (SigeweardZ) 1006)

'I have written down the deaths of all the apostles [. . .] you can read and improve yourselves on this, if you are friendly to your own souls.'

(all from *DOE* s.v. *ārǣman*, sense 1 'to rise; *up arǣman* 'to rise up', sense 2 'to raise, improve (oneself)')

There is thus only one attestation of OE *ārǣman* in which it is used for intransitive translocational motion 'up', namely that of (97a) with the FIGURE of the rising sun (see the discussion of OE *rīsan* that follows).

OE *rīsan*

The situation is similar for *rīsan* 'to rise', mostly found in its prefixed form *ārīsan*. Where it is used for literal motion, it mostly refers to non-translocational motion, i.e. 'stand up, get up', as in (98a–b), just like what we saw in the group of the PATH^{down} verbs (*hnīgan, hyldan, āsīgan* in 5.1.3.2).

(98) a. Theodosius [. . .] **aras** þa of þære flora, and of þam wacan sæcce þe he lange onuppan [. . .] wæs sittende (cf. Sept.dorm. 344 *surrexit ex humo et de sacco quae habebat sub se*). (LS 34 (SevenSleepers) 736, DOE s.v. *ā-rīsan*, sense A.1 'to get up from sitting, lying, kneeling, etc., to stand up')
'Theodosius [. . .] got up from the floor and from the soft sackcloth upon which he had long been sitting.'

b. on morgen, ðа he **aras**, he nam þone stan [. . .] (cf. Gen: *surgens ergo mane*). (Gen 28:18, DOE s.v. *ā-rīsan*, sense A.3 'to get up from sleep or rest')
'In the morning, when he got up, he took the stone [. . .].'

The only instances in the *DOE* attestations that are translocational are those that refer to the rising sun, moon, or stars or to smoke, mist, and the like, as in (99a–c) (see also (97a) with *ārǣman* above).

(99) a. up **aras** ***seo beorhte sunne*** riht on eastende þære heofon. (ByrM 1 2.1.264, DOE s.v. *ā-rīsan*, sense A.8 'of the sun, moon, stars, the light of dawn: to rise, come above the horizon')
'The bright sun rose up right on the east end of the sky.'

b. he [. . .] het his cnapan ðа hwile hawian to ðære sæ gif ænig ***mist*** **arise** of ðam mycclum brymme (ÆLS (Book of Kings) 142, DOE s.v. *ā-rīsan*, sense A.9 'of mist, smoke, steam, etc.: to go up, ascend on high, mount, etc.')
'He [. . .] told his servant to look to the sea in the meantime, if any mist arises from the vast sea.'

c. ðа of ðære stowe ***steam*** up **aras** swylce ***rec*** under radorum. (El 802, DOE s.v. *ā-rīsan*, sense A.9 'of mist, smoke, steam, etc.: to go up, ascend on high, mount, etc.')
'Then steam rose up from this place like smoke under the firmament.'

These uses are very similar to those of *ascenden* in Middle English (see 9.5). The preference of *ārīsan* for FIGURES whose MANNER OF MOTION would actually be hard to describe—In which MANNER does the sun move above the horizon?—seems to be symptomatic for a language which otherwise tends to have MANNER OF MOTION in the verb.

5.1.3.6 Verbs primarily expressing PATH^through

The only candidate for PATH^through, *pæððan* 'to tread a path, to traverse', is only found in poetry (*TOE* s.v.), and the notion of 'to tread, to go on' (i.e. NEUTRAL MOTION) seems to be more prominent in it than that of 'traverse': The verb is derived from 'the Germanic base of *path*, n.', and its equivalents in other Germanic languages also mean 'to tread', 'to go, walk' (*OED3*, s.v. *path*, v.). With *pæððan*, the GROUND is only found expressed in an object, without a preposition, as in (100).

(100) Ic mearcpaþas træd, moras **pæþde** (BT, Exon. 485, 8 ; Rä. 71, 10)
'I treaded the paths, I treaded the moors.'

This looks like what the *DOE* (e.g. svv. *faran*, *fēran*) calls an 'accusative of way', and Mitchell (1985: 580) the 'accusative of extent' (cf. also Kisbye 1972: 65). Therefore, the idea of 'traverse' is likely to be only contextual (see 2.2.3); note also the parallelism with *træd* in (100).

5.1.3.7 Conclusion

This closer inspection of the Old English PATH verbs has shown that most of them are not pure PATH verbs: Some either also have a MANNER component (*wīcan*) or a GROUND component (*gryndan*, *lendan*, *scipian*, to some extent also *innian*). The 'away' verbs *feorrian* and *fyrsian/feorsian* are mostly restricted to glosses and carry the additional meaning of 'far'. Some PATH verbs are restricted to brief descents (*(ā)līhtan* 'to alight, descend' for alighting from horses, chariots, and trees), in others the GROUND argument is profiled as PATIENT or AFFECTED (*forlǣtan*, *grētan*). Still others are more commonly used to describe a change in posture and not in location (*hnīgan*, *hyldan*, *āsīgan*; *ārīsan* and *ārǣman* for animate FIGURES). Some are rare in general (*arǣman*, *genēahian*, *gryndan*) or are rarely used to describe intransitive motion (*innian*). The verbs *gelendan*, *gelīhtan*, *sīgan*, and *gewītan* are all found also in contexts where there is no PATH 'to', 'down', or 'away' involved, which suggests that their PATH meaning is not as prominent as their general MOTION meaning. Also *pæððan* appears to be better classified as a NEUTRAL MOTION verb.

This leaves us with only *(ge)nēahlǣcan* and (rare) *genēahian* 'to approach' as pure PATH verbs in Old English. We can therefore conclude from the discussion of the Old English verbs primarily expressing PATH that there are very few pure PATH verbs. This fully meets the expectations for a language that typically expresses PATH outside the verb (see 5.4 on satellites).

5.1.4 Verbs not primarily expressing MOTION

The fact that PATH is typically expressed outside the verb also allows non-MOTION verbs to occur in motion descriptions, since the meaning 'directional motion' is carried by the construction, and the meaning of the verb will be interpreted in line with the constructional meaning. These verbs are discussed in sections 5.1.4.1–4. Yet in the group of non-MOTION verbs used to describe motion, there are also verbs which do not owe their interpretation to a motion construction, but, for instance, to metonymy, such as OE *sēcan* 'to seek' (see section 2.3). These will be discussed in section 5.1.4.5.

5.1.4.1 'come to be, become': OE *weorðan*

That a MOTION interpretation is created through the construction is most evident with *weorðan*, a copula verb meaning 'to come to be, become'.[86] The verb usually implies a change of state and comes with what Goldberg & Jackendoff (2004: 537) for Present-Day English call a 'property resultative' as in (101a–b).

(101) a. Đa hwile ðe he ðær stod, he **wearþ** ***færinga geong cniht***, and sona eft ***eald man*** (Blickl. Homl. 175, 2, BT s.v. *weorþan*, sense II 'to become, be made, be')
'While he [Simon the Sorcerer] stood there, he suddenly became a young boy and then again an old man.'

b. Sume **wurdon** ***to wulfan*** (Bt. 38, 1, BT s.v. *weorþan*, sense III.c.1 'to become, turn to')
'Some became/turned into wolves.'

(102) Gif hi ***on treowum*** **weorþaþ** (Met. 13, 36, BT s.v. *weorþan*, sense IV 'implying movement: to come, get')
'If they [the tamed birds] get into the trees.'

When combined with a 'spatial resultative', however, as in (102), the composite meaning of verb and construction is read as 'to come, get to'—the change of state is construed as a change of position.

5.1.4.2 'break/burst'

While copula verbs are no longer allowed in the Present-Day English IMC (Rohde 2001), we can also identify some subgroups among the Old English non-MOTION verbs which, according to Rohde (2001: 355), are still possible in the Present-Day English IMC: OE *berstan* 'burst' and *brecan* 'break' can be seen as verbs expressing 'disruption of material integrity', like their Present-Day English continuations *burst* and *break*.

[86] Petré & Cuyckens (2009: 335) argue that OE *weorðan* 'to be, become' has its ultimate origin in a motion verb.

(103) a. he dælde þam earmum mannum, þe ***fram þam herge*** **burston** (O aburston; cf. GREG.MAG. Dial. 4.23.1 *a Langobardorum depraedatione fugientibus cuncta ... largitus est*). (GD 4 (C) 23.293.7, DOE s.v. *berstan*, sense 6.b 'to break away from, escape from')
'He gave [everything he could find in the garden] to the poor men who had burst [i.e. escaped] from the host.'

b. geseah [. . .] stream ***ut*** þonan **brecan** ***of beorge*** (Beo 2545, DOE s.v. *brecan*, sense 4c *brecan of / ut of* 'to burst from or burst out of sth.')
'He saw [. . .] a stream breaking out from there of the mountain.'

Their motion reading in (103a–b) arises through the combination with the satellites *fram þam herge* 'from the host' and *ut of beorge* 'out of the mountain'; the participant role of 'burster'/'breaker' can be construed as a moving FIGURE. The r-relation between verb semantics and constructional semantics would be one of MEANS—e.g. the stream is moving out of the mountain by means of breaking it—or, alternatively, one of RESULT.

5.1.4.3 'labour, toil/fight'

The verbs OE *hīgian* 'to labour, strive, endeavour', *winnan* 'to labour, toil, work', and *feohtan* 'to fight' in an IMC receive the interpretation 'to make one's way', 'to fight one's way'. The meaning of the verb is in a relation of MEANS to the constructional meaning (go by means of labouring/fighting), and the participant role of 'labourer, struggler, fighter' fuses with the argument role of MOVING FIGURE. This use is not common in Present-Day English, except in a *way*-construction, as in the translations of (104b–c) (on the English *way*-construction, see Goldberg 1996; on its development, see Israel 1996 and Traugott & Trousdale 2013: 76–91).[87]

(104) a. he **higode** ***to þære stowe*** (GD 2 (C) 0029 (1.99.19), BTS s.v. *hīgian*, sense C.α ~*tō* of movement, expanded with *DOEC*)
'He strove to the place [where St Benedict was hiding].'

b. Hwæt is ðæt wundor, ðæt geond ðas woruld fareþ [. . .], **winneþ** oft ***hider***? (Salm. Kmbl. 568 ; Sal. 283, BTS s.v. *winnan*, sense III.b 'to win, make one's way')
'What is that wonder that travels around this world [. . .] often works [its way] hither?'

c. þa hie swiðost tweode hwæðer hie aweg comen, þa gecwædon hie þæt hie sume hie beæftan wereden & sume ***<þurh> ealle þa truman ut*** **afuhten**, gif hie mehten. (Or 5 7.121.27, DOE s.v. *ā-feohtan*, sense 2 *ut afeohtan* 'to fight one's way out')

[87] Although it sometimes occurs in Present-Day English as well, cf. *I went to the concert and I **fought** through the crowd* (The Strokes, 'Heart in a Cage') or *They **toiled** up the hill.*

'When they strongly doubted whether they would get away, then they said that some of them would defend them behind, and some fight [their way] out through all the troops, if they could.'

The motion use of OE *feohtan* has been discussed in detail in section 2.2.3 above. The contextual semantics of OE *winnan* and *hīgian* in this construction are comparable to the ones of *travel* in Middle English, which also originally means 'to labour, toil, work' (cf. French *travailler* 'to work'; see the discussion in 2.2.3 and 6.2.3.2).[88] The laboriousness of the way in (104a), for instance, is highlighted in the subsequent lines, which describe the FIGURE's extensive search for the place where St Benedict is hiding.[89] Both verbs are still possible in this use in Middle English, but whereas the motion reading of ME *winnen* appears to continue to be contextual (see 6.2.5.3), 'to go quickly, hurry' has apparently become a decontextualized meaning of ME *hien* (< OE *hīgian*), as can be seen by the sheer number of attestations in this sense in the *MED* (s.v. *hien*) and in uses such as (105):

(105) þe first it gas, þe toþer it **hise** / þe thrid quil es won to rine, / And fleis þe furth wit-vten blin (a1400(a1325) *Cursor* (Vsp A.3) 21278)
'The first [wheel], it goes, the second one hurries, the third wheel is wont to run, and the fourth flies without stopping.'

ME *hien* occurs in a scalar context together with the motion verbs *go*, *run*, and *fly*, ranked by speed in (105), and does not depend on a satellite to receive a motion meaning.

5.1.4.4 Other non-MOTION verbs in the IMC

OE *slēan*

For Present-Day English, Rohde (2001: 355) identifies 'verbs of forceful impact' like *to smash* or *to ram* as possible in the IMC. OE *slēan* 'to strike' can be seen as a 'verb of forceful impact' as well, and similar to the Present-Day English counterparts, the meaning it receives through the intransitive motion construction is 'to move rapidly, rush, dash', see (106).

(106) He on scip astah and **sloh *ut on ða sæ*** (Ap. Th. 6, 6., BT s.v. *slēan*, sense III 'to move rapidly, rush, dash, break, take a certain direction')
'He mounted on a ship and dashed out onto the sea.'

---

[88] See also Ihrig (1916: 157) for a few Modern English dialect verbs which seem to have taken the same semantic development from 'do anything with difficulty, labor' to 'walk with difficulty, travel'. The *OED2* comments that the motion use of *win* 'depends on that of ON. *vinna*' (*OED*, s.v., sense 12). It is of course plausible that the strong contact with another satellite-framing language may have further strengthened the tendencies already present in Old English.

[89] *þa sohte þone Godes wer geond ealle þa stæþhlypan þara munta & geond þa holenesse þara dena & geond þa seaþas & dicas þara landa. Þa gemette he hine lutian in anum scræfe* 'The priest then searched

Usually, OE *slēan* refers to situations where someone strikes someone or something (e.g. *Gif man oþerne mid fyste in naso slæhð* (LawAbt B14.1) 'If someone strikes another one on the nose with the fist'). When the verb is used in the IMC, by contrast, its usual participant structure appears to be shaded, and its meaning component of 'forcefulness' appears to go into the motion reading achieved through the interplay with the constructional semantics.

OE *fācian*

A more difficult case is the verb *fācian*. It is only attested twice all in all, once with a direct object, in (107a), in a sense given by the *DOE* as 'to desire to obtain', and once in combination with an *on*-prepositional phrase, in the sense 'to get to, reach' (107b).

(107) a. he him selfum **facade** Mæcedonia anweald (Or 3 11.82.10, DOE s.v. *fācian*, sense 1 'to desire to obtain (something *acc.*)')
'[Pyrrhus] himself desired to obtain power over Macedonia.'

b. þonne þær east andlanges þære ceapstræte oð cyninges burghege on þone ealdan mylengear þæt þær andlanges þæs ealdan mylegeares oð hit **facað** on þæm ifihtan æsce. (Rec 22 4, DOE s.v. *fācian*, sense 2 *facian on* ? 'to get to, reach')
'Then east along the market street until the king's city hedge on the old mill-yair, then along the old mill-yair, until it gets to the ivy-covered ash tree.'

The context in example (107b) is a description of the estate boundaries of a certain Ealhswið. It is an instance of 'fictive motion' (cf. Talmy 2000a: chapter 2), i.e. it describes a stationary scene in which the FIGURE (the boundary in (107b)) is an unmoving object, such as the fence in PDE *The fence goes from the plateau to the valley* (2000a: 99). I first had intended to eliminate *fācian* from the list of motion verbs for the reason that it is only attested in a context of fictive motion. But since the expression of fictive motion usually functions along the same lines as the one of 'real' motion (see, e.g., Matlock 2004), *fācian* was retained. For the sense development of *fācian* in (107b), the *DOE* refers to Modern English *fetch* as a parallel: In nautical language, the verb *fetch* can be used in a similar way, as in (108).

(108) We tacked and stood over for Cape Elizabeth, under which we **fetched** at half past 5 pm. (1778 J. Cook *Jrnl.* 26 May (1967) III. i., OED s.v. *fetch*, sense 10.e 'to take a course; to reach a specified position, bring one's vessel up.')

---

through all the precipices of the mountains and through all the hollows of the valleys and through all the pits and ditches of the lands. Then he found him lying in a cave.'

OE *(a)redian, niman, healdan*

While in the verbs mentioned so far, the combination with a directional element will have played the major role in yielding the motion reading, the verbs *(a)redian* 'prepare', also 'find', *healdan* 'to hold, keep', and *niman* 'take', by contrast, are more likely to have become possible in this construction (109c–d) via their combination with *weg* or *stīg* 'way', as in (109a–b)—this is also what the *DOE* says on the origin of *ā-redian* in the intransitive motion construction: 'with *weg* understood' (see also Matsunami 1985 and the more detailed discussion in 6.2.5.7).

(109) a. þæt ðu ðe bet mæge **aredian** ***to rodorum rihte stige*** (Met 23.7, DOE)
'That you may all the better prepare/find the right way to heaven.'

b. Dysig bið se wegferenda man, se ðe **nimð** ***þone smeðan weg*** (Hml. Th. i. 164, 8, BT)
'Stupid is the travelling man who takes the smooth way.'

c. lædst me hidres & þidres on swa þicne wudu ðæt ic ne mæg ***ut*** **aredian** (Bo 35.100.4, DOE)
'You lead me hither and thither into a wood so thick that I cannot make/find my way out.'

d. Ge ne gemunon [. . .] ***þonan*** ge **noman** (Bt. 16, 1, BT)
'You don't remember [. . .] where you came from.'

Perhaps a similar process may have been involved for the motion uses of *healdan* 'to hold', as in (110): With respect to these, *BT* refers to a parallel use of ON *halda*, which Cleasby-Vigfusson (s.v. *halda*, sense IV, e.g. *Hrútr hélt suðr til Eyrar-sunds*) describes as elliptical: '*liði, skipi, för, stefnu*, etc. understood'.

(110) Hi **heoldon** ***ofer sæ to Flandran*** (Chr. 1075; Erl. 214,9, BT s.v. *healdan*, sense B.VI 'to hold out, last, hold on, continue, hold with')
'They held [i.e. continued] over the sea to Flanders.'

Thus, at least in Old Norse, combinations such as 'hold the ship to X' / 'hold the direction to X' / 'hold the journey to X' may have given rise to the motion use of ON *halda*. The same may have happened in Old English, or the use may have been taken over from ON *halda* to OE *healdan* (cf. Durkin 2014: 181).

5.1.4.5 Motion meaning through metaphor and metonymy

While in the non-MOTION verbs described in the preceding subsection the constructional semantics are involved in the motion interpretation, there are other non-MOTION verbs where this cannot be the case: The verbs *sēcan* 'seek', *nēosian*, 'seek, visit', and *metan* 'measure' stick out because they are most frequently used—in the case of *metan* and *nēosian* even only used—with a direct object, not with a directional prepositional phrase or adverb. The motion use of *metan* 'to measure' (111) must be figurative, either metonymical or metaphorical (with 'measuring a distance' construable as similar to 'travelling a distance').

(111) Ferdon forð þonon feþelastum ferhþum fægne, **foldweg mæton**, cuþe stræte (Beo. Th. 3271, B.1633, BT s.v. *metan*, sense III 'to measure by paces, to traverse, pass over'; exanded with *DOEC*)
'They went forth thence with footsteps, with joyful minds, measured [i.e. traversed] the earthway, well-known roads.'

Note that the three attestations given by *BT* for the motion use of *metan* all come from poetry. The *OED3* (s.v. *mete*) refers to the completely parallel use of *to measure* (112), a synonym of *metan* that has been borrowed in the Middle English period. This use of *measure* is equally marked as 'Chiefly *poet.*':

(112) With equal pace ***the Temples Nave*** they **measure**! (1682 N.O.tr. N. Boileau-Despréaux *Lutrin* iii. 25, OED3 s.v. *measure*, v., sense 6a 'to travel over or traverse')

The motion uses of the 'seek' verbs, by contrast, also occur in non-literary texts such as the *Chronicles* (113a).

(113) a. Þæt wæron þa ærestan scipu Deniscra monna þe ***Angelcynnes lond*** **gesohton** (Chr. 787; P.54,7, BTS s.v. *gesēcan*, sense II.2 'to go to a place')
'Those were the first ships of the Danish men that sought [i.e. went to] the land of the English.'

b. ðæt hi sunnandagum ***Godes cyrican*** georne **secan** (Blickl. Homl. 47, 28, BT s.v. *sēcan*, sense II.2 'to go or come to: to seek a place, visit, resort to')
'That they should diligently seek [i.e. go to] God's church on Sundays.'

c. Wæs his gewuna ðæt ***Norþanhymbra mægþe*** **sohte** and **neosode** (Bd. 3, 23; S. 554,7, BT s.v. *nēosian*, sense II 'to seek, visit a place')
'It was his habit that he sought and visited the country of the Northumbrians.'

For the 'seek' verbs, *sēcan* (113a–c) and *nēosian* (113c), this use must be based on metonymy, since seeking and visiting a place or person is contiguous with going to that place or person. It is probable that this contextualized motion meaning of *sēcan* had become conventionalized, since the verb is also attested with PATH satellites (114), a combination that would be unusual with the meaning 'seek'.

(114) þa þæt folc ***þider*** **sohte** ***to þam micclan screfe*** (Hml. S. 35, 331–334, BTS, sense II.2 'to go or come to: to seek a place')
'When the people "sought" [i.e. went] thither to the great cave.'

Thus, it seems that *sēcan* in the meaning of MOTION also changed its complementation pattern to the corresponding construction. While in Old English, attestations like (114) with *sēcan* in the intransitive motion construction are still rare, in Middle English the verb is widely attested with PATH satellites (see, e.g., the large number of examples in the *MED* s.v. *sechen*, sense 11).

Finally, we have the verb *gedōn* 'to do', whose use in the meaning 'arrive' is only attested in the *Chronicles*, such as in (115a–b):

(115) a. Foron þa up be Temese oþþæt hie **gedydon** æt Sæferne, þa up be Sæferne. (ChroA, 893.72)
'Then they went up by the Thames, until they "did" at Severn, then up by Severn.'

b. & foron anstreces dæges & nihtes þæt hie **gedydon** on anre westre ceastre on Wirhealum seo is Legaceaster gehaten. (ChroA 893.91)
'And travelled continuously, day and night, [so] that they "did" in a deserted town in Wirral which is called Chester.'

Apart from these two attestations, it occurs in annal 895 (twice), 914, and 917 in MS A of the *Anglo-Saxon Chronicles*.[90] All of these attestations are very similar: The verb—always in the form *gedydon* 'they did'—is always used in a subordinate clause introduced by *(oþ) þæt* '(until) that', preceded by a main clause which describes motion (with *faran* or *gān*), like in (115a–b). Translated literally, the structure thus is 'they travelled so that they did at/in X'. The prepositional phrase following *gedydon* usually starts with *æt* 'at', *on* 'in', or, once, with *innan* 'in'. The prepositional phrases therefore seem to be locative rather than directional; hence, it is unlikely that *gedydon* simply serves as a pro-verb to avoid the repetition of *foron* 'they went'. How this use of *gedōn* as 'to arrive' is motivated is not clear, but since the verb is semantically light, it may be used for all kinds of meanings.[91]

5.1.4.6 Conclusion: motion uses of non-MOTION verbs

To sum up this section, there are two major ways for non-MOTION verbs to be used in a motion meaning: The MOTION reading can be brought about through the interaction of the verb meaning with the intransitive motion construction (5.1.4.1–4, e.g. *berstan, brecan, feohtan, hīgian, weorþan, winnan*), or it can be a figurative extension of the meaning of the verb (see 5.1.4.5 for the 'seek' and 'measure' verbs) without an intransitive motion construction involved. For *sēcan* 'to seek', we have seen that the complementation pattern of the verb has apparently adapted to its metonymical motion use in that, apart from GOAL-GROUND being expressed in the direct object, *sēcan* is increasingly found also with PATH satellites.

---

[90] Among the other manuscripts, the verb is attested in the same structure in the annals for 894 in manuscripts BCD (=893 in A), 896 in BCD (=895 in A), 915 in BC (=914 in A).

[91] In my own Bavarian dialect, for instance, it is also not uncommon for speakers to ask *Wo duast'n hi?* lit. 'Where are you doing to?', meaning 'Where are you going to?' This is, of course, a combination with a directional, but the motion context is similar.

5.1.5 Combination of motion verbs with PATH satellites

The fourth purpose of this study of the Old English motion verb inventory was to allow a first assessment of the combinability of MANNER verbs with PATH satellites (see 4.1.1). Various studies have shown that this combinability is a typical characteristic of satellite-framed, MANNER-salient languages; verb-framed languages, by contrast, do not combine MANNER verbs with PATH satellites to the same extent (see 2.1.4.1–2).

In the overview of the motion verb inventory in section 5.1. (Table 5.1), all the verbs for which at least one attestation with a PATH satellite could be found are printed in italics (see also the attestations in italics in Appendix A). The majority of MANNER verbs (81 verbs, i.e. 76 per cent) are attested with such a satellite. Some examples are given in (116a–e).

(116) a. and he þanon **cleacode** swiðe earhlice ***to porte*** (LS 34 (Seven Sleepers) 446, DOE s.v. *cleacian*)
'And from there he hurried very timidly to town.'

b. An ormæte heort [. . .] **ræsde** ***into þam wudu*** (Hml S. 30.30, BTS s.v. *rǣsan*)
'A giant hart [. . .] rushed into the forest.'

c. þattor **smeh** ***geond ealne þone lichoman*** (Hml. S. 31, 952, BTS s.v. *smūgan*)
'The poison crept through the entire body.'

d. se ongann **flogettan** & **fiþercian** ***ymb his ansyne*** (GD 2 (C) 2.100.18, DOE s.v. *flogettan*)
'It started to fly and flutter around his face.'

e. Hie **swumman** ***ofer to ðæm eglande*** (Nar. 10, 29, BT s.v. *swimman*)
'They swam over to the island.'

It is also noteworthy that in some of these MANNER verb and PATH satellite combinations, the corresponding Latin passage that is given by the dictionary entries is of a different nature:

(117) a. Seo nædre ongan **slincan** ***in þæt scræf*** (*coepit serpens in speluncam* ***venire***) (Gr. D. 211, 16, BTS s.v. *slincan*)
'The snake started to crawl into the cave.'

b. Witodlice ðu becymst to Romebyrig, ***ofer sæ*** ðu **seglast** (*equidem Romam ingressurus es, mare* ***transiturus*** (Gr. D. bk. 2, c. 15)) (Hml. Th. ii 168, 31, BTS s.v. *seglan*)
'Truly, you will go to Rome, you will sail over the sea.'

In (117a), the neutral Latin *venire* 'come' is translated with the MANNER verb OE *slincan* 'to crawl'. In (117b), the Latin future participle *transiturus* 'one who will be traversing', i.e. a form of *transire* (literally 'over-go') is rendered with *seglast ofer* 'you will sail over'. Hence, in both examples, the Old English translation adds MANNER information that is not present in the Latin exemplar. This is yet another indication that Old English is a MANNER-salient language.

Differences between the Latin and the Old English translation do not only involve different kinds of verbs, as in (117a–b), but also different structures, as in (118a–b):

(118) a. He **scet** ***innan sæ*** (***misit se*** *in mare*) (Jn. Skt 21.7, BT s.v. *scēotan* 'to shoot, move rapidly, dart, run, plunge, rush')
'He [i.e. Petrus] rushed into the sea.'

b. Geseah ic balzamum ***of ðæm treowum ut*** **weallan** (*video opobalsamum arborum ramis* ***manans***) (Nar. 27, 23, BT s.v. *weallan*, sense I 'to well, bubble forth, spring out, flow')
'I saw balsam flow out of the trees.'

In (118a), where the literal rendering of the Old English would be 'he shot into the sea', the Latin has the reflexive construction *misit se in mare* 'he cast himself into the sea' (on reflexive constructions, see the end of section 7.2.2). In the Latin sentence in (118b), the SOURCE-GROUND (*arborum ramis* 'the branches of the trees') is joined to the verb without any preposition—the PATH information 'from' is conveyed by the ablative case. In Old English, by contrast, a preposition is needed to introduce the SOURCE-GROUND. To do this, the simple *of* (or *fram*) would have sufficed; however, the translator adds an additional *ut* 'out': in Old English, PATH is typically expressed outside the verb root.

Thus, the majority of the Old English MANNER verbs (76 per cent) are attested in combination with PATH satellites. What about the 24 per cent which lack such an attestation? First of all, the lack of such an attestation obviously does not automatically mean that the verb in question was not possible in this kind of construction—it might just not be recorded. This scenario is not unlikely, in view of the fact that many of the verbs have very few attestations in the extant Old English material (see 3.1.1). For instance, according to the *DOE*, for the verbs *drēopian* 'to drop, drip', *flotian* 'to float', and *fēþan* in the meaning 'to go on foot', which are all not attested with a PATH satellite, we only have two attestations each.[92] Thus, it might just be coincidental that there is no attestation with a PATH satellite.

Second, some MANNER verbs denote a MANNER OF MOTION that is only rarely used as a means to get somewhere, such as the 'dance' verbs *frician*, *s(e)altian*, and *tumbian*: Since FIGURES rarely dance into, out of, or across somewhere, there is usually no need to use these verbs with a directional element. Yet, the fact that the Old English 'dance' verbs do not occur in an intransitive motion construction recalls the implicational scale by Croft et al. (2010: 223; see 2.1.3),

[92] The *DOE* entries include information on how often a word is attested. Since as yet, only the *DOE* fascicles A–G have been finished, this information is only readily available for the verbs listed above. A search in the *Dictionary of Old English Corpus* (*DOEC*) for the verbs from the remainder of the alphabet which lack attestations in the intransitive motion construction, however, suggests that many of them (e.g. *hlēapettan*, *hoppetan*, *hradian*, *s(e)altian*, *slidrian*, *treppan*, *þocerian*) are similarly infrequent.

who place 'dance across' at the highest point of the formal scale. This means that 'dance across' is the complex motion event least likely to be encoded with a satellite-framing construction, owing to the fact that dancing across somewhere is not very typical or natural (2010: 224). Croft et al. (2010: 213) report that in Icelandic, for instance, while it is possible to express 'dance across' with a satellite-framing construction, as in (119a), it is equally possible to use a construction with a finite 'neutral verb of motion' and the 'dance' verb in a participle (119b) (their examples).

(119) a. Ég **dansaði** ***yfir götuna.***
'I danced across the street.'

b. Ég **fór** ***dansandi yfir götuna.***
'I went dancing across the street.'

However, in the light of the recorded material from Old English, the lack of satellite-framing constructions for the 'dance' verbs can in no way be taken as an indicator that Old English is less satellite-framing than Present-Day English: If people in general rarely dance into, out of, or across somewhere, this is probably all the more true for the characters featured in Old English texts—chiefly kings, warriors, saints, monks, etc. Unsurprisingly, in none of the extant texts there seems to have been a need to write about somebody moving somewhere in a dancing MANNER: None of the Old English 'dance' verbs is attested in a description of translocational motion, not even in a construction such as (119b); they are only attested as bare verbs (as in (120a–b)) or as single glosses.

(120) a. & þa ða þære Herodiadiscan dohtor inneode & **tumbode**, hit licode Herode (Mk (WSCp) B8.4.3.1, 14.6, DOEC)
'And when Herodias' daughter went in and danced, it pleased Herodes.'

b. & cweþað, we sungon eow & ge ne **fricudun**; we cwiþdun & ge ne weopun (Mt (WSCp) B8.4.3.1, 11.17, DOEC)
'And said: We sang to you and you did not dance; we lamented and you did not weep.'

To sum up, the high percentage of Old English MANNER verbs (76 per cent) that have attestations with a directional element agrees with the pattern that is expected for a satellite-framed MANNER-salient language. Languages that predominantly follow the verb-framing pattern tend to avoid the combination of MANNER verb and PATH satellite, particularly in the expression of boundary crossing motion (see 2.1.4.1–2; cf., e.g., Slobin & Hoiting 1994; Aske 1989; Narasimhan 2003; as well as the implicational scale in Croft et al. 2010: 223). As we have seen in this section, this does not seem to be the case for Old English. What is more, among the attestations that combine MANNER verb and PATH satellite, there are also several which describe boundary crossing motion (e.g. 116b, 117a, 118a–b).

Combinations with PATH satellites are attested for a similarly high percentage of the NEUTRAL MOTION verbs (90 per cent (43)) and for the non-MOTION verbs (80 per cent (12)). For the PATH verbs, by contrast, combinations with PATH satellites are rarer (58 per cent (11)). This is because some of them are only used with direct object GROUNDS (or often rather PATIENTS/AFFECTED) (*grētan, forlǣtan; genēahian, pæþþan*) or are only attested without GROUND complementation (predominantly those that carry GROUND information in their derivational base: *gryndan, innian, scipian; wican*); see the discussion of the individual verbs in 5.1.3. Section 5.2.3.4 analyses the complementation patterns quantitatively on the basis of the usage study.

5.1.6 Old English inventory: conclusion

Based on the *Thesaurus of Old English* in combination with the dictionary definitions and attestations, this section has shown that for describing INTRANSITIVE MOTION Old English has a large number of verbs at its disposal. Among them, MANNER verbs make up by far the largest group (107 verbs). The fact that some of these MANNER verbs are derived from other MANNER verbs may be taken as an indicator that MANNER OF MOTION was a semantic domain that was under expansion already in the Old English period (5.1.2). The next largest group in the inventory is the one of NEUTRAL MOTION verbs (48 verbs). Verbs primarily expressing PATH are only the third largest group (19 verbs), and, most pertinent for our question in chapter 9, the only ones that qualify as 'pure PATH verbs' among them are *(ge)nēahlǣcan* and *genēahian* 'to approach', as a closer look at these verbs in section 5.1.3 has shown.

Even though this study was based on a thesaurus, there is a smaller group of fifteen verbs which do not primarily express MOTION. In some of them, the motion reading is probably due to metonymy (the 'seek, visit' verbs), but in others ('become', 'break, burst' or 'fight', 'struggle') it crucially depends on their being used with a directional element, i.e. in an intransitive motion construction (5.1.4). Most of the MANNER and NEUTRAL MOTION verbs are attested in such a construction as well (76 and 90 per cent of the MANNER and NEUTRAL MOTION verbs respectively (section 5.1.5); see also the italicized verbs in Table 5.1 and the attestations in italics in Appendix A).

This inventory is thus a first indicator that with Old English we are dealing with a rather prototypical strongly satellite-framing and manner-conflating language. This will be put to test in the next section with a study of how MOTION is typically described in texts, to see how widespread the different verb groups and patterns are in terms of token frequencies.

5.2 *Motion verb use*

As shown in section 5.1, Old English has a large range of motion verbs, including many that express MANNER. We have also seen that the majority of them is attested with a directional element in an intransitive motion construction (see the verbs in italics in Table 5.1 and the attestations in italics in Appendix A). However, as already discussed above, the lexicon is only one side of the coin—the crucial question that lies at the bottom of the motion verb typology is: What is the typical, characteristic pattern in usage? What is frequent? Thus, as a second approach to motion expression in Old English, we will move to the texts and examine how motion is typically talked about (for the texts and the method, see section 4.2). The aims of the usage studies have been outlined in section 4.2.1 as follows:

(a) How frequent are the different types of motion verbs (MANNER, PATH, NEUTRAL MOTION, non-MOTION)?
(b) What are the typical complementation patterns; in particular: Are MANNER verbs usually combined with PATH satellites? What are the typical complementation patterns of PATH verbs?

The results for Old English with respect to these questions will be presented in the following subsections. Section 5.2.1 will show the frequencies of the different types of motion verbs; a discussion of the results with respect to style will follow in section 5.2.2. Section 5.2.3 examines the use of PATH verbs more closely; section 5.2.4 addresses the question of complementation patterns.

5.2.1 Frequency of the different semantic groups

The first aim of the usage study was to investigate how frequently the different kinds of verbs—MANNER, PATH, NEUTRAL MOTION, and non-MOTION—are used. The result is given in Table 5.2 for each of the texts separately; the semantic classification is based on the one from the inventory (Table 5.1 and Appendix A). Since there are three extremely frequent NEUTRAL MOTION verbs—*cuman*, *faran*, and *fēran* account for between 43 and 80 per cent of all motion descriptions—these are represented as an extra category here. Table 5.2 shows the results as a mosaic plot, where the different width of the columns represents the different number of motion descriptions that were found

TABLE 5.2. Frequency of the different verb groups in motion descriptions from Old English texts (highly frequent NEUTRAL MOTION verbs *cuman*, *faran*, *fēran* represented separately).

	CHROA		CHROE		OHTWU		SAINTS		ALL TEXTS	
MANNER	13%	37	5%	8	50%	15	17%	14	13%	74
PATH	6%	17	2%	3	0%	0	6%	5	4%	25
cuman, faran, fēran	62%	177	80%	128	43%	13	48%	39	64%	357
Other NEUTRAL MOTION	9%	25	9%	14	3%	1	24%	20	11%	60
Non-MOTION	10%	28	5%	8	3%	1	5%	4	7%	41
Total motion verbs		284		161		30		82		557

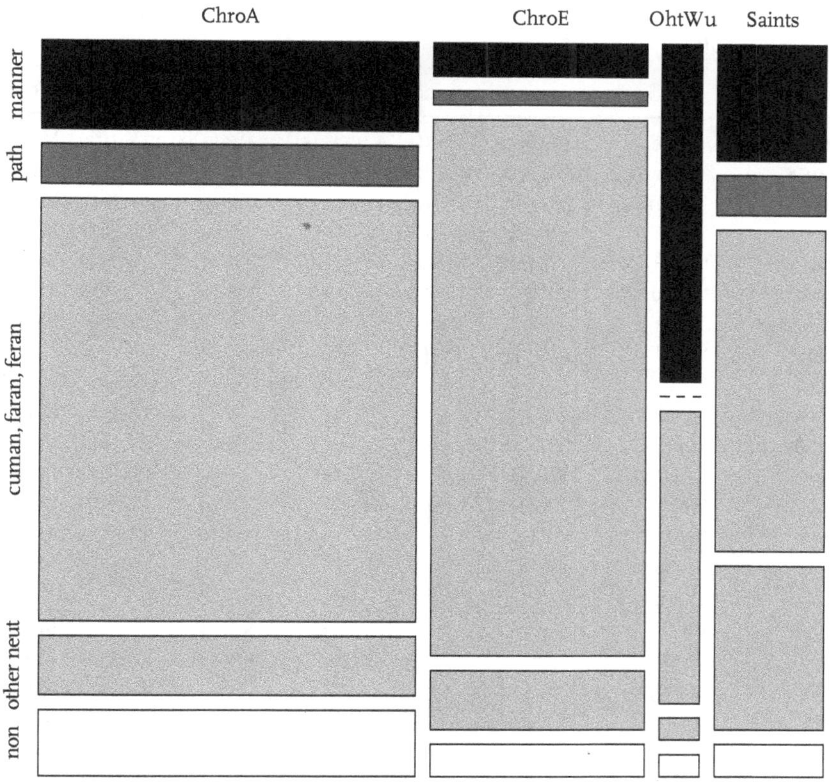

FIGURE 5.2. Frequency of the different verb groups in motion descriptions from Old English texts (highly frequent NEUTRAL MOTION verbs *cuman, faran, fēran* represented separately).

in the respective texts. Section 5.2.2 will discuss the frequency of *cuman, faran,* and *fēran* with regard to stylistic questions.

After the NEUTRAL MOTION verbs, particularly *cuman, faran,* and *fēran* (see the light grey areas in Figure 5.2), the second most frequent group of verbs in the Old English motion expressions (between 5 and 50 per cent, see Table 5.2 and the black areas in Figure 5.2) are the MANNER verbs (obviously very prominent in *Ohthere and Wulfstan*, a text which is much about sailing, with 32 per cent *siglan* 'sail'). The third most frequent group (between three and ten per cent, see Table 5.2 and the white areas in Figure 5.2) are non-MOTION verbs, and the fourth are PATH verbs, ranging between zero and six per cent (see the dark grey areas in Figure 5.2).[93] The frequencies of all the individual verbs are listed in Table 5.3.[94]

---

[93] Except for the *Saints' Lives*, where PATH verbs rank third and non-MOTION verbs are the least frequent group.

[94] Just like in the analysis of the inventory (5.1), the verbs are summarized under their non-prefixed forms; only if a verb occurred with just one specific prefix is this given in brackets (e.g. *(æt)berstan*).

TABLE 5.3. Verbs in motion descriptions from Old English texts (*Chronicles, Ohthere and Wulfstan, Lives of Saints*).

VERB	MEANING	CHROA	CHROE	OHTWU	SAINTS	TOTAL	GROUP
cuman	'come'	21% (59)	38% (61)	17% (5)	34% (28)	27% (153)	$Mo^{neutral}$
faran	'go, travel'	38% (110)	15% (25)	20% (6)	2% (2)	26% (143)	$Mo^{neutral}$
fēran	'go, travel'	3% (8)	26% (42)	7% (2)	11% (9)	11% (61)	$Mo^{neutral}$
gān	'go'	4% (11)	5% (8)	—	7% (6)	4% (25)	$Mo^{neutral}$
rīdan	'ride'	7% (20)	—	3% (1)	4% (3)	4% (24)	Ma^{other}
wendan	'wend, go'	2% (7)	2% (3)	—	7% (6)	3% (16)	$Mo^{neutral}$
(for)lǣtan	'leave'	4% (12)	2% (3)	—	—	3% (15)	Pa^{away}
sēcan	'seek'	4% (11)	1% (1)	3% (1)	2% (2)	3% (15)	non
siglan	'sail'	(1)	—	32% (10)	—	2% (11)	$Ma^{air/fluid}$
(be)stelan	'steal'	2% (7)	—	—	—	1% (7)	Ma^{other}
(ge)dōn	'do'	2% (6)	—	—	—	1% (6)	non
willan	'will, wish'	1% (3)	2% (3)	—	—	1% (6)	non
bēon	'be'	1% (4)	—	—	1% (1)	1% (5)	non
flēon	'flee'	1% (4)	—	—	1% (1)	1% (5)	Ma^{flee}
hweorfan	'turn, change, go'	1% (4)	1% (1)	—	—	1% (5)	$Mo^{neutral}$
feallan	'fall'	—	—	—	5% (4)	1% (4)	Ma^{fall}
būgan	'bow, bend'	—	1% (1)	—	2% (2)	1% (3)	$Mo^{neutral}$
ærnan	'gallop'	—	—	10% (3)	—	1% (3)	Ma^{speed}
irnan	'run'	(1)	1% (1)	3% (1)	—	1% (3)	Ma^{speed}
lendan	'land'	1% (2)	—	—	1% (1)	1% (3)	Pa^{toward}
(ge)wītan	'set out'	1% (3)	—	—	—	1% (3)	Pa^{away}
climban	'climb'	—	1% (2)	—	—	(2)	Ma^{other}
cyrran	'turn, go'	—	1% (1)	3% (1)	—	(2)	$Mo^{neutral}$
folgian	'follow'	(1)	—	—	1% (1)	(2)	$Mo^{neutral}$
fundian	'set out, depart, hasten'	—	1% (1)	—	1% (1)	(2)	Ma^{speed}
gangan	'go'	—	—	—	2% (2)	(2)	$Mo^{neutral}$
healdan	'hold'	—	1% (2)	—	—	(2)	non
hlēapan	'leap, jump'	—	1% (2)	—	—	(2)	Ma^{jump}
(ge)nēalǣcan	'approach'	—	—	—	2% (2)	(2)	Pa^{toward}
(a)rīsan	'rise'	—	—	—	2% (2)	(2)	Pa^{up}
rōwan	'go by water, row'	1% (2)	—	—	—	(2)	$Ma^{air/fluid}$
sīþian	'go, travel'	—	—	—	2% (2)	(2)	$Mo^{neutral}$
wealwian	'wallow, roll'	—	—	—	2% (2)	(2)	Ma^{roll}
(æt)berstan	'break away'	(1)	—	—	—	(1)	non
brecan	'break'	(1)	—	—	—	(1)	non
crēopan	'creep, crawl'	—	1% (1)	—	—	(1)	$Ma^{slow/imp}$

TABLE 5.3. (Continued)

VERB	MEANING	CHROA	CHROE	OETWU	SAINTS	TOTAL	GROUP
(tō)dǣlan	'divide, separate'	(1)	—	—	—	(1)	non
efestan	'to hurry'	—	—	—	1% (1)	(1)	Ma[speed]
fēolan	'make/ force one's way'	(1)	—	—	—	(1)	Mo[neutral]
fōn	'take'	—	1% (1)	—	—	(1)	non
(ūt)rǣsan	'rush, hasten'	(1)	—	—	—	(1)	Ma[speed]
(tō)scēotan	'shoot'	—	1% (1)	—	—	(1)	Ma[speed]
(tō)sciftan	'divide'	—	1% (1)	—	—	(1)	non
(a)settan	'move, convey oneself'	(1)	—	—	—	(1)	Mo[neutral]
sincan	'sink'	—	—	—	1% (1)	(1)	Ma[fall]
(be)stealcian	'move stealthily'	—	—	—	1% (1)	(1)	Ma[other]
(a)styrian	'stir, move'	—	—	—	1% (1)	(1)	Mo[neutral]
(æt)swimman	'swim'	(1)	—	—	—	(1)	Ma[air/fluid]
weorþan	'come to be'	(1)	—	—	—	(1)	non
[no verb]					1% (1)	(1)	non
Total		100% (284)	100% (161)	100% (30)	100% (82)	100% (557)	

Among the verbs found in the usage study, there are only five which were not found in the inventory study (section 5.1), all of them non-MOTION verbs. This shows that the Old English dictionaries and thesauri are a good source for finding out which verbs can be used to express motion. The verbs newly found in the usage study will be briefly presented in the rest of this section: *(tō)dǣlan* and *(tō)sciftan*, both meaning 'to divide', *fōn* 'to grasp, take', *willan* 'to want', and *bēon* 'to be'. Furthermore, there was one motion description without a motion verb (121).

(121) þa gewende se ærendraca ardlice aweg, and gemette be wæge þone wælhreowan Hingwar mid eallre his fyrde fuse ***to Eadmunde*** (AELive (Edmund) 94)
'Then the messenger went quickly away, and met on the way the bloodthirsty Hingwar with all his army quickly to Edmund'

In (121), the FIGURE is Hingwar, and PATH + GROUND is expressed in the prepositional phrase *to Eadmunde*. There is no verb, only the adverb *fūse* 'quickly' (cf. *DOE* s.v. *fūs*; see also Bammesberger 2003: 32), i.e. the literal rendering is '(the messenger met) Hingwar quickly to Edmund', meaning 'Hingwar hurrying to Edmund', as in the translation by Skeat (1900: 321). Thus, we have an intransitive motion construction consisting only of FIGURE and a satellite expressing PATH + GROUND, with no verb, a good example of how the presence of a PATH satellite is enough to convey a motion meaning.[95]

(tō)dǣlan and *(tō)sciftan*

I analysed (122a) below as elliptical, with the verb *tōdǣlde* 'divided' still understood from the previous clause. In principle, though, the structure in (122a) could also be the same as what we just saw for (121), with no verb but only FIGURE (*ōþer dǣl* 'the other part [of the army]') and PATH(+GROUND) (*ēast/tō Hrofesceastre*) expressed.

(122) a. Her **todælde** se foresprecena here on tu, oþer dæl ***east*** oþer dæl ***to Hrofesceastre*** (ChroA, R885.1)
'Here the abovementioned army divided in two, one part east, the other part to Rochester.'

b. Ac se cyng let **toscyfton** þone here ***geond eall þis land to his mannon*** (ChroE, R1085.10)
'But the king let the army divide around all the land to his men.'

In (122b), by contrast, the verb is clearly *tōscyfton* 'divide, separate', and PATH + GROUND is expressed in the satellites *geond eall þis land* and *tō his mannon*.

fōn

The *DOE* entry for OE *fōn* 'grasp, take' gives the meaning of the collocation *fōn on fultum* as 'come to the aid of someone'. Since in (123), *fōn* in this collocation combines with the PATH + GROUND satellite *to þam cynge heora hlaforde*, a motion reading arises, supported by the preceding context of the king sending for the Englishmen:

(123) þe cyng [. . .] Sende þa æfter Englisce mannan [. . .] Ac Englisce men swa þeah **fengon *to þam cynge heora hlaforde*** on fultume (ChroE, 1087.42)
'The king [having become aware of a coup against him] then sent for the Englishmen [promising them benefits which, however, were not to last long]. But the Englishmen nevertheless came to the king their lord to help.'

willan

As discussed in 2.2.4, verbs with modal meanings can fill the verb slot of the Old and Middle English motion construction as well. The texts analysed here contained a few instances of OE *willan* 'to will, want' combined with PATH +

[95] Cf. the existence of a similar structure in colloquial German, such as in *Ich also schnell in den Park* 'So me quickly to the park'.

GROUND satellites, such as *of his rice* 'from his kingdom' and *into þam mynstre* 'into the minster' in (124a–b):

(124) a. & þa salde se here him foregislas & micle aþas, þæt hie ***of his rice* uuoldon** (ChroA, R878.20)
'And then the army gave him hostages and many oaths that they wanted [i.e. wanted to go] from his kingdom.'

b. þa sona on morgen comen ealle þa utlaga mid fela scipe, & **woldon *into þam mynstre*** (ChroE R1070.20)
'Then soon in the morning, all the outlaws came with many ships and wanted [i.e. wanted to go] into the minster.'

This yields a reading of modally broken motion (cf. Berthele 2007: 242), where the motion event is only intended, but not actually accomplished (yet).

bēon

The inventory study has already identified the copula *weorþan* 'to become, come to be' as a verb being used to express motion. In the texts analysed for the usage study, the copula *BE* turned up as a motion verb as well, in attestations such as (125a–b) (cf. also sense B.1.f 'the complement is an adverb, prepositional phrase or conjunction indicating direction: to be, come, go, be on one's way' in the *DOE* s.v. *bēon*).[96]

(125) a. þa ða he ***hamwerd* wæs** (AELive (George), R179)
'When he was homeward [i.e. was going homeward, was on his way home].'

b. þa he þa **wæs *þiderweardes*** (ChroA, R894.39)
'When he was thitherwards [i.e. was going there, was on his way there].'

In all of the six times that *BE* was found in motion uses in the texts, it is combined with an adverbial satellite containing the element *-weard(s)* '-wards'. This appears to highlight the directional meaning of the satellite and crucially conveys the idea that the GOAL-GROUND (*ham*, *þider*) is not reached yet. Such a strong directional is probably needed to coerce a stative verb like *BE* into a motion reading—in the Middle English usage study (6.3.1), it occurs in a motion use in combination with *towards*.

After this brief presentation of the 'new' motion verbs that had not been identified by the inventory study, the following sections will discuss the results of the usage study, starting with stylistic issues.

[96] On the similar use of the copula in varieties of German (such as *der ist auf den Baum hinauf* 'He is up to the tree'), cf. Berthele 2007.

5.2.2 Discussion: MANNER verbs, neutral MOTION verbs, and style

One of the aims of the usage study was to investigate whether MANNER verbs are frequently used in motion descriptions. To have an idea what is meant by 'frequently used', it should be recalled that for Present-Day English, Slobin (2004a: 231) finds MANNER verbs in 45 per cent of motion descriptions in the frog story narrations he examines (see 2.1.2) and in 40 per cent of motion descriptions in novels. In Cappelle's study of the *BNC*, MANNER verbs only account for around 34 per cent of the motion descriptions in Present-Day English fictional and for around 20 per cent in biographical texts (2012: 187). Slobin (2004a) does not address the remaining verbs, but it is rather likely that the bulk of them will be *come* and *go*, which are the Present-Day English equivalents of the highly frequent Old English *cuman* and *faran/fēran*: Cappelle (2012: 185) reports a good 40 per cent of PDE *come* and *go* in the motion descriptions from the *BNC* subcorpora of fictional and biographical texts.

In the Old English texts (apart from *Ohthere and Wulfstan* with 50 per cent MANNER), by contrast, MANNER verbs do not reach comparable frequencies, but range between 5 and 17 per cent. MANNER verbs are indeed used more frequently than PATH and non-MOTION verbs, but—at least in the *Chronicles* and in the *Lives of Saints*—they are vastly outnumbered by the three very neutral motion verbs *cuman* 'come' and *faran/fēran* 'go, travel'.

Yet, the three verbs *cuman* 'come' and *faran, fēran* 'go, travel', as we have seen in 5.2.1, are conspicuously more frequent in the *Chronicles* (62 per cent in ChroA and 80 per cent in ChroE) than in the other two texts (43 per cent and 48 per cent, respectively). In view of the difficulties of finding a 'colloquial' Old English text (see 3.1.2.1), these differences in frequency must be discussed with respect to stylistic issues. We will focus on the style of the *Chronicles* in particular, because (a) it is in the *Chronicles* that the neutral verbs *cuman, faran,* and *fēran* are most frequent, and (b) the number of motion descriptions in them is more representative.

The style of the *Chronicles* has been discussed in 4.2.2.1. The preponderance of the neutral, unspecific motion verbs *cuman, faran,* and *fēran* is quite likely related to this repetitive, formulaic style, for which (126) is an example. Lexical and structural variation is clearly not a major goal here:

(126) 918 Her on þysum geare **com** micel sciphere ***hider ofer suþan of Lidwiccum*** . . .

919 Her on þys gere Eadweard cyng **for** mid fierde ***to Bedanforda*** foran to Martines mæssan . . .

920 Her on þys gere foran to middum sumera **for** Eadweard cyning ***to Mældune*** . . . (ChroA, 918–20)

'918. Here in this year came a large ship-army hither over the south from the *Lidwicces* . . .

919. Here in this year went King Edward with troops to Bedford before Martinmas . . .

920. Here in this year before Midsummer went King Edward to Maldon . . ..'

This idea is supported by the fact that whenever the annals become more narrative and the 'story' more dramatic, the rarer motion verbs appear in larger numbers. I would like to illustrate this with the episodes of Cynewulf and Cyneheard (ChroA, annal 755, see (127)) and the 'tumult at Glastonbury' (ChroE, annal 1083.1–28, see (128)).

(127) [. . .] & þa ongeat se cyning þæt, & he ***on þa duru* eode**, & þa unheanlice hine werede, oþ he on þone æþeling locude, & þa **utræsde *on hine***, & hine miclum gewundode. & hie alle on þone Cyning wærun feohtende oþ þæt hie hine ofslægenne hæfdon; & þa on þæs wifes gebærum onfundon þæs cyninges þegnas þa unstilnesse, & þa ***þider* urnon** [. . .] hie simle feohtende wæran oþ hie alle lægon butan anum Bryttiscum gisle, & se swiþe gewundad wæs. ða on morgenne gehierdun þæt þæs cyninges þegnas þe him beæftan wærun þæt se cyning ofslægen wæs, þa **ridon** hie ***þider***, [. . .] & þa ***þær to* eodon**; & þa gebead he him hiera agenne dom feos & londes gif hie him þæs rices uþon, & him cyþdon þæt hiera mægas him mid wæron þa þe ***him from* noldon**; & þa cuædon hie þæt him nænig mæg leofra nære þonne hiera hlaford, & hie næfre his banan **folgian** noldon, & þa budon hie hiera mægum þæt hie gesunde ***from* eodon**; [. . .]. & hie þa ymb þa gatu feohtende wæron oþþæt hie ***þær inne* fulgon**, & þone æþeling ofslogon [. . .]

'[. . .] and then the king became aware of this [i.e. the fact that he is besieged by Cyneheard] and he **went *to the doors*** and defended himself bravely until he looked at the etheling [=Cyneheard], and then he **rushed *out on him*** and wounded him severely. And they were all fighting against the king until they had slain him. And then, because of the behaviour of the woman, the king's thegns became aware of the commotion, and they **ran *thither*** [. . .] they kept on fighting until they all lay [dead], except one British hostage, and he was severely wounded. Then in the morning, the king's thegns that had been behind him heard that the king was slain, they **rode *thither*** [. . .] and then **went *thereto*** and he [=Cyneheard] offered them their own choice of money and land if they granted him the kingdom, and they [=Cyneheard's people] told them that their kinsmen were with him, who **did not want** [to go] ***from him***. And then they [the king's people] said that no kinsman was dearer to them than their lord and they would never **follow** his murderer, and they wished their kinsmen to **go *away*** unharmed; [. . .] and they were fighting around the gates until they **forced their way *therein***, and killed the etheling.'

In the episode of Cynewulf and Cyneheard in (127), we find the motion verb types *gān* 'go', *fēolan* 'to force one's way', *folgian* 'to follow', *irnan* 'to run', *rǣsan* 'to rush', *rīdan* 'to ride', and *(ne) willan* 'to (not) want [to go]'. Conspicuously, in this episode the neutral *cuman, faran*, and *fēran* do not occur at all. In ChroA overall, by contrast, they account for 62 per cent of all the motion descriptions.

The 'tumult at Glastonbury' episode is given in (128), with all motion verbs in bold and their PATH (+GROUND) satellites in bold italics.

(128) Anes dæges þe abbot **eode** ***into capitulan***. & spræc uppon þa munecas. & wolde hi mistukian. & sende æfter læwede mannum. & hi **comon** ***into capitulan*** on uppon þa munecas full gewepnede. & þa wæron þa munecas swiðe aferede of heom. nyston hwet heom to donne wære. Ac **toscuton**. sume **urnon** ***into cyrcean***. & belucan þa duran into heom. & hi **ferdon** æfter heom ***into þam mynstre***. & woldon hig utdragan. þa ða hig ne dorsten na ***utgan***. Ac reowlic þing þær gelamp on dæg. þæt þa Frencisce men bræcen þone chor & torfedon towærd þam weofode þær ða munecas wæron. & sume of ðam cnihtan **ferdon** ***uppon þone uppflore***. & scotedon adunweard mid arewan toweard þam haligdome. swa þæt on þære rode þe stod bufon þam weofode sticodon on mænige arewan. & þa wreccan munecas lagon onbuton þam weofode. & sume **crupon** ***under***.

'One day the abbot **went** ***into the chapter-house***, and spoke to the monks, and wanted to maltreat them, and sent after laymen. And they **came** ***into the chapter-house***, against the monks, fully armed, and the monks were much afraid of them and did not know what to do. But they **rushed** ***in different directions***. Some **ran** ***into the church*** and locked the door unto them. And they **came** after them ***into the church*** and wanted to tear them out, so they did not dare to **go** ***out***. A rueful thing happened that day: the Frenchmen broke the choir and threw toward the altar where the monks were. And some of the men **went** ***onto the upper floor*** and shot down with their arrows onto the sanctuary, so that many arrows stuck in the cross that stood upon the altar. And the poor monks lay around the altar, and some **crept** ***under***.'

In this episode, the motion verbs *crēopan* 'creep', *gān* 'go', *irnan* 'run', and *to-scēotan* 'rush in different directions' (lit. 'to-shoot') all occur closely together, and the percentage of *cuman, faran*, and *fēran* drops to 38 per cent, as opposed to their 80 per cent in ChroE overall. In these narrative passages, the text seems to 'zoom in' on the action, conveying an immediacy that is matched by the more specific and colourful rarer motion verbs. With respect to this 'zooming' effect, it is probably not a coincidence that this annal contains the expression *Anes dæges* 'one day' (cf. German *eines Tages*): In contrast to the usual *her on þysum geare* 'here in this year' (126), which suggests more of a bird's-eye view, *anes dæges* signals 'zooming in on close-up'.

In most of the other annals analysed for the present study, by contrast, the events appear to be told rather in formulaic terms from a retrospective, summarizing distance (as in (126)), a distance for which the more aloof neutral verbs *cuman*, *faran*, and *fēran* seem to be stylistically more adequate. After all, they convey only the most important information (MOTION) and thus are more 'classically objective' than those that conflate MANNER.

5.2.3 Discussion: PATH verbs

Centrally, this study is interested in whether the pure PATH verbs borrowed in Middle English from French and Latin are a novelty in the native system. The inventory study in 5.1 has suggested that they are: We have seen that apart from *(ge)nēalǣcan*, *genēahian* 'approach', the Old English verbs predominantly expressing PATH are not pure PATH verbs (5.1.3). The analysis of the texts now shows that they are also rather infrequently used: PATH verbs account for only a minority, namely four per cent (25) of all the motion descriptions (see Table 5.2). Crucially, however, they only include the following types:

(129) PATH verbs in the Old English texts:
forlǣtan (15), *lendan* (3), *gewītan* (3),
arīsan (2), *genēalǣcan* (2)

The most frequently used PATH verb is *(for)lǣtan* 'leave, abandon' with 15 occurrences, thus accounting for 60 per cent of the motion descriptions with PATH verbs. The problematic status of OE *(for)lǣtan* as a MOTION verb has been addressed in detail in 5.1.3.1—its direct object is often profiled as PATIENT/AFFECTED that is abandoned, left behind for good, rather than more neutrally as SOURCE-GROUND. This is also the case in (130a–c) from the texts analysed here.

(130) a. þa mynte Laurentius ðe þa was ercebiscop on Cent þæt he wolde suð ofer sæ & ***eall*** **forlæton**. ac him com to on niht se apostolus Petrus & hine hetelice swang, for ðan ðe he wolde ***Godes hyrde*** **forlæton** (Chro A 616.5)
'Then Laurentius, who then was archbishop in Kent, thought that he wanted [to go] south over the sea and abandon everything. But the apostle Peter came to him at night and chastised him fiercely, because he wanted to abandon God's flock.'

b. & foron to Mældun, & ymbsæton þa burg, & fuhton þær on, oþ þam burgwarum com mara fultum to utan to helpe, & **forlet** se here ***þa burg***, ***& for fram*** (Chro A 921.49)
'And [the Vikings] went to Maldon and besieged the town, and fought on it, until more help came to the townsmen from outside. And the army abandoned the town and went from [it].'

c. Se cyng syððan sende here to Dunholm. & let besittan þone castel. & se biscop griðode and ageaf þone castel. & **forlet** ***his biscoprice***. & ferde to Normandige. Eac manige Frencisce men **forleton** ***heora land***. & ferdon ofer sæ (Chro E 1087.83–6)

'The king then sent an army to Durham, and let them besiege the castle. And the bishop agreed and gave up the castle, and relinquished his bishopric and went to Normandy. Also many Frenchmen abandoned their land and went over sea [where they received land].'

The idea of 'giving up, leaving behind for good' is present in all of the examples in (130a–c), whether bishop Laurentius ponders abandoning Britain and God's flock after king Eadbald has renounced the Christian religion (130a), the Vikings give up the siege of Maldon (130b), or whether the bishop of Durham gives up his bishopric and moves to Normandy, and the Frenchmen leave France to receive land in Britain (130c). Note also that the separate mention of *& for fram* 'and went from [it]' in (130b) can be taken as another indicator that *(for)lǣtan* is not a pure PATH verb, or else the same state of affairs would be expressed twice.

OE *lendan, gewītan,* and *arīsan* are also problematic PATH verbs (see the discussion in 5.1.3), so the only pure PATH verb found in the 557 motion clauses of the usage study is *genēalǣcan* 'to approach', used twice, once bare (131a) and once with a PATH + GROUND satellite (131b).

(131) a. **genealæc** nu ærest (AELive (George) R.33)
'Now approach first.'

b. þa hæðenan **genealæhton** ***to þam halgan Oswolde.*** (AELive (Oswald) R.155)
'The heathens approached to the holy Oswald.'

In view of their rare occurrence, what we can say about the complementation patterns of PATH verbs is necessarily not very representative—*forlǣtan* always takes a direct object, and the others are either used bare or with an adverbial or prepositional satellite.

We can conclude that pure PATH verbs are not only rare as types in the Old English motion verb inventory (see 5.1.3), but that they are also extremely rarely used: In the 557 motion descriptions analysed for the usage study, there were only 25 instances with a verb from the group primarily expressing PATH, and of those, only two instances (of the verb *(ge)nēalǣcan*) can be considered to represent a pure PATH verb. Thus, in the Old English period, there is no real native 'forerunner' for the pure PATH verbs *enter, ish/issue, descend, avale, ascend,* and *mount,* which are borrowed in the Middle English period and investigated in chapter 9.

5.2.4 Complementation patterns

The discussion of the Old English usage study will end with this final section on complementation patterns. In the inventory study, we have seen that most motion verbs are attested at least once with a PATH satellite: 76 per cent of the MANNER verbs, 90 per cent of the NEUTRAL MOTION verbs, 80 per cent of the non-MOTION verbs, and 58 per cent of the PATH verbs (see 5.1.5). The interest here now is to see whether in actual usage, the combination, particularly of

TABLE 5.4. Complementation of the different motion verb groups in Old English texts.

	AT LEAST 1 SATELLITE		NPGROUND		BARE VERB		TOTAL
MANNER	82%	(61)	12%	(9)	5%	(4)	74
PATH	28%	(7)	60%	(15)	12%	(3)	25
cuman, faran, fēran	90%	(321)	1%	(5)	9%	(31)	357
Other neutral	92%	(55)	3%	(2)	5%	(3)	60
Non-MOTION	63%	(26)	37%	(15)	—	—	41
Total	84%	(470)	8%	(46)	7%	(41)	557

MANNER verbs, with PATH satellites is frequent as well, as is expected to be the case in a satellite-framing, MANNER conflating language (2.1.4.1–2). Table 5.4 gives an overview of the complementation patterns of the different verb types in the texts analysed here.

The result is that most motion descriptions contain at least one PATH satellite: 84 per cent overall, 82 per cent of those with MANNER verbs, and around 90 per cent of those with neutral MOTION verbs. The lower percentage (63 per cent) in the non-MOTION verbs is due to metonymic uses of *gesēcan* (10 times) and *sēcan* (5 times) 'to seek', which take a direct object (NPGROUND) (e.g. *Claudius [. . .] Bretene lond gesohte* 'Claudius [. . .] sought [i.e. came to] Britain'). The even lower percentage of satellites in the PATH verbs (28 per cent) is due to the fact that *forlǣtan* 'to leave, abandon', which takes a direct object (NPGROUND), accounts for 60 per cent of the motion descriptions with PATH verbs. A handful of motion clauses also feature a direct object encoding NPGROUND in combination with MANNER verbs (12 per cent) or neutral MOTION verbs (1 per cent for *come/fare* and 3 per cent for the others). These include the transitive use of *flēon* 'to flee' and accusatives of way as in (132a–b), but also transitive uses of prefixed motion verbs whose prefixes do not carry spatial meaning any longer, as *ge-* and *of-* in (133a–b) (see section 5.4). The function of these prefixes is rather transitivizing (cf. Petré & Cuyckens' 'non-predicative prefix constructions' (2008)) or aspectualizing, and they are therefore not counted as satellites in the present study.

(132) a. þa Walas **flugon** ***þa Englan*** swa fyr (ChroA 473.1)
'The Welsh fled [i.e. fled from] the Englishmen like fire.'

b. And þonne **rideð** ælc ***hys weges*** (OhtWu 1.21.4)
'And then each one rides his ways.'

(133) a. se [. . .] ***þæt feoh*** **gearneð** (OhtWu 1.20.36)
'The one who [. . .] "ge-rides" [i.e. reaches by riding] the money [deposited outside the town].'

b. & þa ne mehte ***hine*** [i.e. Æþelwold] mon **ofridan** (ChroA 901.12)
'Then they could not "of-ride" him [i.e. overtake him by riding].'

Other prefixes, by contrast, such as *oð-* 'away', *tō-* 'apart', and *æt-* 'away' in (134a–c) are clearly spatial. These function as PATH satellites. As realization of PATH satellite, however, these pure prefixes are definitely marginal in the motion descriptions of the texts analysed here: There are only 11 instances of them in the 557 clauses. The far more prototypical PATH satellites are prepositional phrases (e.g. *to þam scipum* in (134c)) and adverbs (e.g. *ut* in (134c)), as shown in Table 5.5.

(134) a. Þa gefengon hie þara þreora scipa tu [. . .] & þæt an **oðwand** (ChroA 897.30)
'Then they captured two of the three ships [. . .] and that one [i.e. ship] got away.'

b. & se Ecgbryht lædde fierd to Dore wiþ Norþanhymbre, & hie him þær eaþmedo budon & geþuærnesse, & hie on þam **tohwurfon**. (ChroA 827.10)
'And Ecgbryht led an army to Dore against the Northumbrians, and they offered him submission and concord there, and on that they turned apart [i.e. separated].'

c. buton þa þe þær ***ut ætswummon to þam scipum*** (ChroA 918.21)
'Except those that there out away-swam to the ships.'

Almost three quarters of the motion descriptions with PATH satellites contain a prepositional phrase, and 43 per cent contain an adverb. (Since often, as in (134c), various—adverbal and prepositional—satellites are stacked onto one and the same verb (see 2.1.4.2), the numbers in Table 5.5 add up to more than 100 per cent.)

We can conclude that for describing motion events in Old English, the most frequent pattern (84 per cent overall) is the use of PATH satellites. Typically, these are prepositional phrases and adverbs; pure prefixes play only a marginal role. Also, MANNER verbs combine with PATH satellites in the majority of cases

TABLE 5.5. Types of satellites in the Old English motion clauses with PATH satellite according to verb type.

	AT LEAST 1 PP		AT LEAST 1 ADVERB		TOTAL WITH AT LEAST 1 SATELLITE
MANNER	66%	(40)	57%	(35)	61
PATH	86%	(6)	14%	(1)	7
cuman, faran, fēran	79%	(252)	40%	(128)	321
Other neutral	62%	(34)	47%	(26)	55
Non-MOTION	65%	(17)	42%	(11)	26
Total	74%	(349)	43%	(201)	470

(82 per cent). That the percentage is lower for non-MOTION verbs and PATH verbs (63 and 28 per cent) is due to the two verbs OE *(ge)sēcan* and *forlǣtan*, which both combine with direct objects—the other motion clauses with non-MOTION verbs and PATH verbs contain PATH satellites as well. Thus, the standard case is for Old English to express PATH outside the verb root, in adverbs and prepositional phrases.

5.2.5 Old English usage: conclusion

The study of motion expression in four Old English texts has shown that PATH verbs are very rarely used: They occur in only four per cent of the motion descriptions (25 instances; see Table 5.2), and, what is more, only two of those 25 instances can be considered to represent a pure PATH verb (*(ge)nēalǣcan*; see 5.2.3). As we have seen in section 5.2.4, PATH is instead expressed in satellites in the vast majority of cases: 84 per cent of the motion descriptions feature at least one PATH satellite (a prepositional phrase in 74 per cent of the cases, an adverb in 43 per cent). Also for MANNER verbs, the standard case is to combine with PATH satellites (82 per cent). Thus, the characteristic pattern clearly is to have the semantic element of PATH expressed outside the verb root.

The picture is less clear-cut with respect to the conflation patterns. In the *Chronicles* as well as in the *Saints' Lives*, the information most frequently expressed in the verb is MOTION only (usually *faran/fēran* 'go, travel' and *cuman* 'come'), without any MANNER or PATH information. MANNER verbs are indeed the second most frequent group, but they lag far behind the NEUTRAL MOTION verbs (Table 5.2). Among the Old English texts analysed in this study, the MANNER verb frequency of 40 to 45 per cent reported by Slobin (2004a: 231) (on the basis of novels and frog story narrations) or of 20 to 34 per cent by Cappelle (2012) (on the basis of biographical and fictional texts) for MANNER-salient Present-Day English is only reached in *Ohthere and Wulfstan* (with 50 per cent MANNER verbs). In the other texts, it ranges between 5 (*Chronicle E*) and 17 per cent (*Saints' Lives*).

A closer look at different passages from the *Chronicles* (5.2.2) suggests that the high frequency of the NEUTRAL MOTION verbs *faran/fēran* and *cuman* as opposed to MANNER verbs is due to the topic and style of the texts: In the passages that display the typical formulaic chronicle style, MANNER verbs are extremely rare. In the more narrative passages which zoom in on the action, by contrast, MANNER verbs become a lot more frequent. Therefore, in view of the fact that the extant Old English texts cannot really be compared with Present-Day English novels and frog story narrations (see 3.1.2.1), a cautious conclusion with respect to conflation patterns in Old English is that MANNER verbs are indeed a comparatively frequent choice, but only if this is justified by the subject matter and the style of the texts.

5.3 *Treatment of Latin* PATH *verbs in Gospel translations*

The usage study in 5.2 has found that PATH verbs are an extremely rare choice in motion descriptions in autonomous Old English texts. The present section will use an additional approach to test the range of application of the PATH verbs that existed in Old English (see 5.1.3): For this, passages that contain PATH verbs in the Late Latin text of the Vulgate will be compared to their rendering in the Old English translation of the gospels (ed. Liuzza 1994). Liuzza (2000: 50) characterizes this Old English translation as 'literal but relatively idiomatic' (see also Lenker 1997: 54–9). Therefore, if the Old English PATH verbs were idiomatic choices for describing MOTION, we should expect them to be freely used in the translation when triggered by a Latin PATH verb in the Bible, since the translation apparently aimed at a literal rendering of the Latin text and at being idiomatic at the same time.

First, however, a word on the 'Latin PATH verbs' might be in order. Latin, of course, is seen as a satellite-framed language itself, with prefixes as PATH satellites (e.g. Talmy 2000b: 103), as in formations such as *in-repere* 'to in-creep'. Schøsler (2008: 127), however, suggests that already in Classical Latin, MANNER information in the verb root was not very frequent, and in Late Latin, it was still rarer (for details on the development of motion expression in Latin and French, see chapter 7). In fact, some of the verbs that were originally formed of a MANNER verb with a PATH prefix are already lexicalized to the degree that they can be seen as PATH verbs in Late Latin, i.e. in the language of the Vulgate (cf., e.g., for *descendere* Kopecka 2009: 417; for *intrare, exire* Stolova 2008: 256). *Descendere* and *ascendere*, for instance, are originally prefixations of *scandere* 'to climb' (*de-scandere* 'to off-climb' and *ad-scandere* 'to on/to-climb'). As Stolova (2008: 257) notes, though, the prefixed forms *de/a-scendere* 'were much more common, especially in prose, and eventually [. . .] replaced the simple form.' In the Vulgate, non-prefixed *scandere* is already very rare: The only attestation I found is *porro David ascendebat clivum Olivarum* ***scandens*** *et flens* (2 Sam. 15:30) 'Then David ascended Mount Olivet, climbing and crying'. Late Latin *ascendere, descendere, introire, ingredi, intrare,* and *exire* will therefore be treated as PATH verbs here.

5.3.1 Material and method

Of course, the term 'the' Bible is a little inappropriate, since many different manuscripts of the gospels in the Latin Vulgate existed at the time the Old English translation was written (cf. the overview in Lenker 1997: 28–41 and Liuzza 2000: 1–49, and references therein). But an exemplary check in the comparisons of manuscripts before the tenth century in Fischer's *Varianten zu Markus* (1989) shows that with respect to the verbs we are interested in here, the various manuscripts do not really differ in their lexical choices, but rather in the specific forms (number, tense, etc.) they employ. Thus, for instance, for Mk 8:11 *et exierunt Pharisaei* 'and the Pharisees came out', all the versions have some form of *exire*: Most have the form *exierunt* 'they exited', some use the participle *exeuntes* 'exiting (pl.)'. Similarly, for Mark 8:26 *si in vicum introieris* 'if you enter into the town', most versions have the perfect subjunctive singular *introieris* (from *introire*), some the plural *introieritis*, some a participle *introiens* 'entering (sg.)', or *intres* (present subjunctive of *intrare*). Hence, for our question, the problem of not knowing precisely which manuscripts served as the basis for this Old English translation is not really problematic in the first place, since it is the presence of the Latin PATH verbs in general, and not their specific word forms, that could be expected to trigger Old English PATH verbs.

So with the help of a digitized version of the Vulgate (ed. Fischer & Weber 1997),[97] I looked for verses in the gospels containing *ascendere, descendere, introire, ingredi, intrare,* and *exire,* i.e. verbs for the PATHS UP, DOWN, IN, and OUT, which are also the PATHS covered by the French loan verbs examined in chapter 9.[98] Furthermore, since the verbs OE *(ge)nēalǣcan* and *genēahian* 'to approach' have emerged as pure PATH verbs in the inventory study (5.1.3.7), I also extracted verses containing L *adpropinquare* 'to approach' to see whether this prompted the Old English translators to use the corresponding Old English PATH verb. These verses were then compared to their translations in the Old English version of the gospels in Liuzza (1994). The results are discussed in the next section.

5.3.2 Results and discussion

The analysis of the verbs used in the Old English translation yields a very clear picture: L *adpropinquare* 'to approach' is translated with OE *genēalǣcan* 'to approach' in each and every instance. For all the other Latin PATH verbs, which, in contrast to *adpropinquare* 'approach', do not have a pure PATH verb equivalent in Old English, the translation mainly uses NEUTRAL MOTION verbs, as shown in the overview in Table 5.6. The following subsections will discuss the renderings of these Latin PATH verbs in more detail.

5.3.2.1 *intrare, introire, ingredi*

For the 'enter' verbs *intrare, introire,* and *ingredi* the translators never use a PATH verb, but mostly *gān* 'go, walk' (92 per cent) together with a PATH satellite, such as in (135).

(135) ***intravit*** *autem rex* (Mark 2:26)
Ða **eode** se cyning ***in***
'Then the king went in.'

[97] Available via *Stuttgarter Elektronische Studienbibel* (SESB) Vers. 3.0.
[98] The search terms *ascen**, *descen**, *intra**, and *ingre** should have captured all instances of *ascendere, descendere, intrare,* and *ingredi*. For the *-ire* combinations *exire* and *introire*, the search terms *exi** and *introi** do not capture the first person singular and third person plural present indicative (*-eo* and *-eunt*), the present subjunctive (*-ea-*) and some participial forms. Forms of *inire* and *egredi* have not been included in the search, but they are not so frequent anyway (*egred** occurs nine times in the Gospels, *egress** 24 times; *init* zero, *inibat* zero, *iniit* zero, *inierunt* twice).

TABLE 5.6. Old English verbs translating Latin PATH verbs in the Gospels.

	INTRARE/ INTROIRE/ INGREDI		*EXIRE*		*DESCENDERE*		*ASCENDERE*		*ADPROPINQUARE*	
cuman	2%	(3)	5%	(5)	24%	(12)	—	—	—	—
faran/fēran	4%	(5)	31%	(31)	24%	(12)	31%	(18)	—	—
gān	92%	(120)	54%	(54)	14%	(7)	22%	(13)	—	—
(a)stīgan	—	—	—	—	24%	(12)	38%	(23)	—	—
genēalǣcan	—	—	—	—	—	—	—	—	100%	(28)
Other	2%	(2)	10%	(10)	12%	(6)	8%	(5)	—	—
Total		(130)		(100)		(49)		(59)		(28)

This is consistent with the findings in the preceding sections: There is no Old English verb that specifically means 'enter'. Apart from *gān*, *faran/fēran* and *cuman* are also used to translate the Latin 'enter' verbs, though much less frequently (see Table 5.6). In one verse, the 'enter' verb is left untranslated, and in another one, the Old English translation uses the modal verb *magan* 'can, may' (136).

(136) *quisque non receperit regnum Dei velut parvulus non* ***intrabit*** *in illud* (Mark 10:15)
Soþlice ic secge eow swa hwylc swa godes rice ne onfehð swa lytling ne **mæg** he ***on*** þæt
'Whoever does not receive God's kingdom like a child, he may not [i.e. may not go] into it.'

The information 'in' is almost always (95 per cent) rendered in the Old English version by a preposition or adverb meaning 'in' (*in, into, innan* in 69 examples, *on*[99] in 57 examples). Hence, prototypical renderings are those in (137a–b).

(137) a. *ipse* ***intravit*** *in quoddam castellum* (Luke 10:38)
se hælend **eode** ***on*** sum castel
'The saviour went into a certain village.'

b. *aut quomodo potest quisquam* ***intrare*** *in domum fortis* (Matt 12:29)
oþþe hu mæg man **ingan** ***on*** stranges hus
'Or how can one go into the house of a strong person.'

In very few cases, the information 'in' is not explicitly expressed in the translation, or a preposition *to* or *through* is used instead. To conclude, in all renderings of verses with Latin *intrare, introire, ingredi*, the Old English version follows the satellite-framing strategy, with PATH being expressed either in adverbs or prepositions.

[99] *In* was largely displaced in its prepositional use by *on* in Old English, particularly in West-Saxon (cf. *OED*, s.v. *in*, prep.).

5.3.2.2 *exire*

Also Latin *exire* is rendered with the NEUTRAL MOTION verbs *gān, faran/fēran,* or *cuman* in most cases: in 90 per cent of the examples. In terms of PATH expression, around two thirds of the translations have a satellite (an adverb or prepositional phrase) with a meaning more or less equivalent to that of L *ex-*: OE *fram* 'from', *of* 'of, from, out of, off', *þanan* 'thence', or *ūt* 'out'. Typical translations are the ones in (138a–b):

(138) a. *iterum* ***exivit*** *ad Iudaeos* (John 18:38)
þa **eode** he eft ***ut*** to þam iudeon
'Then he went out again to the Jews.'

b. ***exirent*** *a Bethania* (Mark 11:12)
hi **ferdon** ***fram*** bethania
'They went from Bethany.'

Compared to the translations of the 'into' verbs, which almost all had a satellite meaning 'in', it is surprising that only two thirds of the *exire*-translations make use of a PATH satellite meaning 'out/away'. Where is the PATH information in the remaining third? Upon closer inspection, those verses with *exire* that lack an 'out/of/from' satellite in the Old English translation actually do not refer to situations in which the notion of going out of somewhere is particularly salient. An example is (139a), where the SOURCE GROUND—from inside a house—is only implicit in the context of the last supper mentioned in the preceding verses. In (139b), the SOURCE GROUND is even less salient, as not even the context allows a clear identification.

(139) a. *et hymno dicto* ***exierunt*** *in montem Olivarum* (Mark 14:26)
And gecwedenum lofe hi **ferdon** on elebergena munt
'And after the praise had been sung, they went to Mount Olivet.'

b. *in crastinum voluit* ***exire*** *in Galilaeam* (John 1:43)
On mergen he wolde **faran** on galilea
'The next day, he wanted to go to Galilee.'

Crucially, none of the verses that lack an Old English *fram, ūt,* or similar particle explicitly specify the SOURCE GROUND in the Latin, i.e. they are all like the examples in (139), and different from those in (138) where the SOURCE is either given next to the verb (*a* ***Bethania*** in (138b)) or explicit in the preceding verses (*introivit ergo iterum* ***in praetorium*** (John 18:33) for (138a)). The lack of a salient SOURCE GROUND in verses like (139a–b) appears to be the reason why the translators left out an 'ex' satellite here.[100]

[100] It is conceivable that the frequent use of L *ex-* in sentences that do not refer to MOTION situations with a salient SOURCE is a precursor of a similar phenomenon in Old French. There, preverbal *en* (< L *inde* 'thence'; often in combination with a reflexive, OF *s'en*) is also extremely frequent with motion

Verses where the Old English translation uses a PATH verb to render *exire* are extremely rare: There is one translation with *forlǣtan* 'leave' (140), and three with *gewītan* 'go, depart' (two of which occur in combination with an *of/from* PATH satellite), although it is not quite clear whether the latter should really be regarded as a PATH verb (see the discussion in 5.1.3.1).

(140) *et* ***exiit*** *ab eo daemonium* (Matt 17:17)
and se deofol hyne **forlet**.
'And the demon left him.'

It is not surprising that *forlǣtan* 'leave' is used in a translation of one of those verses that refer to a demon leaving a person (140), rather than in one of those referring to somebody going out of a town, such as (138b): As we have seen in sections 5.1.3.1 and 5.2.3, OE *forlǣtan* often implies that something (or somebody) is left behind, abandoned for good. *Forlǣtan* would therefore probably be less idiomatic when used to refer to a situation where it is likely that the FIGURE will come back or where the SOURCE GROUND is not affected by the departure.

In another two instances, *exire* is rendered with a MANNER verb: *flōwan (ūt)* 'flow (out)' in (141a), the 'default' MANNER OF MOTION (Slobin 1997: 459) for liquid FIGURES, and *springan (ofer)* 'spring (over)' in (141b).

(141) a. ***exivit*** *sanguis et aqua* (John 19:34)
þar **fleow** blod ***ut*** and wæter
'There flowed out blood and water.'

b. *et* ***exiit*** *fama haec in universam terram illam* (Matt 9:26)
and þes hlisa **sprang** ofer eall þæt land
'And the report sprang over all the land.'

Hence, also for the Old English versions of the verses with *exire*, we can conclude that hardly any PATH verbs are used, but that PATH is expressed typically in satellites.

5.3.2.3 *ascendere, descendere*

Ascendere and *descendere* are also most often translated with NEUTRAL MOTION verbs: *cuman*, *faran/fēran*, *gān*, and *(ā)stīgan*. OE *(ā)stīgan* is a verb that is very often used for upward and downward motion alike (see the discussion in 3.2.2.1), so it may be surprising that the translators did not choose it more frequently to render *ascendere* and *descendere*—for both Latin verbs, the translation with *(ā)stīgan* is outnumbered by the renderings with *cuman*, *faran/fēran*,

---

descriptions, also in those where the SOURCE is not clear, or at least not salient. Scholars of Old French suggest that OF *en* often does not have any spatial meaning ('from there'), but has rather acquired an inchoative function (e.g. Buridant 2000: §328.2; Marchello-Nizia 2002: 215; for more details, see the discussion in 7.1.2). It is possible that *ex-* in the biblical *exire* in examples such as (139a–b) has a similar function, based on the metonymy of initial PATH being contiguous to the beginning of the MOTION event.

and *gān*. This may perhaps have to do with the fact that in the gospels, *ascendere* and *descendere* (like ἀναβαίνω 'to ascend' and καταβαίνω 'to descend' in the Greek version) are often used to describe motion from or towards Jerusalem—probably because of its elevated location—or down to Capernaum.[101] It is possible that in these cases, adding the information 'up'/'down' seemed unnecessary to the translators: They only chose to use *āstīgan* in one single verse out of 25 of this kind. In the rest of them, which all use *faran/fēran* and *wendan*, also 'up'/'down'-satellites are conspicuously lacking, see for instance (142).

(142) ***descendebat*** *ab Hierusalem in Hiericho* (Luke 10:30)
ferde fram hierusalem to hiericho
'Went from Jerusalem to Jericho.'

Yet also in those contexts where the upward or downward direction is more salient, i.e. motion down from or up to mountains, trees, roofs, heaven, or down to hell or from the cross, *(a)stīgan*, although its use rises markedly (18 out of 40, e.g. (143a–b)), is slightly outnumbered by the NEUTRAL MOTION verbs, such as in (144).

(143) a. *cum autem* ***descendisset*** *de monte* (Matt 8:11)
þa se hæl[a]nd of þam munte ***nyþer*** **astah**
'When the saviour descended down from the mountain.'

b. ***ascendit*** *in arborem sycomorum* (Luke 19:4)
stah ***up*** on an treow
'Went up on a tree.'

(144) *si rex Israhel est* ***descendat*** *nunc de cruce et credemus ei* (Matt 27:42)
Gyf he israhela cyning sy **ga** nu ***nyþer*** of þære rode and we gelyfað hym
'If he be the king of Israel, he shall now go down from the cross and we believe him.'

Like for *exire* above, we also find a few attestations where *descendere* is rendered with a MANNER verb, *ēf(e)stan* 'to hurry', in the verses referring to Zacchaeus hurrying down from the tree.

(145) *Zacchee festinans* ***descende*** (Luke 19:5)
Zacheus **efst** to þinum huse
'Zacchaeus, hurry to your house.'

The comparison with the Late Latin verse in (145) looks like one of the textbook juxtapositions of a PATH-conflating and a MANNER-conflating language: While in the original, PATH is expressed in the verb, and MANNER in the adverbial participle *festinans* (L *festinare* 'to hurry'), in the translation, MANNER is found

[101] Although this 'up' and 'down' here might also be related to social factors, or just be conventional, such as in Bavarian, a German dialect that makes heavy use of directional adverbs (e.g. *auf Passau* ***obe*** 'down to Passau', *auf Falkenberg* ***auße*** 'out to Falkenberg', *auf Minga* ***aufe*** 'up to Munich'), and where the choice of the adverb is not always determined by topography (cf., e.g., Rowley 2007).

in the verb, and PATH—although not *down from the tree*, but *to your house*—in a prepositional phrase.

5.3.2.4 Different rhetorical styles?

Before we conclude, let us have a brief look at Luke 8:33, where two Latin verbs are reduced to only one Old English verb:

(146) ***Exierunt*** *ergo daemonia ab homine et* ***intraverunt*** *in porcos* (Luke 8:33)
þa **eodon** hig ***of*** þam men ***on*** þa swyn
'Then they [i.e. the demons] went from the men into the pigs.'

The comment on this verse in Liuzza (2000: 69) is: 'the translation omits the subject *daemonia* and the verb *intraverunt*. The latter looks like an error'. However, according to Slobin's studies, as we have seen in section 2.1.4.2, satellite-framing languages tend to accumulate PATH phrases on one single verb (*eodon* + *of þam men* + *on þa swyn*), while verb-framed languages would prefer to segment the PATHS and use a separate verb for each PATH segment. Thus, the omission of *intraverunt* is probably less of an error, but rather an accomodation to the 'rhetorical style' (Slobin 1996b: 201–4; Berman & Slobin 1994: 118–9) of Old English as a satellite-framed language.

5.3.3 Conclusion

This comparative study focussed on the Old English translation of the Late Latin verbs *intrare, introire, ingredi, exire, ascendere, descendere*, and *adpropinquare* in the Gospels. We found that for the translation of L *adpropinquare* 'to approach', the Old English pure PATH verb *genēalǣcan* 'to approach' is used consistently. For the other Latin PATH verbs, Old English does not have a pure PATH verb equivalent. In a majority of cases, these Latin PATH verbs are not rendered with PATH verbs in the Old English translation, but with NEUTRAL MOTION verbs, rarely also with MANNER verbs (OE *ēfestan* 'hurry', *flōwan* 'flow', *springan* 'spring') and non-MOTION verbs (*magan* 'can, may'). The PATH information conveyed by the Latin verbs is usually given in satellites in the Old English version. In contexts where the PATH information expressed by the Latin verb is not salient in the MOTION event that is described, it tends to be left out in the Old English translation.

We have not investigated the Old English renderings of L *abire* 'go away', *egredi* 'exit', and *transire* 'go across'. Yet, the Latin–Old English word list in Liuzza's edition of the gospels indicates that for these verbs the picture is quite similar: For L *abire* 'go away', the word list gives OE *cyrran* 'to turn', *faran, fēran, gān*, and *ūtgān*, and for L *egredi*, it has OE *fēran, ūtgān, ūtgangan*, with the latter two having PATH in the particle *ūt*; for L *transire* it lists OE *āgān, faran, fēran, gān, gangan*, and *gewītan*, none of which is a PATH verb that would mean 'go across'. Hence, like the studies of the Old English inventory and usage in the preceding sections, the Gospel translations also support the diagnosis that in Old English (with the exception of *genēalǣcan* 'to approach'), PATH is typically not expressed in the verb itself, but in a satellite.

5.4 PATH *satellites*

What we have seen so far is that the Old English motion verb inventory has only a very limited number of verbs that express PATH alone (5.1), and that the PATH verbs that exist are also used very infrequently (5.2 and 5.3), compared to NEUTRAL MOTION verbs (such as *faran*), MANNER verbs (such as *rīdan*), or even verbs that do not primarily express MOTION on their own (such as *brecan* or the modals). In this section, then, we will turn to PATH satellites. As argued in section 2.1.3, I use a wider notion of the term 'satellite', like many other recent studies, to include not only prefixes and adverbs, as in Talmy (2000), but also prepositional phrases. The typical PATH satellite in Old English is (or, in the case of prepositional phrases, contains) a free morpheme (following Thim 2012, these will be referred to as 'particles' in this section), which can be realized in different positions, either as an adverb before the verb as what looks like a prefix (written together (147a) or apart (147b)), as an adverb after the verb (147c), or in a prepositional phrase before (147d) or after the verb (147e).

(147) a. þæt hi na mihton ***incumen*** (ChroE, 1070)
'So that they could not in-come.'

b. sona þa ***inn*** **eode** an þæs cyninges þegna (ÆLive (Oswald), R87)
'Soon there in-went one of the king's theigns.'

c. & þa oðre ða dura bræcon þær adune & **eodon *inn*** (ChroE, 1083)
'And then the others broke down the doors and went in.'

d. and syððan þæs sumeres mid ferde ***into Wealan*** **for** (ChroE, 1121)
'And then during the summer he went with an army into Wales.'

e. & þanon **wendon *in Wihtland*** (ChroE, 1001)
'And thence they went to the Isle of Wight.'

I use the particle *in* as an example in (147a–c, e) because it can occur in all these positions and functions (Hiltunen 1983a calls this kind of particle 'prepositional adverb'). Other particles cannot be used as preposition (Hiltunen's 'phrasal adverbs', such as *forþ*), but only as adverbs before or after the verb, as in (148a–b).

(148) a. Ðæt mæden ða ***forð*** **eode** mid cynelicum reafe ymbscrid (ApT 40.9, in Hiltunen 1983a: 121)
'Then this girl stepped forward, clothed in royal garments.'

b. **Cuman** nu myccle hundas ***forð*** & hine abitan beforan þyssym casere (BlHom 181.19, in Hiltunen 1983a: 200)
'Now let great dogs come forth and bite him before the emperor.'

There are many studies on the position of these particles in Old English, and the changes it underwent on the way to Middle English (e.g. Hiltunen 1983a, 1983b; Fischer et al. 2000: 180–210; Fischer & van der Wurff 2006: 190–3; van Kemenade & Los 2003; Elenbaas 2007; Thim 2012). For our question, this issue of position is no great concern, because the particle is outside the verb root, whether it is pre- or postverbal. As argued in detail by Thim (2012), its positional changes are a consequence of the changes in word order.

Word order in Old English is typically (but by no means exclusively) verb-second in main clauses and verb-final in subordinate clauses. In early Old English, verb-final is still more widespread in main clauses as well. (For details, see e.g. Denison 1993; Traugott 1992; Thim 2012: 93–100). In the Middle English period, SVO spreads in both main and subordinate clauses (e.g. Fischer 1992). The position of the particle reflects this: Both phrasal adverbs such as *adun* 'down', *aweg* 'away', *forþ* 'forth', *niðer* 'down', *up* 'up', *ut* 'out', and prepositional adverbs in adverbial function such as *in* 'in', *ofer* 'over', *on* 'on, in', *to* 'to', *þurh* 'through', and *under* 'under' were most frequently preverbal in early Old English, but they became predominantly postverbal in late Old English main clauses, and, in early Middle English, also in dependent clauses and with non-finite verb forms (Hiltunen 1983a: 105–25, 160–7). Denison (1981), for instance, shows that already in the *Orrmulum*, an early Middle English text from around 1200, as few as six per cent of all particles are preverbal (1981: 174; quoted in Claridge 2000: 85). Roughly speaking, as word order shifts to VO, particles become postverbal.[102] Particle position thus is mainly 'epiphenomenal to basic word order', as Thim (2008: 299) puts it.

However, this apparently neat situation of PATH being expressed in free morphemes is a bit more complicated at more detailed examination: First, in addition to these particles, Old English also has prefixes which are invariably bound to the verb (see 'prefixed verbs' in Figure 5.3 below). An example is the prefix *ā-*, which—at least originally—meant 'away, out' (from Proto-Germanic **uz* 'out'). Most of these prefixes are spatial in origin; in Old English, however, they typically have aspectual or transitivizing functions, and 'frequently there does not seem to be any meaning difference at all between the simplex and the prefixed form' any longer (Kastovsky 1992: 377; for aspectual functions of OE *ā-* see also Broz 2013). They are discussed in more detail below.

Second, in some formations even an element such as OE *ofer* 'over' that is otherwise a particle, and hence free and stressed, has developed into an inseparable, unstressed prefix (cf., e.g., Kastovsky 1992: 375–6; Hiltunen 1983a: 25–7, 158, 185; see 'inseparable particle' in Figure 5.3 below). An indicator for inseparability is that *ne* 'not' or, in the case of an inflected infinitive, *to* do not intervene between the particle and the verb (Mitchell 1978: 246; Hiltunen 1983a: 27), as in the following examples (both from Mitchell 1985: §1973):[103]

---

[102] A more accurate phrasing would probably be 'verbs become pre-partical', since it is actually the position of the verb that changes, cf. Thim (2012: 115).

[103] If the particle and the verb are separable, the position of *ne* 'not' and *to* is between the two, such as in *ge **in** ne **eodun*** (Lk(WSCp) 0504(11.52), DOEC) lit. 'you in not went' or in *nyd hi **inn** to **farenne*** (ÆCHom II, 26, 0064 (216.100), DOEC) lit. 'urge them in to go'.

(149) a. gefera, gefera, gemyne þæt ðu ure gecwedrædenne 7 geferrædenne to longe **ne oferbrec** (*Or* 242.6)
'Comrade, comrade, remember that you do not violate [lit. over-break] our agreement and our companionship too long.'

b. hio bið micle ðe ieðre **to oferfeohtanne** (*CP* 277.24)
'It is much easier to conquer [lit. over-fight].'

These lexicalized combinations of prefixed particle and verb, however, often do not preserve the 'original locative meaning' of the particle (in (149a–b), for instance, *ofer* is no longer spatial), and the meaning of the combination is often metaphorical (Kastovsky 1992: 375), as it is in (149a–b), where *oferbrecan* means 'violate' instead of 'over-break', and *oferfeohtan* means 'conquer' instead of 'over-fight'. Further examples of such combinations of motion verbs with inseparable particles are *ofercuman* 'to overcome' or *oferstīgan* 'to surpass' (cf. Hiltunen 1983a: 158). These lexicalized, inseparable combinations are hardly ever used to describe LITERAL MOTION. Generally in Old English, 'the more metaphorical the meaning of a combination [of verb and prefixed prepositional adverb], the more likely it is to be inseparable' (Hiltunen 1983a: 185). An overview of the synchronic situation in Old English just sketched is given in Figure 5.3.

The typical Old English PATH satellite is the prepositional phrase or the separable particle (see the results of the usage study in Table 5.5 (section 5.2.4), where the satellite contains at least one prepositional phrase in 74 per cent of the cases and at least one adverb in 43 per cent). The inseparable particles and the prefixes are a less frequent means of conveying PATH information; they will be discussed in what follows. We will start with the pure prefixes and, after placing them in a diachronic context, address the issue of the inseparable particles. The account is mainly based on Thim's diachronic study of phrasal verbs (2012).

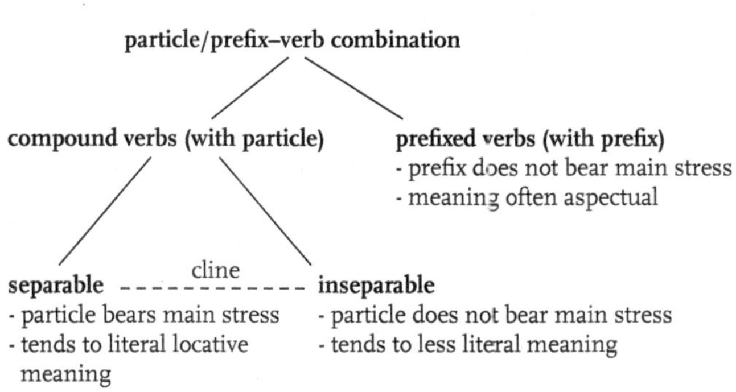

FIGURE 5.3. Old English particle/prefix-verb combinations.

According to Thim (2012; see also Dietz 2004), the Old English verbal prefixes which are always bound and—in contrast to the particles—also unstressed are the following:

(150) **Old English verbal prefixes** (Thim 2012: 148, 151)[104]

a-	< Proto-G. **uz* 'out'
æt-	< Proto-G. **at* 'at, to'
an-/on-	< Proto-G. **anda* 'away (towards)'
be-	< Proto-G. **βi* 'by'
ed-	< Proto-G. **iþ*, **ið* 'back (again)'
fer-/for-	< pre-OE **fur* 'away'
fræ-	< pre-OE **fra* 'forward, away'
full-	< Proto-G. **fulla* 'full'
ge-	< Proto-G. **γa* 'near, at, with'
mis-	< Proto-G. **missa* 'wrong'
oþ-	< Proto-G. **unþa* 'away'
samod-	< pre-OE **samuð* 'together'
te-	< pre-OE **te* 'apart'
tŏ-	< pre-OE **tuz* 'apart'
un-/on-	< Proto-G. **und/anda* 'away (towards)'

Similar lists in the literature show slightly different inventories of prefixes mainly because they are based on different criteria. Thus, Kastovsky (1992: 377–81) does not include as prefixes those that have a corresponding free form (e.g. *æt, of, on, to*), and de la Cruz 1975 (and following him, also Hiltunen 1983a) only treats *to-*, *a-*, *on-*, *ge-*, *for-*, *of-*, and *be-* as 'pure prefixes'. Those either have no etymologically related free form (the free morpheme *to* that is cognate with German *zu*, for instance is a homonym of the prefix *to-*, cognate with German *zer-*, and therefore not etymologically related to the other *to*), or, in the case of *for-*, *of-*, and *be-*, have 'markedly isolated themselves from the latter on functional grounds', as de la Cruz argues, which is why they should be seen as separate morphemes (1975: 47). For instance, while the particle *be* has meanings such as 'near', 'along', 'by' (*DOE*, s.v. *be*), the prefix *be-* can have a transitivizing function (e.g. *sprengan* 'sprinkle, intrans.', *besprengan* 'besprinkle, trans.'; see also Petré & Cuyckens 2008), an intensifying or a privative meaning (e.g. *beniman* lit. 'be-take', i.e. 'deprive') (*BT* s.v. *be-*; see also Hiltunen 1983a: 48).

The verbal prefixes in the list in (150) not only have in common that they are invariably bound, but also that in many cases the spatial meaning that most of them originally had (see the etymologies) has given way to aspectual meanings or transitivizing functions, so much so that sometimes they even have

[104] However, on the claim that all these verbal prefixes are unstressed, see Minkova, who, in the context of dismissing the traditional view of functional stress shifting (that prefixes are stressed in nouns since these formations existed before the fixation of stress in Germanic; that they are unstressed in verbs, because these are innovations from after the fixation of stress) as too simplistic, points out that *ed-* and *and-* carry stress not only when prefixed to nouns, but also when prefixed to verbs (with the examples of *ándswàrian* 'to answer' and *édniwian* 'to renew', in *Maldon* 44, *Bwf* 340, *MB* 11 39, *Juliana* 485, where, due to the alliterative verse, stress on the prefix cannot be refuted) (Minkova 2008; 2012;

no clearly recognizable meaning any longer (e.g. Hiltunen 1983a: 48–51; van Kemenade & Los 2003; Thim 2012: chapter 5.1–2).

Yet the development of the prefixes must have started in the same way as that of the separable particles, in that they too have their origin in spatial adverbs and prepositions (e.g. de la Cruz 1975: 49; Dietz 2004; van Kemenade & Los 2003; Thim 2012).[105] For one thing, this is a very common grammaticalization process for pre-verbs, not only in the Germanic or the Indoeuropean languages in general, but crosslinguistically (Booij & van Kemenade 2003: 4): Free spatial adverbs evolve into elements that are fused to the verb and acquire aspectual semantics. Another indicator that these prefixes used to be free morphemes originally is that in Gothic, the Germanic language attested earliest (in the *Wulfila* Bible translation from the fourth century), other morphemes (such as an interrogative marker) can occur between the prefix and the verb. Thus, even though the prefixes are already bound in Gothic, their former separability is still reflected in the fact that clitic particles can intervene between them and the verb (van Kemenade & Los 2003: 98; Thim 2012: 166–81). The fact that, compared to Old English, in Gothic a higher number of prefixes still have corresponding adverbs or prepositions has also been cited as evidence for this (Dietz 2004: 568; see also Thim 2012: 169).

Thus, Old English pure prefixes must be regarded as representing a rather advanced stage of the widespread development from spatial adverbs to bound elements: They have become invariably bound to the verb and have acquired aspectual or other meanings, which often replaced their original spatial meaning. According to Thim's overview (2012: 152), literal spatial meanings in OE are still available (next to the more abstract ones) only for the pure prefixes in (151), but no longer for the ones in (152).

(151) *æt-* (originally 'at, to', then also 'away', see Dietz 2004: 573–9), *ed-* 'back', *oþ-* 'away', *samod-* 'together', *to-* 'apart'

(152) *a-, an/on-, be-, fer-/for-, fræ-, ge-, te-, un-/on-*.

However, this does not entail that the prefixes in (151) *always* have literal spatial meanings: *æt-*, for instance, is found in combinations where it apparently does not add anything to its base—*æt-slīdan* only means 'to slide, slip, fall', just like the simplex *slīdan* 'to slide, slip, fall'; the same holds for *æt-stæppan* 'to go, tread' (cf. *DOE* s.vv. and Appendix A). Furthermore, even in combinations where the Present-Day English equivalent given by the *DOE* would suggest a literal spatial meaning of *æt-*, such as in *æt-yrnan* 'to run away', this spatial meaning need not always necessarily be conveyed by the prefix. In (153a), for instance, it would be present in the context as well, namely in *ut* 'out'. In others, such as (153b), however, it is indeed carried by *æt-* alone.

---

on stressed preverbal *ed-* see also Bülbring 1902: §77). Dietz (2004: 565) argues that these stressed verbal prefixed forms are denominal formations.

105 Even *ge-* is thought to be of 'locative/circumstantial' origin in the end, probably being a cognate of Latin *cum* 'with, together' (Booij & van Kemenade 2003: 4; de la Cruz 1975: 49).

(153) a. Þa teah hy hine be his claðum & cwæð to him slap mid me þa **ætarn** he ***ut*** & forlet his wæuels on hyra handa. (*Gen* (Ker) B8.1.2, 39.12, DOE s.v. *æt-yrnan*)
'Then she pulled him on his clothes and said to him "Sleep with me." Then he ran out and left his cloak in her hands.'

b. Ðeah þræla hwylc hlaforde **ætleape** & of cristendome to wicinge weorþe (*WHom* 20.3 B2.4.2C, 101, DOE s.v. *æt-hlēapan*)
'If any servant escapes from (lit.: away-runs) the lord and turns from Christendom to the Danes.'

The only clearly spatial prefixes that were found in the analysis of more than 550 Old English motion descriptions in section 5.2.4 (in which 470 had at least one PATH satellite) are *æt-* 'away' (two occurrences), *oþ-* 'away' (three occurrences), and *tō-* 'apart' (six occurrences). This confirms that for the expression of PATH, prefixes hardly play a role any longer in Old English.

The particles (*adun, æfter, forþ, fram, in, of, ofer, under, up, ut,* etc.) by contrast, which as we have seen above are still flexible in their position, are very frequent in the motion descriptions analysed in section 5.2. They are still at an earlier stage of what is, however, essentially the same development that the prefixes have taken. This is convincingly advocated, for instance, by van Kemenade & Los (2003), Dietz (2004), and Thim (2012), who also stresses that the only thing that is different with the particles is that their development into bound morphemes is forestalled (a) by the fact that word order is becoming more and more fixed to SVO in Middle English, which entails that the particles become restricted to postverbal position, and (b) by the fact that they are stressed (2012: 164). This, however, concerns only the formal side of their development. Semantically, Thim points out the familiar path from predominantly spatial particle semantics in Old English to more aspectual and also non-compositional uses later on, such as we have seen in section 4.2.3 (examples (69a–b)) with lexicalized OE *forþferan* 'to die' (2012: 156, 176–83; see also, e.g., de Smet (2010: 84–96) on *out* and *forth*). Since this development is already under way in Old English, there are (a) also instances of particles whose meaning is no longer clearly spatial, and (b) also some combinations of particle and verb in which the particle has become inseparable (see the overview in Figure 5.3 above).

In sum, therefore, the typical realizations of PATH satellites in Old English are adverbs and prepositional phrases. The prefixes that originally carried PATH information and have become inseparably bound to the verb have lost their PATH meaning to a great extent, although they are still available as PATH satellites to a limited degree. This is supported by the results of the usage study in section 5.2, where the only prefixes with clearly spatial meaning that were found in the motion descriptions are *æt-* and *oþ-* 'away', as well as *tō-* 'apart'. These, moreover, only occurred in 11 out of the 470 motion descriptions with satellites, which suggests that prefixes hardly play a role as satellite-realizations any longer in Old English motion descriptions.

CHAPTER 6

Talking about MOTION in Middle English

ANALOGOUS TO THE STUDY of Old English motion verbs in chapter 5, in the chapter on Middle English, we will also first look at the inventory of verbs that can be used to express INTRANSITIVE MOTION (6.2), and then study Middle English motion description on the basis of text excerpts (6.3). There will be no separate section on satellites, since their further development has already been addressed in section 5.4 of the preceding chapter. The translation of French and Latin PATH verbs into Middle English will be discussed in the scope of chapters 8 and 9. Before we come to the studies of the Middle English inventory and usage, section 6.1 will give a brief overview of the language contact situation with French in the Middle English period, which had a tremendous impact on the vocabulary of the English language.

6.1 *French influence on Middle English*

The Middle English period is commonly taken to begin around 1100, i.e. a few decades after the Norman Conquest. The period boundaries between Old and Middle English are not based on the lexicon, but rather on the decline of the (inflectional) endings, with Old English being characterized by Sweet (1888: §581) as the 'period of *full* endings (*mōna, sunne, sunu, stānas*)', and Middle English as the 'period of *levelled* endings (*mōne, sunne, sune, stǫnes*)'. While the beginning of Middle English also coincides roughly with the beginning of the Norman rule in England, the French influence on the lexicon, which was a late consequence of the Norman Conquest,[106] becomes noticeable only from the late thirteenth century onwards (see, e.g., Lutz 2002a: 162), i.e.

[106] Although of course, contact between speakers of French and English already existed before the Norman Conquest, such as in the course of the Cluniac reform (see, e.g., von Mengden 2012: 27–9), through the Normans brought to the English court by Edward the Confessor (1042) after his exile in Normandy (see, e.g., Skaffari 2012: 1673), and, obviously, also through trade and other contacts.

in the ‘middle’ of the Middle English period. Thus, with regard to the lexicon, in the first two centuries of the Middle English period, the language is not very different from Old English, but can still be characterized as Anglo-Saxon (Lutz 2002a: 156–61; for a concise discussion of the ‘end of Old English’, see also von Mengden 2012: 32; on periodization in general, see Curzan 2012).[107]

As is well known, the lexicon of today’s English contains a very sizable French element, prominent even in the core vocabulary: Scheler (1979: 72–3), for instance, found 38 per cent French loans in the ‘General Service List’, a list of about 4,000 frequent words in Present-Day English; the *World Loanword Database* (cf. Haspelmath & Tadmor 2009 and http://wold.clld.org) identified approximately 35 per cent loans from French and Latin in a list of roughly 1,460 basic meanings (Grant 2009: 381–3). Note that although verbs are usually much less ‘borrowable’ than nouns, Grant (2009: 370) found that a good third of the verbs in this list of basic meanings are not native to English, but have been borrowed, most of them from French. More recently, Durkin (2014: 37) has shown that around half of the 1,000 most frequent words in the *British National Corpus* (BNC) are loans from French and/or Latin. The bulk of these loanwords that have become part of the core vocabulary of English came into the language in the Middle English period (Durkin 2014: 39). Later loans tend to stay more restricted to intellectual, cultural, and technical vocabulary.

It is particularly from the beginning of the fourteenth century onwards that the English lexicon sees this massive influx of French words (cf. the figures in Durkin 2014: 259–63). This is due to the fact that late medieval England was a strongly multilingual society (with the most important languages being English, French, and Latin, but also Flemish and other languages being involved, see, e.g., Machan 2012). Latin was mainly used as ‘a language for specialized written purposes’ (Vezzosi 2012: 1708), i.e. as a ‘High Language’ (Schendl & Wright 2011: 18), and Latin loans are therefore more typical cultural borrowings, belonging to more ‘learned’ registers (cf. the results in Scheler 1977: 83–4, also the discussion in Lutz 2002a: 147–8). This is different for French, as shall be briefly discussed in the following paragraphs.

Immediately after the Conquest, the new aristocracy of England, of course, spoke their Northern French dialects (not exclusively Norman French, since William also had Picard, as well as Breton and Flemish, followers). Their descendants became native speakers of English, or native bilinguals very soon (e.g., Machan 2012: 523; Townend 2006: 67), but many members of the English literate classes acquired French as a second language, probably in early childhood (Ingham 2010b; 2012), French being a language of prestige and

[107] One difference between the vocabulary of Old English and early Middle English, though, is the higher number of loans from Scandinavian which are attested in the Middle English period. These have been adopted into the language in the Old English period (from the Viking settlers), and their late appearance is rather due to the respective provenance of the extant texts, with most Old English texts being written in the West-Saxon dialect.

the language of education, business, and so forth.[108] This also led to habitual code-switching in late Medieval England, as attested, for instance, in mixed-language business writing (e.g. Wright 2000; 2011).

Naturally, during these centuries of di- or triglossia, particularly after the loss of Normandy in 1204, the French spoken in England developed into a separate variety of French, usually called Anglo-Norman or Anglo-French (see the discussion in Wogan-Browne 2009: 1–10), which became increasingly different from the varieties of French spoken on the continent (e.g. Rothwell 1991: 175).[109] Anglo-French shows several specific features that can be explained as a function of its close contact with Middle English (see, e.g., Ingham 2010a; see also 6.2.3.2 below), as it 'evolved in parallel and in constant contact with Middle English on the soil of England' (Rothwell 1998: 159).

Middle English, in turn, was heavily influenced by Anglo-French as well. Continental French also featured as a contact variety for Middle English, but its influence was apparently more through the written medium (e.g. Skaffari 2012: 1672), and therefore different in kind from the everyday contact with spoken Anglo-French. The traditional overestimation of Continental French as a source of loanwords in English, and the concomitant underappreciation of Anglo-Norman as a donor language has been repeatedly pointed out by Rothwell (e.g. 1993a, 1993b, 1998), such as in the following:

> [I]n literally thousands of cases forms and meanings were adopted [. . .] into English from Insular, as opposed to Continental, French. The relationship of Anglo-French with Middle English was one of merger, not of borrowing, as a direct result of the bilingualism of the literate classes in mediaeval England (Rothwell 1991: 174).

Rothwell has shown for a number of words that they cannot have entered English through Continental French, but must have been 'homegrown' words or usages, adopted from Anglo-Norman. Among them are, for instance, *coro(u)ner*, *duty*, *dues*, *record*, and *recorder*, which are used in meanings in English and Anglo-Norman that are not attested in Continental varieties of French (Rothwell 1993a: 311–13; 1998: 147). While the adoption of the former words, as terms from the legal system, can be explained through the fact that the legal system after the Conquest was French (see also the discussion of the replacement of Anglo-Saxon legal terms in Lutz 2002a: 148–50), there are other Anglo-Norman words (or specifically Anglo-Norman semantic developments

[108] The compulsory quote with respect to French as the prestige language in the thirteenth century is Robert of Gloucester's *Vor bote a man conne frenns me telþ of him lute* (7542, ed. Wright 1887) 'For, unless a man knows French, people think little of him.'

[109] Later Anglo-Norman, as attested from the thirteenth century onwards, had often been discarded as an imperfect learner variety (e.g. Pope 1956: §1077), but has recently regained scholarly attention (cf. the volumes by Short 1993, Wogan-Browne 2009, Ingham 2010, numerous articles by Rothwell, and, above all, the *Anglo-Norman Dictionary*).

of French words) in English which do not belong to typically superstratal areas of influence. These include, among others, *to sack* in the meaning 'to expel, drive out' (AN *saker*), *to puke* (AN *pucher*) (Rothwell 1993a: 315–17), and *flour* (Rothwell 1998: 156).

Thus, in many cases, 'the *form* of a French word was taken into Middle English, but its *meaning* reflects the French of London rather than that of Paris' (Rothwell 1998: 162). An example of this kind of *faux amis* is French *donjon*/ English *dungeon*: Originally, the word refers to a 'structure high above the ground' (ibid.). In Anglo-Norman, only the meaning components of 'strength' and 'security' remain, and the word starts to be used for strong and secure structures 'below ground'. We will return to Anglo-Norman as a source for loanwords in English in section 6.2.3.2 and elsewhere below, in a discussion of a number of French-borrowed motion verbs which equally have most likely come from the English-influenced Anglo-Norman variety, rather than from Continental French.

To conclude, this section has shown that, starting in the late thirteenth century, the English vocabulary changes tremendously due to the long-term coexistence of (Anglo-)French and English: English takes up a wealth of (Anglo-) French words, which, crucially (a) also involve many verbs and (b) often become part of the everyday vocabulary. Therefore, if we want to investigate the 'native' system of talking about MOTION, into which the French PATH verbs are borrowed (*enter, descend, mount,* etc.), we cannot be content with investigating the Old English situation, but need to take into account the Middle English situation as well. This will be done in the following sections, starting with the motion verb inventory.

6.2 *Motion verb inventory*

The material and method of the study of the Middle English motion verb inventory is explained in section 4.1. The present section presents the general results (6.2.1), i.e. the different kinds of motion verbs that have been identified in the Middle English inventory (MANNER, PATH, NEUTRAL MOTION, and non-MOTION; see 2.3). In section 6.2.2 we will see that the proportion of French and Latin loans is much higher in the PATH verbs as compared to the other groups of motion verbs. The subsequent sections will take a closer look at the different verb groups: Section 6.2.3 will discuss the origins of the new Middle English MANNER verbs, with a particular focus on those borrowed from French, as some of them, such as ME *frushen*, are not motion verbs in French and only start to be used to express MANNER OF MOTION in Middle English (or Anglo-Norman). Section 6.2.4 investigates the factors that enable verbs that primarily express CAUSED MOTION, such as *casten* 'to throw, cast', to be used in the intransitive motion construction as MANNER verbs and shows that these also include some French loans that were not used for INTRANSITIVE MOTION in the donor language. Section 6.2.5 presents the non-MOTION verbs in the inventory, and section 6.2.6 shows that, like in Old English, also in Middle English most MANNER verbs have attestations with PATH satellites, as expected in a satellite-framing, manner-conflating language. The PATH verbs found in the inventory will not be treated in a separate section here, since the pure PATH verbs *entren, ishen/issuen, (a)mounten, ascenden, avalen,* and *descenden* will be discussed in detail in chapter 9. Section 6.2.7 will sum up the findings of the Middle English inventory study.

6.2.1 General results

An overview of the results is given in Table 6.1. Like in the Old English inventory study, the verbs that are attested in an intransitive motion construction—again the vast majority—are printed in italics. Appendix B, which can be found on the companion website to this book [www.oup.com/us/motionandtheenglishverb/], also gives the meanings, attestations, etymologies, and prefixes for the single verbs.

As shown in Table 6.1 and Figure 6.1, MANNER OF MOTION verbs form the largest group in the Middle English inventory as well, accounting for a good half of all the verbs that can be used to express MOTION (258, i.e. 55 per cent). Different from Old English, though, the group of verbs primarily expressing PATH is slightly larger (72, i.e. 15 per cent, compared to ten per cent in Old English), which is mostly due to borrowings from French and Latin, as we will see in the following section. The group of verbs that do not primarily express MOTION is also slightly larger: It contains 46 verbs, i.e. it accounts for ten per cent of the overall motion verbs (compared to eight per cent in Old English); the group of NEUTRAL MOTION verbs has 89 members and thus accounts for 19 per cent of the Middle English motion verbs (compared to 25 per cent in Old English).

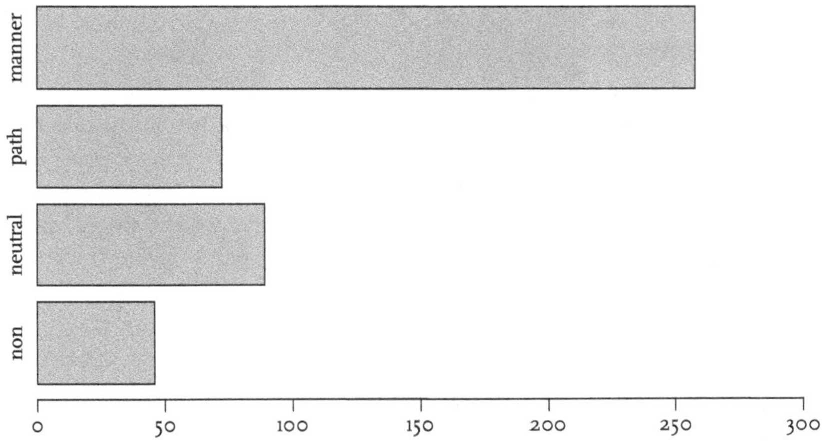

FIGURE 6.1. Types of verbs in the Middle English motion verb inventory.

TABLE 6.1. Middle English motion verb inventory.*

a) **Verbs primarily expressing MANNER OF MOTION (Ma) (55% (258))**

1. **Speed (Ma^{speed})** (91): *blenchen, breiden, breithen, bronten, brushen* (v.1), *busshen* (v.2), *cacchen, casten, chacen, clingen, dashen, dingen, driven, dushen, efesten, ferken, fiken,* fisken, *flatten, flingen,* floden, *flusshen, friken, frushen, fusen,* gadden, galopen, *girden* (v.2), *glenten, gushen, halen, hasten, hien, hurlen, jacen,* jumpen, *lashen, launcen, launchen, lepen, lopen, louken* (v.2), *lushen, platten* (v.1), *poperen, pressen, priken, rapen* (v.1), *rappen* (v.2), *rasen* (v.3), *rashen, reken* (v.1), *remen* (v.3), *rennen* (v.1), *rennen* (v.2), *resen* (v.2), rillen, *routen* (v.4), *rushen, sailen* (v.2), *shaken, sheten, shokken* (v.1), *shunten, skillen* (v.2), *sleten, smiten, sparren* (v.2), *speden, sperren, stertelen, steten, straken* (v.1), *streken* (v.1), *swengen, swepen, sweien* (v.2), *swingen, swippen, thirlen, thringen,* treunten, *wainen* (v.2), *walopen, werpen, whippen, whirren, wincen, winden* (v.1), *wingen, (a)wrenchen*
2. **Slow or impeded motion (Ma$^{\text{slow/impeded}}$)** (26): *blonderen, clambren,* clocken, *craulen, crepen, gog(e)len,* falteren, halten, hirplen, *hobelen, hurtelen, hurten, lurken, raumpen, rothelen* (v.2), *slinken, smuȝen, snaperen, sniken, spurnen, stakeren,* stamperen, *stumpen, stomblen, swafren, waveren*
3. **Implying movement through air or fluid (Ma$^{\text{air/fluid}}$)** (10): *fleten* (v.1), *flien, flikeren, floteren, floten, hoven* (v.1), *rouen* (v.1), *seilen, swalteren, swimmen*
4. **Falling (Ma^{fall})** (23): *deven* (v.2), *dippen, disten, diven, dompen, doppen, douken, drepen* (v.2), *droppen, duven, fallen, helden, louten* (v.1), *plumben, plumpen, plungen, resen* (v.3), *senchen, sinken, tumben, tumblen, wadelen, welten* (v.2)
5. **Gliding (Ma^{glide})** (10): *flouen, gliden, leken,* ridelen, *sien, sliden, slideren, sliken, slippen, stremen*
6. **Jumping (Ma^{jump})** (9): *hippen, hoppen, sauten* (v.2), scopen (v.1), *skippen, sprenten, springen, spritten* (v.2), *sterten*
7. **Rolling (Ma^{roll})** (9): *relen* (v.1), *rollen* (v.2), trenden, *turnen, waltren, walwen, wharven,* whirftlen, *whirlen*
8. **Fleeing, escaping (Ma^{flee})** (7): *escapen, flemen* (v.1), *flen* (v.1), *scapen* (v.1), *skeuen* (v.2), *smolten,* stoken (v.1)
9. **Other manner of motion (Ma^{other})** (73): *abaten, airen,* amblen, balen, *benden* (v.1), *blusteren, bustelen, butten* (v.1), claveren, *climben, cliven, crouden* (v.1), dauncen, enterchaungen, foten, *frounten, gengen, getten, icchen, lithen* (v.1), *marchen* (v.2), prauncen, *raiken,* railen (v.5), *riden, roilen* (v.1), roiten, *rokken, romblen, romen,* scalen (v.2), *scateren,* shiggen, *shoggen, shouven, skulken, slenchen,* spacen, *spinnen,* stampen, *stalken, stelen, steppen, striden, swangen, swarmen, threngen, thresten, thricchen, throngen, throuen, trailen,* trampen, tramplen, *trappen* (v.3), *treden,* tricen, trippen, *trotten, waggen, walken, wallen* (v.1), *waven,* wawen, *weiven* (v.2), *welken* (v.2), wigelen, *wosen, wrien* (v.2), *wringen, writhen, yeten* (v.3), *yongen* (v.2)

b) **Verbs primarily expressing PATH (Pa) (15% (72))**

1. **Towards (Pa^{towards})** (11): *ap(p)rochen, ariven, felen* (v.3), *ilengen, lenden,* londen, *neighen* (v.1), *neighlechen, neighlen, prochen* (v.1), *riven* (v.3)

2. **In (Pa^{in})** (4): borden (v.2), *entren*, havenen, *shippen*
3. **Away (Pa^{away})** (21): aloinen, *declinen, departen, ferren, firsen, forloinen*, leten, leven (v.1), *outraien, parten, receden, remeven, remuen, rimen, sceren, severen, sheden, swerven, twinnen* (v.1), *voiden, witen* (v.4)
4. **Down (Pa^{down})** (11): *avalen, descenden, enclinen*, grinden (v.2), grounden, *lighten* (v.2), *lightnen, seien* (v.2), *setlen, sien* (v.1), *valen*
5. **Up (Pa^{up})** (6): *amounten, ascenden, heien, mounten, risen, wecchen*
6. **Through (Pa^{through})** (3): *crossen*, overthwerten, traversen
7. **Forward (Pa^{forward})** (3): avaunten, *forthen, furtheren*
8. **Around/in circle (Pa^{circle})** (4): compassen, envirounen, *ringen* (v.1), virounen
9. **Along (Pa^{along})** (1): costeien
10. **Back (Pa^{back})** (6): *arreren, bounten, rebounden, recuren, returnen, resorten*
11. **Out (Pa^{out})** (2): *ishen, issuen*

c) **NEUTRAL MOTION verbs (Mo) (19% (89)):** *beien, beren* (v.1), *bouen* (v.1), *cairen, carien, charen* (v.1), *comen, diverten, drauen, dressen, dwelen, erren* (v.1), *faren, flecchen, flitten, folwen, fonden, founden* (v.1), *gangen*, gaten, *glauncen, gon* (also *yede*), *helden, hitten, hoken, journeien, kerven, leden, malskren, marken* (v.2), *mepen, meven, passen, picchen, pilgrimagen, proceden, purseuen, putten, quecchen, raken* (v.2), raungen, *raven* (v.2), *rechen* (v.1), *recchen* (v.1), *reisen* (v.2), *rengen, repairen, routen* (v.8), *scrithen, setten, seuen* (v.1), *sheren, shiften, shrinken, silen* (v.3), *sithen* (v.2), *skelten, skiften, slingen, sparplen, spreden, steiren, stien, stiren, stonden* (v.1), *stragelen, straien, strecchen, stressen, striken, temen* (v.1), *ten* (v.1), *terven, tighten* (v.1), *tiren* (v.2), *tracen, trinen* (v.2), *trollen, waden*, wageren, *walten, wandren, wæilien, weien* (v.1), *wenden*, wendren, *weven* (v.2), *willen* (v.2), woren

d) **Verbs not primarily expressing motion (non-MOTION) (10% (46)):** *amen, ap(p)lien, atlen, auntren, boillen, bounen, breken, bresten, busken, chesen, coveren* (v.2), *dighten, fon, fongen, geten* (v.1), *greithen, holden* (v.1), *kennen* (v.1), *labouren, leiken, listen* (v.1), *menen* (v.1), *meten* (v.1), *monen* (v.2), *moten* (v.2), *mouen* (v.3), *nimen*, pakken, *profren, purposen, releven*, richen (v.1), *sechen, shapen, swinken, taken, thinken, threpen, travailen, trussen, weden, wilnen, winnen, wisen, worthen, yeven*

* For meanings, attestations, etymologies, etc., see Appendix B.

6.2.2 Etymology

Before we come to the discussion of the results with respect to the questions that the inventory study aims to address—size of manner verb inventory, non-MOTION verbs in motion uses, and combinability of motion verbs with PATH satellites (see 4.1.1)—, it is interesting to take a quick look at the proportion of French and Latin loans within the four groups of verbs (Table 6.2). In the following, French and Latin are treated together (the dark areas in Figure 6.2),

TABLE 6.2. Proportion of French/Latin loans in the different types of Middle English motion verbs.

	MANNER	PATH	NEUTRAL	NON-MOTION
French/Latin	18.6% (48)	51.4% (37)	27% (24)	23.9% (11)
Other	81.4% (210)	48.6% (35)	73% (65)	76.1% (35)

because in some verbs, both have played a role as donor languages (e.g. in ME *receden*, from Latin *recedere* and French *receder*, cf. also Durkin 2009: 166–7). Note, however, that the number of French loans among the motion verbs greatly exceeds that of Latin origin (see the etymology column in Appendix B).

The light areas in the mosaic plot in Figure 6.2 subsume verbs of Old English and of other Germanic etymology, as well as verbs of unknown origin (such as the 'probably imitative' *hurlen, jumpen, lashen, lushen*). From a Middle English point of view, the light areas in Figure 6.2 will thus more or less represent the 'native' motion verbs: The Scandinavian loans cannot really be seen as newcomers anymore, since they will have entered the language in the Old English period already and therefore have probably long been nativized (cf., e.g., Käsmann 1961: 16). Furthermore, it is often impossible to say precisely which of the two languages—native Old English or Old Norse—a verb comes from.[110] The loans from other Germanic languages which have been borrowed only in the Middle English period (e.g. *blusteren* and *shokken* from MLG) have been put into this category as well, because also there it is frequently unclear whether a verb is a previously unattested native verb or a loan from Low German.[111] Verbs for which the *MED* gives a mixed Germanic-Romance etymology, such as *turnen* (OE *turnian* & OF *torner*, in Ma^{roll}) are also put in this category. The different widths of the columns in the mosaic plot in Figure 6.2 represent the different sizes of the groups of motion verbs.

What is immediately striking in Figure 6.2 is that French and Latin loans account for slightly more than half of the verbs primarily expressing PATH (51.4 per cent), while in the other groups, their proportion only ranges from roughly a quarter (in the NEUTRAL and non-MOTION verbs) to 18.6 per cent (in the MANNER verbs) of the total. Particularly if we bear in mind that most of the native PATH verbs are not pure PATH verbs (e.g. *shippen, witen*, see the discussion in section 5.1.3), this is an indication that PATH verbs in English are really a rather 'French phenomenon'.

110 Cf. the phrasings 'Prob. ON' or 'ON & OE' from the *MED* etymologies in Appendix B (e.g. for *blenchen, rennen, swippen, whirren* in Ma^{speed}). On the topic of identifying Scandinavian loans, cf. Durkin (2014: 191–211).

111 Cf. '?OE, also cp. MLG', 'Prob. ON & cp. MLG', and similar phrasings from the *MED* etymologies in Appendix B (e.g. for *plumpen* in Ma^{fall}, *slippen* in Ma^{slide}, *trappen* in Ma^{other}).

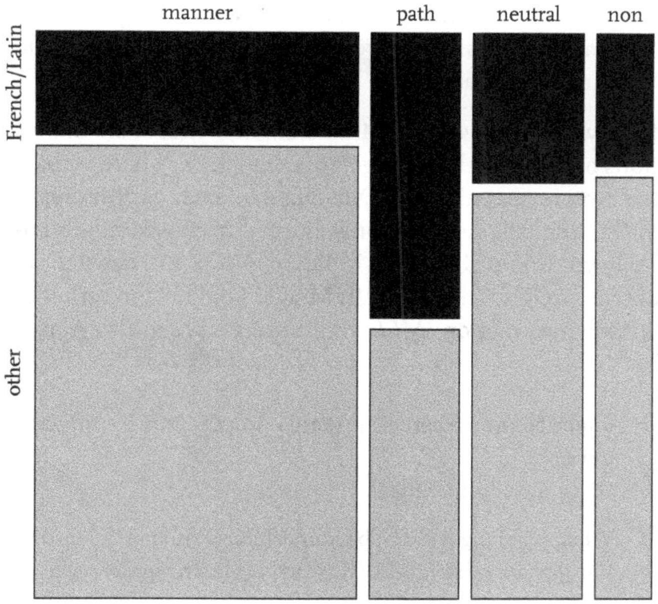

FIGURE 6.2. Proportion of French/Latin loans in the types of Middle English motion verbs.

The following two sections, then, will discuss the new MANNER verbs and their origins (6.2.3) with a particular focus on primarily CAUSED MOTION verbs used as INTRANSITIVE MOTION verbs (6.2.4). They will also show that several among the 18.6 per cent loans in the Middle English MANNER verbs have only received their MOTION meaning in Middle English (or in Anglo-Norman, probably through Middle English contact influence, cf. section 6.1).

6.2.3 Verbs primarily expressing MANNER OF MOTION

As we have seen above (Table 6.1), MANNER verbs make up around half of the Middle English motion verb inventory (55 per cent, 258 verbs). In this, the situation is comparable to the Old English inventory, in which MANNER verbs accounted for 57 per cent (107 verbs). Different origins of new MANNER verbs in Middle English have been discussed in Fanego (2012: 44–50) (see also 2.1.4.3): semantic extensions from existing verbs, new Middle English formations, and borrowings from other languages. Therefore, these sources of new MANNER verbs are only briefly repeated here in section 6.2.3.1, with examples from my own data. One source of new MANNER verbs, however, namely borrowing from French, will be discussed in more detail in section 6.2.3.2. A subtype of Fanego's 'extensions from existing verbs', namely using CAUSED MOTION verbs as INTRANSITIVE MOTION verbs, will be the subject of section 6.2.4.

6.2.3.1 Origins of new Middle English MANNER verbs I: semantic extensions, word formation, borrowing from other Germanic languages

Among the new MANNER verbs in Middle English, first of all, there are semantic extensions of native verbs (Fanego 2012: 44–5), i.e. where verbs do not have intransitive motion meanings in Old English, such as ME *dippen* from OE *dyppan*. In Old English, *dyppan* is only attested in caused motion uses, meaning 'to dip, lower into or immerse in liquid' (*DOE* s.v., sense 1), as in (154a), and 'to baptize' (*DOE* s.v., sense 2). In Middle English, the verb is additionally used for intransitive motion (*MED* s.v., sense 3a–b, see also App. B Ma^fall^), as in (154b).

(154) a. **duppe** þanne a feþer on ele (PeriD 62.43.2, DOE s.v. *dyppan*, sense 1 'to dip etc.')
'Then dip a feather in oil.'

b. What quik þing þat it be þat **duppeþ** ***þerynne***, anon it lepeþ vp aȝen ((*a*.1387) Trev. *Higd.*(StJ-C H.1) 1.117, MED s.v. *dippen*, sense 3.a 'to plunge (into water), dip')
'Whatever living thing plunges into it [the Dead Sea], it leaps up again immediately.'

In these cases, we can, of course, never be sure whether Old English also had an intransitive motion use which just happens not to be attested. Another example for a new manner verb in Middle English is ME *hien* 'to go quickly'. As discussed in 5.1.4.3, it originally means 'to labour, strive, endeavour' and is attested only once in a contextual motion meaning in Old English. In Middle English, this appears to have become a decontextualized, coded meaning of the verb.

Next to these semantic extensions of existing verbs, there are new Middle English formations (Fanego 2012: 45). Some of them are formed with the suffixes *-er* and *-le* (< OE *-rian*, OE *-lian*) already mentioned in 5.1.2 on Old English MANNER verbs, for instance ME *gog(e)len* 'to stagger' (cp. Wel. *gogi* 'to shake'), ME *hurtelen* 'to stumble' (from ME *hurten* 'to stumble'), ME *poperen* 'to ride rapidly' (from ME *poppen* 'to strike'), or ME *stamperen* 'to walk clumsily, stagger' (from ME *stampen* 'to stamp, tread heavily'). The origin of others, according to the *MED*, is 'probably imitative', such as ME *flusshen* 'to move rapidly, rush, dart, spring', *hurlen* 'to rush violently or quickly', *jumpen* '?to walk quickly, run, ?jump', *lashen* 'to flow, gush', and *lushen* 'to make a dash, rush, charge' (all in Ma^speed^).

Thirdly, as also mentioned in Fanego (2012: 46–8), there are MANNER verbs borrowed from Scandinavian (many of which will have come into the language during the Old English period) and from other Germanic languages such as Middle Dutch or Middle Low German, with whose speakers there were strong commercial relations including during the Middle English period (see, e.g.,

Hendriks 2012: 1663–5). As mentioned above (footnote 111), it is often hard to tell whether a verb is native (it might just not be attested in Old English) or borrowed from another Germanic language—note the frequent mixed Germanic etymologies in the etymology column in Appendix B. For the MANNER verbs ME *blusteren* 'to stray blindly or blunderingly', *shokken* 'to move rapidly, rush', and *trampen* 'to tread heavily, stamp' the *MED* gives exclusively a Middle Low German origin. Like English, all the Germanic languages will most likely be equally manner-salient and satellite-framing, so it is not surprising that the contact with those languages will have further increased the English MANNER verb inventory.

The fourth source for Middle English MANNER OF MOTION verbs is borrowing from French (Fanego 2012: 49–50). In the following section, I want to take a closer look at those verbs (see also 6.2.4.4 below).

6.2.3.2 Origins of the new Middle English MANNER verbs II: borrowing from French

MANNER verbs borrowed from French deserve an extra subsection, because it is often perceived as extraordinary that MANNER verbs should be borrowed from a verb-framed language. In Fanego (2012: 49), for instance, this is explained with earlier French still being a satellite-framed language:

> [T]he large number [of manner verbs] that English borrowed from French may initially seem surprising, considering the alleged status of French as a V-framed language. However, [. . .] recent research (cf. Kopecka 2006; Brucale, Iacobini, & Mocciaro 2011) on the expression of motion events has tended to refine this characterization, showing that earlier French was in fact predominantly S-framed, as indeed was its parent language, Latin. Prior to its typological shift in the direction of V-framed languages, French appears to have had an important number of manner verbs used to describe displacements of various kinds. [examples with ME *flatten* 'to rush' < OF *flatir*, ME *sauten* < OF *sauter*, and ME *travailen* < OF *travaillier* follow] (Fanego 2012: 49)

However, it has to be stressed again that the mere fact that a language is classified as verb-framing or PATH-conflating does not entail the claim that it does not *have* MANNER OF MOTION verbs: As we have seen in section 2.1.2, what verb-framing or PATH-conflating entails is that MANNER verbs are not as frequently *used* as PATH verbs are, particularly not in combination with (boundary-crossing) PATH satellites (2.1.4.1). The size of the MANNER verb lexicon also tends to be smaller in verb-framed languages (see 2.1.4.3), but, of course, they do have some MANNER verbs. Thus, French MANNER verbs can be adopted into Middle English without any problem.

Note also that in this, the Middle English use of a verb need not exactly mirror its use in the donor language. ME *chacen* and *cacchen*, for instance, are adoptions from OF/AN *chac(i)er* 'to chase, pursue, hunt', i.e. from a verb that is mostly used transitively, such as in (155a). According to Tober–Lommatzsch (*TL*), OF *chacier* is sometimes also used as an intransitive verb (155b), in the sense 'ride fast' (*AND*, *DMF*, and Godefroy do not list such a use)—though note that *TL* does not have any attestation with a PATH satellite. Finally, OF *chacier* can also be used figuratively in the sense 'to strive for something, pursue something', with the GOAL expressed in a prepositional phrase with *a* (155c).

(155) a. turmente out e vent cuntraire Ki a la cité ***le* chaça** (*Rou* I 546, TL s.v. *chacier*)
'There was a thunderstorm and a contrary wind, which chased him to the city.'

b. puis prist fort a **chacier**, Cel jour passa maint tertre et maint desruble fier (*Jub. NRec.* I 98, TL s.v. *chacier*)
'Then he started to ride fast [lit. chase] strongly; this day, he passed many a hill and many a wild cliff.'

c. Se jou voloie ***a ce* chacier**, Vos nel me poriiez vëer (*Guil. D'A.* 2510, TL s.v. *chacier*, sense 'sich um etwas bemühen, nach etwas streben')
'If I want to pursue this [lit. chase to this], you cannot prevent me from doing so.'

With examples such as (155a–b) as an input, it is not astonishing that in Middle English, the verb *chacen* is also used in the meaning 'to ride fast', but with PATH satellites (156). Thus, it is not surprising at all that MANNER verbs should be borrowed from Old French and then be used with PATH satellites in Middle English.

(156) Comynge fro a cuntre þat men called Ierico, ***To a iustes in iherusalem*** he **chaced *awey*** faste. (*c*.1400(*c*.1378) *PPl.B* (LdMisc 581) 17.51)
'Coming from a country that people call Jericho, he hurried away quickly to a joust in Jerusalem.'

Another aspect in the issue of borrowed MANNER verbs from French is that it is important to avoid the fallacy of simply equating the contact language 'French' with medieval French as spoken on the continent: As discussed in section 6.1 above, for much of the Middle English period, the most influential contact variety of French was Anglo-Norman, i.e. the French spoken in England. This variety of French is in many ways different from continental varieties, not least because it was influenced by English as well. To repeat the quote from Rothwell above, often 'the *form* of a French word was taken into Middle English, but its *meaning* reflects the French of London rather than that of Paris' (Rothwell 1998: 162). This also seems to be true for the verbs AN *fruisser*/ME *frushen*, AN *travailler*/ME *travailen*, and, to some extent, AN *laborer*/ME *labouren*, which all have motion uses in Anglo-French and Middle English, while their continental French counterparts OF *froissier*, OF *travailer*, and OF *laborer* do not, as will be shown in the following.

We will start with AN *fruisser*/ME *frushen*. The verb means 'to break, smash', and in continental French, according to *TL* and *DMF*, it is used transitively 'to break something', as in (157a), or intransitively 'to break' (157b):

(157) a. Maisuns ardent, **fruissent** celiers (*Rou* I 354, TL s.v. *froissier*)
'They burn houses, they break cellars.'

b. Et sa lance de freisne **froisse** (*Clig.* 3769, TL s.v. *froissier*)
'And his lance of ash-wood breaks.'

For Anglo-Norman, by contrast, the *AND* also lists the use of AN *fruisser* with *de* in the meaning 'to break off from, to splinter away from', as in (158a).[112] The same use is found in Middle English with ME *frushen* (158b), for fire breaking out of a cross, i.e. moving out of the cross by breaking it.

(158) a. Quant nostre nef fud depecee, E par piecez degetee, Nus dous enfanz, ne sai cument, Venimes sur un tablement, Ki ***de la nef* fruissé** esteit (*S Clem* 6649, AND s.v. *fruisser*)
'When our ship was shattered and broken into pieces, we two children, I don't know how, came onto a plank which had broken from the ship.'

b. A fyre **fruscheth *out of his croys.*** (*a*.1500(*a*.1400) *Chev.Assigne* (Clg A.2) 332, MED s.v. *frushen*, sense 3)
'A fire burst out of his cross.'

In Middle English, this use (158b) is to be expected: As we have seen in section 5.1.4.1 verbs meaning 'to break' (e.g. OE *brecan, berstan*; ME *breken, bresten*) can be used in an intransitive motion construction, so it is not surprising that this should also be possible for French loanwords meaning 'to break'. It is conceivable that the parallel use in Anglo-Norman in (158a) is due to the influence of English on native bilingual speakers, or on speakers with L2 Anglo-Norman: Various studies on present-day languages have shown this kind of L1 and L2 transfer with respect to motion expression. Larrañaga et al. (2012), for instance, report that the combination of *correr* 'run' with *en el banco* 'in(to) the bank' is used in descriptions of a bank robber running into a bank by English learners of Spanish across three levels of proficiency, a combination not normally used by speakers of L1 Spanish (cf. 2.1.4.1), but typical for English (for similar findings with English learners of French, cf. Harley 1989, Harley & King 1989, and Treffers-Daller & Tidball 2015).[113]

The picture is similar for AN *travailler*/ME *travailen* and AN *laborer*/ME *labouren*, as already addressed briefly in 2.2.3 in relation to the discussion of

112 Of course it is possible that such a use in continental French is just not recorded in Tobler–Lommatzsch and the *DMF*.

113 Other studies on transfer from satellite-framed L1s on verb-framed L2s are Cadierno 2004 and Cadierno & Ruiz 2006 (L1 Spanish, L2 Danish), Navarro & Nicoladis 2005 (L1 English, L2 Spanish), and Bernini et al. 2006 (Germanic L1s, L2 Italian). Hohenstein et al. 2006 have also found influence in the other direction with L1 (Spanish) being influenced by L2 (English).

verb meaning.[114] As we have seen above, both verbs originally mean 'to toil, work hard'. Their motion use seems to be an 'English' development: For *laborer*, no such sense is attested according to *TL*, Godefroy, *FEW*, or *DMF*; also for *travailler*, motion uses are effectively absent in continental varieties of medieval French.[115] In Anglo-French, by contrast, *travailler* is widely attested in motion uses (see Rothwell 1993b: 34, and *AND* s.v. *travailler*), such as in (159a–b).

(159) a. Anne est si grosse enzeynte qe ele ne poet en propre persone **travailler** ***en voz partyes*** (*Corr Lond* 245.529, AND, s.v. *travailler*, v.n. sense 10 'to travel')
'Anne is so very pregnant that she cannot herself travel to your place.'

b. si le dit W. ne **travaille** point ***de lieu en lieu***, mesque demourre . . . (*GAUNT1* i 347, AND, s.v. *travailler*, v.n. sense 10 'to travel')
'If the abovementioned W. does not travel from place to place, but stays [. . .].'

What is more, the motion meaning of AN *travailler* is not only attested for the verb, but also for the nominal and adjectival uses of its participle forms, as in (160a–b), which point to its more decontextualized status:

(160) a. a graunde damage de eux [. . .] et desease a toutz lez **travaillantz** de Somers' vers Londres (*Lett & Pet* 327.27, AND s.v. *travailler*, p.pr. as s. sense 3 'traveller')
'[Service is no longer held in a church in Knoyle in the diocese of Salisbury] to the great detriment of [the inhabitants], and inconvenience to all the travellers from Somers[et?] toward London.'

b. 'je vien tout droit de Venyse' . . . 'j'en ai grant joye de vous, que vous estez si bien **travaillee**' (*Man lang* 92, AND s.v. *travailler*, p.p. as a. 6 'well-travelled')
'I come straight from Venice.' [. . .] 'I'm very glad for you that you are so well travelled.'

[114] I classified *travailen* and *labouren* as non-motion verbs, for the reasons explained in section 2.2.3, but they may nonetheless serve as illustration for the importance of taking into account that Anglo-Norman as a donor language cannot be equated with continental varieties of French, because Anglo-Norman itself is influenced by English.
[115] The sense is not listed in Godefroy (s.v. *travaillier*, nor in the nouns *travaille, travaillement*, and *travailleur*). For the noun *travail*, *TL* has two attestations in the sense 'Reise, unstetes Wandern' ('journey, vagrant wandering'), but both could be read as 'effort, hardship' as well, such as *mes ne veil mïe conter Ses jornées ne raconter, Ne son travail ne met a some* (Méon II 435, 279) 'I do not want to tell more, neither to retell his days, nor to sum up his travel/hardship'. One of these two attestations is from Anglo-Norman anyway (Marie de France, *Lai de Milun*, 512). In the entry for the verb *travaillier*, *TL* gives two attestations of participle forms in the sense 'reisen' ('travel'): *Par tantes terres ai esté travillans* (Mon. Guill. 5079) 'Throughout so many countries I have been travelling', and *Ma dame, je vous prie Que vous voeliés aler et soyes* ***travellïe*** *jusques a Moradin, qui tant a seignorie* (Godefr. de Bouillon 33502) 'Madame, I beg you that you would like to go and be travelled to Moradin, who has

For AN *labourer*, like for ME *labouren* (see 2.2.3), this use does not seem to be so widespread as for AN *travailler*/ME *travailen*, but the *AND* has two attestations in a motion reading (161a–b):

(161) a. puis qe **laborer** a pee, que pur asperité del chemyn que pur loignetaigne del lu, a moy surdroit graund desese et grevance (*SAMPS1* 420.85b, AND s.v. *laborer*, v.n. sense 3 'to make the physical effort of travelling')
'[I need a horse] since travelling on foot, because of the bad condition of the way and the distance of the place, great trouble and distress would arise for me.'

b. Nostre dit soverain seignur [. . .] si ***avant*** **labora**, q'en brieve temps [. . .], y gaina la ville de Hareflu (*Rot Parl1* iv 94., AND s.v. *laborer*, *laborer avant* '(mil.) to advance (through battle)')
'Our abovementioned sovereign lord [. . .] worked so forwards that in a short time [. . .] he won the town of Harefleur there.'

It is rather likely that the semantic development of AN *travailler* toward the motion meaning 'travel' (and the scattered use of AN *laborer* in a motion reading) is due to the contact influence of Middle English on Anglo-Norman. In Old and Middle English, verbs meaning 'to labour, toil, struggle' can be used in an intransitive motion construction (see 5.1.4.3 for OE *winnan*, *hīgian*, 6.2.5.3 for ME *swinken*, *winnen*). Thus, one scenario is that speakers of Middle English started to do the same in their L2 Anglo-Norman with AN *travailler*. Another scenario is that Middle English speakers started to use ME *travailen* in a motion sense and that this use was taken over into Anglo-Norman. The earliest attestations of this use of ME *travailen*/AN *travailler* in the two languages are from approximately the same time, and thus do not reveal which scenario is more likely: According to the *AND*, the earliest attestation of intransitive *travailler* in a motion use in Anglo-Norman is the one in (162a) from the late thirteenth century *Mirror of Justices*.[116] In

so much authority'. The very same attestation, though, is given in Godefroy under the entry *travaillié*, part. passé, in the sense 'accablé de travail, de fatigue' ('exhausted, tired'), which yields a far more convincing reading of the passage as 'Madame, I beg you that you would like to go—even though you may be tired—to Moradin', i.e. with a concessive *et*. The sense 'to exhaust through a voyage' (*'fatiguer par un voyage'*) of OF *travillier* for which the *FEW* refers to *Gerbert de Metz* (a Champagne *Chanson de Geste* from the twelfth century) is a caused motion use, such as in *De longues terres voz ai ci* ***travilliez*** (GerbM 8979, ed. Taylor 1952) 'From distant lands I have *travelled* you hither'.
The *DMF* gives the sense 'travel' as 'regional', belonging to the Picard and Anglo-Norman dialects (s.v. *travailler*, sense B.3 Région. (Picardie, anglo-normand) 'Se fatiguer, faire des efforts en voyageant, voyager'). The attestations given for this sense in the *DMF* entry are from the *Manière de langage* (*Man. lang.* G. (1396)), a book for learning French (from England), and from the Picard Froissart, who had lived at the English court for several years and is known to have retained some Anglo-French expressions (cf. Molencki (2012: 153–155)).

[116] The *AND* gives one earlier attestation of *travailler* in a motion use, but with a reflexive pronoun (*vus*), i.e. in a caused motion structure. It is from the early 13th century Anglo-Norman *Vie de St Jean*

Middle English, according to the *MED*, the earliest attestation of *travailen* in a motion use (162b) is from *c*.1300, i.e. only slightly later. Furthermore, since there is only one single attestation of ME *travailen* prior to that (*c*.1275), it seems that the motion use is basically there from the start in Middle English.

(162) a. en taunt qe les justices funt les jurours **travailler** ***es plus foreins marchies des conties***, ou mieux vaudreit qe justices **travaillassent** ***de hundred en hundred*** qe de travailler taunt de people. (AND s.v. *travailler*, expanded with Whittaker 1895: 178)
'For the justices cause the jurors to journey to the extreme boundaries of the counties, whereas it would be better that the justices should journey from hundred to hundred and not labour so many folk.' [trans. Whittaker 1895: 178]

b. ȝwy ne bi-holde ȝe þe heiȝe temple..Of sonne and Mone and steorrene al-so, ***fram þe este to þe weste*** Þat **trauaillieth** [Hrl: Wendeþ] and neuere werie ne beoth ne neuere ne habbeth reste? (*c*.1300 *SLeg.Kath.*(LdMisc 108) 24, MED s.v. *travailen*, sense 5.a.a 'to make a journey, travel, go')
'Why do you not look at the high temple [. . .] of sun and moon, and also stars, which travels from the east to the west and is never weary, and never has rest?'

Whichever scenario has led to it, it seems evident that AN *travailler*/ME *travailen* 'to travel' is a use that has developed in England and has originated either in Middle English, or in Anglo-Norman due to the influence of Middle English (cf. the discussion of the contact influence in section 6.1). Therefore, simply regarding ME *travailen* as a motion verb borrowed from Old French, as in the quote at the beginning of this section, means unnecessarily 'making a leap across the Channel in search of a "borrowing" from mediaeval French', to quote Rothwell (1991: 174). Quite to the contrary, it appears that AN *travailler*/ ME *travailen* only *became* a motion verb through English influence. As we will see below (sections 6.2.4.4 and 6.2.5; see also 6.3.1 (example (202)) on ME *hosteien*), there are more French forms among the Middle English motion verbs whose intransitive motion use is an English development and is often not even attested in Anglo-Norman.

Before, however, in section 6.2.4, we will take a closer look at another type of verb that 'extends' to an intransitive motion use as MANNER verb in Middle English: verbs that are otherwise primarily used for CAUSED MOTION, such as ME *werpen* or *casten* 'to throw'.

---

l'Aumônier, a translation of the Latin *Vita S. Joannis Eleemosynarii* (*Ne vus ennuit, . . ., De* ***vus a Rume travaillier****: N'irrez pas sul, od vus irrai* 'Please [lit. that it annoys you not to] trouble yourself to Rome. You won't go alone, I will go with you.' (*S Jean* ANTS 6578; *AND* s.v. *travailler*). According to Urwin (1981: 63), the use of *travailler* here is modelled on the respective passage in the Latin text, which has a passive infinitive: *ne dedigneris* ***fatigari usque ad reginam civitatem*** 'you will not object to be troubled to the royal city.' (PLD 73, 379A).

6.2.4 CAUSED MOTION verbs used for intransitive motion

As discussed in 2.2.3, the term MOTION verb in the inventory studies covers verbs which are (a) mostly used for intransitive motion (such as ME *crepen* 'to creep'), (b) used for both intransitive and caused motion (such as ME *hasten* 'to hasten'), and (c) verbs mostly used for caused motion, but also found in intransitive motion uses (such as ME *werpen* 'to throw'). Hence, what the verbs in the category 'MOTION verb' (MANNER, PATH, and NEUTRAL MOTION) have in common is that they give access to a MOTION EVENT frame.[117] What is interesting, though, is that there are a range of verbs in the Middle English motion verb inventory which are most often used for CAUSED MOTION EVENTS, and only sporadically for INTRANSITIVE MOTION EVENTS. These are, for instance, the abovementioned *werpen* 'to throw' as well as *casten* 'to cast, throw', *smiten* 'to throw, hit' (all in Ma^{speed}), or *throuen* 'to hurl, throw' (in Ma^{other}). Typically, they are used to express caused motion events, as in (163a–d), where the FIGURE, expressed in the direct object, is printed in bold italics:

(163) a. Đe child ***it*** **warp** dun to de grund. (*a.*1325(*c.*1250) *Gen.& Ex.*(Corp-C 444) 2640, MED s.v. *werpen*, sense 2.a 'to throw (somebody or something), fling, hurl, etc.')
'The child threw it [the crown] down to the ground.'

b. Ho..**Kest** vp ***þe cortyn*** & creped with-inne. (*c.*1400(?*c.*1390) *Gawain* (Nero A.10) 1192, MED s.v. *casten*, sense 11.f ~ *up* 'lift (something)')
'She cast up the curtain and crept within [i.e. into the bed].'

c. A geaunt..**smot** to grounde ***Cleodalis*** (*c.*1330(?*a.*1300) *Arth.& M.*(Auch) 9884, MED s.v. *smiten*, sense 1.b.a ~*(a)doun* 'to knock (somebody or something) down, etc.')
'A giant knocked Cleodalis to the ground.'

d. Horn **þreu** ***him*** ouer þe brigge. (*c.*1300(?*c.*1225) *Horn* (Cmb Gg.4.27) 63/1076, MED s.v. *throuen*, sense 1.a 'to hurl (somebody or something), fling, throw, cast')
'Horn threw him over the bridge.'

However, the same verbs are also attested in intransitive motion uses, such as (164a–d), where the FIGURE is not expressed in the object, but in the subject. The intransitive motion uses are a lot less frequent than the caused motion ones above.[118]

---

[117] This section is a slightly condensed version of Huber (2013).

[118] In *PPCME2*, for instance, *casten* has more than 150 attestations in caused motion use, but only one in an intransitive motion use; *werpen* has more than 60 attestations in caused motion use but only three intransitive motion attestations; *smiten* has more than 20 caused motion attestations and only three in intransitive motion use.

(164) a. ***Þai*** wan in wightly, **warpit *to þe yates,*** Barrit hom full bigly with boltes of yerne. (*c*.1540(?*a*.1400) *Destr.Troy* (Htrn 388), MED s.v. *werpen*, sense 1a 'to hasten, rush, etc.')
'They went in boldly, threw [i.e. hastened] to the gates, barred them very firmly with iron bolts.'[119]

b. ***To Arthur-is tombe he*** **caste.** (*a*.1500(?*a*.1400) *Gawain* (Nero A.10), MED s.v. *casten*, sense 18 'to move quickly, rush, dash, leap, etc.')
'He threw [i.e. moved quickly] to Arthur's tomb.'

c. ***Into þe mede þai*** **smiten** wiþ rape, ȝif þai miȝt ouer þe water scape (*c*.1330(?*a*.1300) *Arth.&M.*(Auch), MED s.v. *smiten*, sense 9.b 'to go quickly, hasten, rush, leap, etc.')
'They smote [i.e. hastened] into the meadow in a hurry, so that they might escape over the water.'

d. He þrompelde atte þrexwolde and **þreuh *to þe grounde.*** (*c*.1390 *PPl.A(1)* (Vrn), MED s.v. *throuen*, sense 1.i 'to fall, hurtle, etc.')
'He stumbled at the threshold and threw [i.e. fell] to the ground.'

While in the caused motion uses there are two participant roles, the AGENT (the 'thrower') and the FIGURE (the 'thrown'), in the intransitive motion uses, there is only one participant, which, in a way though, has both roles. This has been argued, for instance, in Kemmer (1993), according to whom intransitive motion events are semantically 'middle' (and not intransitive), because they involve actions that 'by definition involve a volitional entity acting on itself to propel itself forward' (Kemmer 1993: 58). Thus, conceptually the semantic roles of Initiator (or agent) and Endpoint (i.e. the body, the figure) of the action can be distinguished, although these two 'refer to a single holistic entity' (Kemmer 1993: 66) in events which are semantically 'middle'. The two kinds of events—caused and intransitive motion—are therefore not so very different from each other.

The sporadic use of caused motion verbs to express intransitive motion (without any reflexive pronoun, i.e. *he threw to the ground* instead of *he threw himself to the ground*) as in (164a–d) has also been observed in Matsunami (1985). However, I think that, instead of simply observing that the reflexive pronoun of caused motion verbs is sometimes dropped, we might also explain why this could happen. This will be attempted in the following (cf. also Huber 2013).

It is possible that, in Middle English, the use of primarily caused motion verbs (CM verbs, such as *casten* 'to throw, cast') in the intransitive motion

---

[119] In the interpretation suggested here and in the *MED*, the MOVING FIGURE is expressed in the subject *Þai*, and the prepositional phrase *to þe yates* 'to the gates' is the PATH satellite. This attestation could also be read as CAUSED MOTION 'they threw the gates shut', as is done in *OED2* (s.v. *warp* v.). In this reading, *þe yates* would be the direct object coding the MOVING FIGURE, and the adverb *to* would constitute the PATH satellite.

construction (IMC) is facilitated by various factors: One factor which may have been involved in the permeability between the two groups of verbs is that caused motion verbs derived from intransitive motion verbs (IM verbs), such as OE *ferian* 'to transport' from *faran* 'to move' sometimes fall together in Middle English due to sound changes and thus get confused (cf., e.g., MED s.v. *faren* v., sense 16). Another factor is the occurrence of CM verbs in structures which are syntactically and semantically extremely similar to those in which typical IM verbs (such as *rennen* 'to run') are used. The boundary between the two types of verbs, which are used for similar events anyway, may become more blurred by this constellation of parallel patterns. The relevant patterns—(a) verb and reflexive pronoun, and (b) *be* + past participle—are presented in the following subsections (6.2.4.1–2). Section 6.2.4.3 argues how this may lead to the sporadic occurrence of caused motion verbs in intransitive motion constructions. Finally, section 6.2.4.4 will link this up with the intransitive motion use of French loan verbs in Middle English, which are not intransitive motion verbs in the donor language (cf. the discussion in 6.2.3.2 above).

6.2.4.1 Similar patterns I: verb + reflexive pronoun

We will start with the combination of the verbs with a reflexive pronoun. As already mentioned above (4.2.3), the reflexive pronouns in Old and Middle English are identical to the object case pronouns, i.e. *me*, *him*, etc., as exemplified in (165).

(165) A wood man most sauage, Which that fedde **hym** most with flessh of man. ((?*a*.1439) Lydg. *FP* (Bod 263), MED s.v. *him*)
'A wild, most savage man, who fed himself mostly with human flesh.'

The combined forms with *-self* (*myself*, *himself*) that English has today are used mainly contrastively in Old and Middle English and only start to take over the function of reflexive pronouns roughly after 1500. Before that, it is overwhelmingly the bare pronoun that is used to refer back to the subject (Keenan 2002: 345–6; König & Siemund 2000: 44, 49; Lange 2007: 3; Lutz 2002b: 46; Mustanoja 1960: 153, 430; Peitsara 1997: 288).

Verbs that primarily express CAUSED MOTION can of course combine with all sorts of objects, including objects that are coreferential to the subject, and thus realized by reflexive pronouns (as in 166), just like in Present-Day English (e.g. *he threw himself into the river*). In these cases, the reflexive pronoun fills a valency slot of the verb: The verb is commonly used with a second participant, which, in examples like (166), is coreferential to the subject.

(166) He..**smiten** ***hem*** amid þe pres So grehound doþ out of les. (*c*.1330(?*a*.1300) *Arth.& M.*(Auch) 9027, MED s.v. *smiten*, sense 9.b refl. 'hurl oneself')
'They [. . .] hurled themselves into the midst of the crowd, like an unleashed greyhound does.'

Different from Present-Day English, however, in Old and Middle English the reflexive pronouns are also used with verbs where they do not fill a valency slot, such as *gon* 'go' and *stalken* 'to walk cautiously' in (167a–b) (see also 4.2.3). The reflexive pronoun is rendered as REFL in the translation, because this use is no longer available in Present-Day English with those particular verbs.

(167) a. Þe werwolf..**goþ *him*** to a gret heiȝ-waye. (*a.*1375 *WPal.*(KC 13) 1846, MED s.v. *him*)
'The werewolf [. . .] goes REFL to a broad highway.'

b. Into the chamber he **stalked *hym*** ful stille. ((*c.*1395) Chaucer *CT. Cl.*(Manly-Rickert), MED s.v. *stalken*, sense a. refl. 'move quietly')
'He stalked REFL very quietly into the chamber.'

In this kind of use, the pronouns are usually described as not having an argument role, since they do not fill a valency slot of the verb. This use thus traditionally is labeled 'pleonastic reflexive' (e.g. Sweet 1892: §1106, Mitchell 1985: §§271–4), and its function has been described as to convey 'the notion that the person denoted by the subject was particularly affected by the result of the action or event' (Visser 1963–73: §328). These pleonastic reflexives occur very often with MOTION verbs.[120]

Syntactically then, the two patterns described in the preceding paragraphs (cf. (166) and (167)) look exactly the same: There is a MOTION verb and a dative pronoun coreferential with the subject. Also semantically there is much similarity, since in both patterns there is a FIGURE that moves itself somewhere.[121] This configuration of inherent semantic relatedness and identical patterns

---

[120] In addition to MOTION verbs, pleonastic reflexives are also frequent with psychological verbs (e.g. *I fear me* 'I am afraid'; *I remembre me* 'I remember') and verbs of social behaviour (e.g. *I bere me* 'I behave'), according to Ito (1998: 60; cf. similarly Lange (2007: 87) using Kemmer's (1993) terms of 'change in body posture', 'non-translational motion', 'translational motion', 'emotion', 'cognition'). In an approach inspired by Fillmore's Case Grammar, Ito argues that the common denominator of these 'essentially reflexive' groups of verbs is their 'built-in self-directedness' (Ito 1998: 61), in that the activities they denote 'involve a relationship between the total self and one of its integral parts' (Ito 1998: 60). In the case of MOTION verbs, it is the Agent who acts upon himself to get his body (the Object) to another location. Thus, even though this object is not an argument of the verb, it can still be assigned the semantic role of Object. This argument is exactly in line with Kemmer's (1993) classification of translational motion verbs as semantically middle.

Moreover, it is particularly with these verbs that the simple pronouns (e.g. *me, him*, i.e. those without *-self*) are attested for a longer time: Even at the time when the *-self* pronouns gradually start to take over, after around 1500, they take over especially with those verbs for which coreferentiality of subject and object is unusual, such as *see* (*he saw him* vs. *he saw himself*). This is not the case for MOTION verbs (just like for psychological verbs and for verbs of social behaviour), where, consequently, the simple reflexive pronoun can still be found well into the Early Modern English period, such as in Shakespeare's "run thee to the parlour" (*M Ado* iii.1.1) (Kisbye 1972: 106–9; Peitsara 1997: 297, 303; Lange 2007: 139–40).

[121] See also the discussion of *hasten* in combination with a reflexive pronoun in 4.2.3: In the case of verbs that are commonly used for intransitive and caused motion alike, it is not even possible to decide whether the reflexive pronoun fills an argument slot of the verb, or whether it is a pleonastic reflexive.

may open up the way for CM verbs to be used intransitively such as in (169b), by an analogy illustrated in (168–169):

(168) a. He **leop *him*** out in-to þe strete (*c*.1300 *SLeg.*(LdMisc 108))
'He leapt REFL out into the street.'

b. Jn to þe water he **lep** þo (*c*.1400(?*a*.1300) *KAlex.*(LdMisc 622))
'Then he leapt into the water.'

(169) a. Beues in to þe sadel ***him* þrew** (*c*.1330(?*c*.1300) *Bevis* (Auch))
'Bevis threw himself into the saddle.'

b. Of his horse he **þreou** (*c*.1275(?*a*.1200) Lay. *Brut* (Clg A.9))
'He threw [i.e. jumped] of his horse.'

Since both intransitive (168a) and caused motion verbs (169a) can be used with a reflexive pronoun and referring to the same kind of situation, it seems plausible that the latter should sometimes occur intransitively (169b) as well, in analogy to intransitive uses as in (168b).

6.2.4.2 Similar patterns II: *be* + past participle

In addition to the shared combination with a reflexive pronoun, there is another pattern shared by intransitive and caused motion verbs: Both can be used as past participles with the auxiliary *be*. In the case of intransitive verbs, this is a perfect periphrasis, such as in (170).

(170) In to the chaunbyr anoon he ***is* i-lope**. (*c*.1460(*a*.1449) Lydg. *2 Merch.* (Hrl 2255))
'He is [i.e. has] immediately run to the chamber.'

The origin of this perfect lies in 'statives built around resultative participles' (McFadden & Alexiadou 2010: 415), i.e. in examples such as (170), *be* is originally a copula, and the participle functions adjectivally as subject complement. This origin is visible in the nominal inflectional endings with which the participles sometimes still appear in Old English (e.g. Fischer & van der Wurff 2006: 140). As argued by McFadden & Alexiadou (2010), the *be*-periphrasis retained the resultative interpretation that is at the origin of this construction until it fell out of use by Late Modern English: *In to the chaunbyr anoon he is i-lope* (170) thus is interpreted as 'he **is** in the chamber as a result of having run there' (on the resultative character of this construction, see also Visser 1963–73: §1898; Denison 1993: 366; Elsness 1997: 340–6; Fischer & van der Wurff 2006: 141). Consequently, McFadden & Alexiadou analyse *be* in examples such as (170) as 'nothing more nor less than the copula, combining here with a resultative participle' (2010: 393). Hence, they point out, it is very much like the German stative passive, such as in *Er ist geheilt* 'he is healed' (ibid.: 404–7). Their view fully corresponds to the present argument that the *be*-perfect and the passive are very close in their resultative character.

Just like the perfect-of-result periphrasis with *be*, the passive also has its origin in a construction with copula and adjectival past participle functioning as subject complement (e.g. Fischer & van der Wurff 2006: 153; Petré & Cuyckens 2009: 327). Similar to German, Old English also has separate auxiliaries to distinguish dynamic (OE *weorðan*, G *werden*, e.g. G *er wird geheilt* 'he is being healed') and stative passives (OE *bēon/wesan*, G *sein*, e.g. G *er ist geheilt* 'he is healed'). Yet, this dynamic/stative difference between OE *weorðan* and *bēon/wesan* + past participle is becoming vague already in Old English (cf. Brunner 1962: 286; Fischer & van der Wurff 2006: 152–3; and Mailhammer & Smirnova 2013, who show that this overlap arises from the combination with the past tense, such as in (172)). Already in this period, *weorðan* is less frequent than *bēon/wesan*, and by the late Middle English period, it is no longer used at all (Fischer & van der Wurff 2006: 153; Mossé 1973: §145; Petré & Cuyckens 2009: 312). Thus, passives with *be* can have both stative and dynamic interpretations, as exemplified in (171), which can be read as 'has been taken' (stative) or 'is being taken' (dynamic).

(171) Alsso it is seyde þat þe Erle off Penbroke **is taken** in-to Brettayn
(*Paston Letters* ed. Davis 1971)
'It is also said that the Earl of Pembroke is being taken / has been taken into Brittany.'

Due to the identical form of the *be*-perfect and the passive, these constructions are notoriously ambiguous in cases where the main verb can be used both transitively and intransitively. This is pointed out by Mitchell (1985: §735), referring to the following Old English example:

(172) Þa Dænescan . . . **wurdon awende** to þære mestre untriuwðe.
perfect: 'The Danes [. . .] had turned to the greatest treachery.'
passive: 'The Danes [. . .] had been turned to the greatest treachery.'

OE *āwendan* is found both in transitive ('to turn somebody') and intransitive ('to turn') uses, and thus there is no way of deciding whether the structure with *bēon* (or the past tense of *weorðan* in (172)) represents a passive or a perfect (cf. also Denison 1993: 346; Brunner 1962: 286). The ambiguous nature of *be* + past participle is also noted by Fischer & van der Wurff, according to whom this ambiguity may be one of the reasons why the *be*-perfect was finally lost: 'a phrase like *she was returned* could mean 'she had returned' as well as 'she had been returned'' (2006: 142).

Both the passive (for the CM verbs) and the perfect periphrasis (for the IM verbs) with *be* therefore have in common that they often highlight a state or a result. Hence, it can be argued, it is less important whether the MOTION is CAUSED or INTRANSITIVE, since the resulting state is the same in both cases—somebody or something has changed position. Observe the following examples, where the result seems to be the most important information: The knights are in the hall (173a), the man is no longer at his place (173b), and Aristotle is ascended to heaven (173c).

(173) a. Ich & þas tweie cnihtes leopen ut of þan fihte, and hider in we **beoð idraȝen.** (*c.*1275 (?*a.*1200) Lay.*Brut* (Clg A.9))
perfect: 'I and these two knights leapt out of the fight and have drawn in here [i.e. into castle Tintagel].'
passive: 'I and these two knights leapt out of the fight and are drawn in here [i.e. into castle Tintagel].'

b. Þat man þat so doth hath cast awey his sheld, and **is remwed** from his place. (?*c.*1425 Chaucer *Bo.*(Cmb Ii.3.21))
perfect: 'That a man who does so has cast away his shield and has removed from his place.'
passive: 'That a man who does so has cast away his shield and is removed from his place.'

c. Aristotille **was mountid** vp to hevene in lyknes to a dowue of fyre. (*c.*1450 *Royal SSecr.*(Roy 18.A.7))
perfect: 'Aristotle had mounted up to heaven in likeness of a dove of fire.'
passive: 'Aristotle had been mounted up to heaven in likeness of a dove of fire.'

In these examples, it does not really matter whether the change of place is the result of some sort of passive process—in that the people are drawn, removed, or raised by some other power or influence or their own will—or of an active one: Whether Aristotle in (173c) is said to 'have risen to heaven' or to 'have been lifted to heaven' is of little consequence and basically amounts to the same thing.

6.2.4.3 CAUSED and INTRANSITIVE MOTION verbs in a family of related constructions

The number of motion verbs that share this potential for ambiguity is substantial: Around a hundred verbs of the altogether 465 verbs from the Middle English inventory (Appendix B) have attestations in both CAUSED MOTION and INTRANSITIVE MOTION uses. Around 30 per cent of the MOTION verbs (i.e. MANNER, PATH, and NEUTRAL MOTION verbs) found in the inventory study have attestations with a reflexive pronoun (cf. column five in Appendix B). Examples for two of them, *flitten* 'to convey (something)' and *wenden* 'to walk, turn', are given in (174–175) with attestations for the four relevant uses: (a) intransitive motion, (b) caused motion, (c) reflexive pronoun, and (d) *be* + past participle.

(174) FLITTEN

a. intransitive motion
I rede we **flitte** Ilkone ***till oure contre.*** (*a.*1450 *Yk.Pl.*(Add 35290), MED s.v. *flitten*, sense 2.b 'to flee, escape from something')
'I advise that everyone of us should flit [i.e. escape] to our country.'

b. caused motion
Thydour..they wold ***hym*** **flytte.** (*a.*1500 *Tundale* (Adv 19.3.1), MED s.v. *flitten*, sense 1.a 'to convey something, etc.')

'They wanted to flit [i.e. take] him there.'

c. reflexive pronoun

Ner to þat place þe prest ***him* flit** (*a*.1425 *NHom.(3) Leg.Suppl.Hrl.* (Hrl 4196), MED s.v. *flitten*, sense 1.d refl. 'to betake onself, turn')

reflexive: 'The priest flitted himself [i.e. betook himself] closer to that place.'

middle: 'The priest flitted REFL [i.e. went] closer to that place.'

d. *be* + past participle

Fro me, to þy sone, my soule **is flet** (*a*.1450 *The tixt of holy writ* (Dgb 102), MED s.v. *flitten*, sense 2.a 'to go, come, depart')

perfect: 'My soul has flitted [i.e. has gone] from me to your son.'

passive: 'My soul is flitted [i.e. has been taken] from me to your son.'

(175) WENDEN

a. intransitive motion

þei weren wont to **weynden *þeder*** eueryche ȝer. ((*c*.1390) *Chart.Abbey HG* (LdMisc 210), MED s.v. *wenden*, sense 2a.a 'to make one's way, travel, etc.')

'They were used to wending [i.e. travelling] there every year.'

b. caused motion

Alisaunder wiþ outen asoyne **Hadde want *his ost to Macedoyne***. (*a*.1425 *KAlex.* (LinI 150), MED s.v. *wenden*, sense 2b.b 'to cause (somebody) to take a certain course, etc.')

'Alexander had wended [i.e. directed] his army to Macedonia without delay.'

c. reflexive pronoun

The wyf [. . .] **wente *hir out*** to pisse ((*c*.1390) Chaucer *CT.Rv.*(Manly-Rickert), MED s.v. *wenden*, sense 4.a 'to go one's way etc.')

reflexive: 'The wife [. . .] wended herself [i.e. directed herself] outside to pee.'

middle: 'The wife [. . .] wended REFL [i.e. went] outside to pee.'

d. *be* + past participle

Qwo may þe hwngri lyon sle þat ***be þe wod* es went**? (*a*.1500 *Ho may þe* (Lin-C 44), MED s.v. *wenden*, sense 2a.b 'to wander, roam')

perfect: 'Who is able to kill the hungry lion that has wended [i.e. has gone] to the wood?'

passive: 'Who is able to kill the hungry lion that is wended [i.e. is directed] to the wood?'

I argue that all these four patterns form a family of constructions, related in various ways, as illustrated in Figure 6.3.

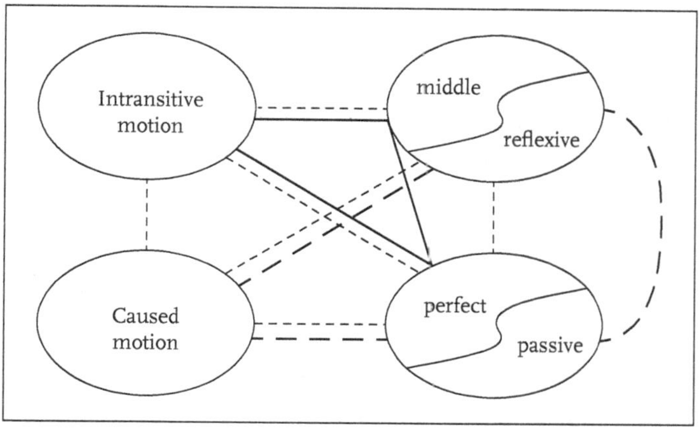

FIGURE 6.3. A family of related constructions in Middle English.

The two circles to the right represent the two pairs of similar patterns described in the preceding sections: They are split in two, because strictly speaking they constitute different constructions, but still 'share' a circle since, as discussed above, syntactically they look identical and semantically, they are very close. The patterns represented by the four circles, then, are related first of all semantically because they all describe situations in which somebody or something moves—represented here by their common frame. Secondly, they are related by the fact that a substantial number of verbs occur in all four of them, as mentioned above—this is illustrated by the thin broken lines. Thirdly, and more specifically, intransitive MOTION verbs, which typically occur in the INTRANSITIVE MOTION construction, often also occur with a non-argument reflexive pronoun (labeled *middle* in the diagram, cf. Kemmer 1993; Lange 2007: 86–9), and as past participles with *be* as resultative perfect, thus linking the INTRANSITIVE MOTION pattern with these two patterns (bold lines). In a similar way (bold broken lines), CAUSED MOTION verbs link the CAUSED MOTION construction to the passive and to the reflexive construction, since they occur in all three of them. The bold links between the INTRANSITIVE MOTION construction and the two patterns on the right disappear in later stages of the language, when the pleonastic reflexive is lost and the perfect is limited to the auxiliary *have*.

I suggest that this closely-knit network of constructions may be held responsible for the sporadic occurrence of Middle English CAUSED MOTION verbs in the INTRANSITIVE MOTION construction: A CM verb also regularly occurs with reflexive pronouns (ME *Beues in to þe sadel* ***him þrew***, PDE *Bevis* ***threw himself*** *into the saddle*) and in the passive construction (ME *a wight* ***is*** *from hire whiel* ***ythrowe***, PDE *someone* ***is thrown*** *from her wheel*). IM verbs occur in the same patterns, except that the pronoun here is strictly speaking not an argument (ME *Ser Darie . . .* ***Trottis him*** *on to Tirgre*, no longer so in PDE *Sir Darius* ***trots*** *on to Tigris*) and the *be*-periphrasis a resultative perfect (ME *folkis . . . þat* ***are furth ronne***, with auxiliary *have* in PDE *people that* ***have run*** *away*). These parallel patterns may lead to a higher degree of permeability between the two groups of verbs.[122] It is fostered further by the abovementioned factor that in Middle English, CM verbs sometimes become homonymous with the IM verbs they originally derived from. Additional 'group pressure' is exerted by the number of verbs that are regularly used to express both INTRANSITIVE and CAUSED MOTION (such as ME *hasten*) and thus occur in all four constructions.

6.2.4.4 French CAUSED MOTION verbs

Before turning to primarily CAUSED MOTION verbs from French that are found in the Middle English INTRANSITIVE MOTION verb inventory, this section will sum up the arguments of the preceding subsections. We have seen that native verbs that are usually found in CAUSED MOTION constructions can sporadically occur in the INTRANSITIVE MOTION construction in Middle English (see also Matsunami 1985). I have argued that an explanation for this phenomenon may lie in the particular Middle English configuration of patterns in which CAUSED MOTION and INTRANSITIVE MOTION verbs are commonly used. Both kinds of verbs regularly appear with a simple reflexive pronoun, which serves as a pleonastic pronoun in the case of INTRANSITIVE MOTION verbs, and as a reflexive pronoun with CAUSED MOTION verbs. With both kinds of verbs, the patterns are basically identical in form (V + object pronoun + directional adverbial) and meaning (a FIGURE is moving itself from somewhere to somewhere). Both kinds of verbs are also often found as past participle with *be*, a pattern that can be interpreted either as passive (CAUSED MOTION) or perfect (INTRANSITIVE MOTION), with both interpretations frequently sharing a resultative meaning. This co-existence of formally and semantically extremely similar patterns is likely to minimize the boundaries between INTRANSITIVE and CAUSED MOTION verbs, which enables the sporadic use of CAUSED MOTION verbs in the INTRANSITIVE MOTION construction.

Once again, verbs borrowed from French take part in this phenomenon as well: ME *butten*, *frounten*, and *halen* are borrowed from French verbs with caused motion senses, such as 'strike (down)' or 'haul, hoist', see (176a–c). In the donor languages, according to the dictionaries (*AND*, *DMF*, *TL* s.vv.), these do not have intransitive motion uses. They only receive such use in Middle English, probably in analogy to native verbs of similar semantics (see also the discussion in 6.2.3.2).

---

[122] It may be possible that such a configuration influenced the development of *went* from an originally causative verb *wendan* (<**wand-jan* 'to make something/somebody wind') to an intransitive motion verb which then yielded the suppletive past tense form of *go*.

(176) a. ME *butten* from OF *bouter* 'thrust, strike'
Þe kniȝt ***donward*** gan **butten** ***Amidward þe hors gutten.*** (*c*.1330(?*a*.1300) *Arth.& M*.(Auch) 5165, MED s.v. *butten*, sense b. 'throw oneself, plunge')
'The knight plunged down into the middle of the horse-guts.'

b. ME *frounten* from OF *afronter* 'clash, strike down, affront'
Ector [. . .] **frunt** ***of hys fol*** flat ***to þe ground.*** (*c*.1540(?*a*.1400) *Destr.Troy* (Htrn 388) 6890, MED s.v *frounten*, sense c. 'to fall, tumble down')
'Hector [. . .] tumbled of his horse flat to the ground.'

c. ME *halen* from OF *haler* 'to haul, hoist'
He..**Hallede** ***out at þe hal-dor*** (*c*.1400(?*c*.1390) *Gawain* (Nero A.10), 458, MED s.v. *halen*, sense 4.a 'to hasten, rush, etc.')
'He rushed out at the hall door.'

The only use in which these verbs may express motion in the donor languages is when they are combined with a reflexive object pronoun, i.e. in a structure that is basically one of CAUSED MOTION, where FIGURE and AGENT are coreferential:

(177) a. [. . .] qe nul ne *se* **boute** ***mesmez avant*** (BOZ *Cont* 152, AND s.v. *bouter, se bouter avant* 'thrust oneself forward')
'[. . .] that nobody thrust themselves forward.'

b. ***Hors del haven se*** **sunt buté** (*S Edm* 1360, AND s.v. *bouter, se bouter hors* 'to leave')
'They have pushed themselves out of [i.e. left] the port.'

It is also possible that Middle English speakers interpreted the French reflexive pronoun in sentences with CAUSED MOTION verbs like AN *hors se sunt buté* 'they have pushed themselves out' (as in (177)) as non-obligatory, since many INTRANSITIVE MOTION verbs can be used with a pleonastic reflexive in Old French as well: AN *courir* 'to run', for example, is found both without (178a) and with a reflexive pronoun (178b).

(178) a. ore je **cure** humblement (SAMPS2 (T) 165v, AND s.v. *coure*, v.n. 1 'to hasten')
'Now I run, humbly.'

b. et puis *se* **courent** fort pour atteindre les chivalx (*Man lang* 78, AND s.v. *coure*, v.refl. 1 'to run')
'And then they run REFL strongly to get to the horses.'

Thus, once again, we have seen that some French loans among the Middle English MANNER verbs are French forms, but have English meanings or, rather, uses (see also 6.2.3.2). Therefore, to say, for instance, that ME *halen* is a MANNER OF MOTION verb borrowed from French means missing some important steps: In French, *haler* is used as a CAUSED MOTION verb ('haul sth./oneself somewhere'). It is adopted into Middle English in this use and is additionally used in the INTRANSITIVE MOTION construction ('haul somewhere').

6.2.5 Verbs not primarily expressing MOTION

In the Old English inventory, we have encountered 15 non-MOTION verbs used to express motion (section 5.1.4). For the Middle English period, there are far more of these (46 of them): Some of the semantic groups identified for Old English are also present in Middle English—sometimes with the very same verbs (e.g. OE *brecan*/ME *breken*, see 6.2.5.2), sometimes with verbs of similar meaning (e.g. OE *winnan*, ME *swinken*, see 6.2.5.3).

6.2.5.1 Adapted complementation patterns

Two of the verbs found in the Old English inventory to be used metonymically[123] to express motion (OE *sēcan* 'to seek, visit' and *metan* 'to measure', cf. 5.1.4.5) are still part of the inventory in Middle English. What is interesting, though, is that, when used to express motion, their complementation pattern has been 'adjusted' to their motion reading: In Old English, *sēcan* had mostly, and *metan* always, been used with GROUND in a direct object, i.e. with its normal valency ('seek somebody'; 'measure something'). In Middle English, by contrast, *sechen* is often found in an intransitive motion construction in its motion use, i.e. with a directional adverbial, such as *unto the chambre, ouer the see*, and *oute of youre sight* in (179a–c).

(179) a. Be nyhte ***unto the chambre*** he **soghte** ((*a*.1393) Gower *CA* (Frf 3) 2.2474, MED s.v. *sechen*, sense 11.a 'to come, go, travel')
'By night, he sought [i.e. went] unto the chamber.'

b. The peple..**soughten** ***ouer the see.*** (*c*.1450(?*a*.1400) *Parl.3 Ages* (Add 31042) 434, MED s.v. *sechen*, sense 11.a 'to come, go, travel')
'The people sought [i.e. travelled] over the sea.'

c. ***Oute of youre sight*** nowe **is** it **soghte.** (*a*.1450 *Yk.Pl.*(Add 35290) 449/25, MED s.v. *sechen*, sense 11.d *ben sought* 'to have gone away')
'It [a vision] is now sought [i.e. gone] out of your sight.'

Note that it is not only in the combination with PATH satellites that the verb has adapted to its now more decontextualized motion reading in Middle English, but also in that its past participle is used with a form of *be* in (179c),

[123] 'Searching X' is contiguous to 'going to X', therefore the use of *sechen* 'search' must be metonymically motivated. The motion reading of *meten* 'measure' could be both metonymical (in going, a FIGURE also 'measures' the distance, e.g. in counting the steps) or metaphorical (measuring a distance as similar to walking the distance).

the auxiliary usually used for MOTION verbs in this periphrastic construction. Normally, *sechen* would be used with the auxiliary *have*.[124]

ME *meten* 'to measure' in its motion reading is now also sporadically attested with PATH satellites in Middle English (cf. *doun of þe mounte in-to a mirk vale* in (180)), and no longer only with GROUND in the direct object as in Old English (see 5.1.4.5):

(180) Þen **metis** he ***doun of þe mounte in-to a mirk vale***. (*c*.1450(?*a*.1400) *Wars Alex.* (Ashm 44) 4803, MED s.v. *meten* (v.1), sense 5.a 'to traverse (a certain distance); go, journey, pass')
'Then he measures [i.e. goes] down from the mountain into a dark valley.'

This change in the complementation patterns of *sechen* and *meten* in motion uses can also be taken as an indication that the intransitive motion construction (i.e. the combination of a verb with a PATH satellite) is by far the most entrenched pattern for motion expression in Middle English (see also 6.2.6 and 6.3.4).

The following subsections (6.2.5.2–5) will address the relations that can hold between the non-MOTION verbs and the intransitive motion construction: Many of the non-MOTION verbs used in the intransitive motion construction in Middle English can be grouped into specific r-relations between verb meaning and construction meaning (see 2.2.2). A good half of the non-MOTION verbs can be explained in this way, i.e. by an interaction of constructional and verbal meaning. For many of the remaining non-MOTION verbs, however, it is more likely that they owe their intransitive motion use to other developments, which will be discussed in 6.2.5.7 (the non-MOTION verbs that will not be discussed in more detail here are ME *weden* 'be or become crazy' (see (30b) in 2.2.4); ME *boillen* 'boil', *leiken* 'to engage in a game or contest, play', as well as ME *ap(p)lien*, whose meanings 'to go, take a course, land' are also present in Latin *applicare* and AN *applier*, from which it is borrowed (cf. footnote 62)).

6.2.5.2 'break/burst'

Just like OE *brecan* 'break' and *berstan* 'burst', also their Middle English reflexes ME *breken* and *bresten* (181a–b) are possible in motion uses (and remain until the present day, see Rohde's 'verbs of disruption of material integrity' (2001: 355)).

(181) a. His Barouns defended þe Tour fast, But þey **brake** [LinI: breken] ***yn*** at þe last. (*a*.1400(?*a*.1350) *Siege Troy*(1) (Eg 2862) 2009, MED s.v. *breken*, sense 12.b 'to break (into a building, place); make forced entry')
'His barons vehemently defended the tower, but at last, they [i.e. the Greek enemies] broke in.'

[124] Cf., e.g., *Seth ye folye* **haue sought**, *folie haue ye founden*. (*a*.1500(?*c*.1450) *Merlin* (Cmb Ff.3.11) 572, MED s.v. *sechen*, sense 12 'in proverbs') 'Since you have sought folly, you have found folly.'

b. A colfox..The same nyght ***thurgh out the hegges* brast *In to the yerd.*** ((*c*.1390) Chaucer *CT.NP.*(Manly-Rickert) B.4408, MED s.v. *bresten*, sense 12.b 'to break (into a building, place); make forced entry; also, ~ *in*')
'The same night, a coal-fox burst through the hedges into the yard.'

The relation of these verbs to the constructional meaning MOTION is most adequately described as one of MEANS: In (181a), for instance, the Greeks move into the tower by means of breaking it; in (181b), the fox moves into the yard by bursting the hedges that enclose it. Alternatively, breaking the tower and bursting the enclosure could be interpreted as the RESULT of the motion.

6.2.5.3 'labour/toil'

Also the verbs meaning 'to labour, toil' can be characterized as related to MOTION by an r-relation of MEANS ('move somewhere by means of toiling'). The 'labour, toil' verbs that were found in Middle English are *labouren*, *swinken*, *travailen*, and *winnen*, as in (182a–d), i.e. two native and two borrowed verbs. The reflex of OE *feohtan* (ME *fighten*) 'to fight' is no longer attested in motion uses, and in ME *hien*, the reflex of OE *hīgian* 'to labour, strive, endeavour', the formerly contextual motion reading has become decontextualized as 'to go quickly' (see Ma^{speed} in Appendix B). ME *swinken* 'to toil' is newly attested in the IMC, just like the French loans *labouren* and *travailen* (see the discussion in 2.2.3 and 6.2.3.2).

(182) a. Sche myth not enduryn so gret jurneys as þe man myth..And þerfor sche labowryd as long as sche myth tyl þat sche fel in sekenes.. Owr Lord sent leeuyn, thundyr, & reyne ny al þe tyme þat þei durst not **labowryn *owtward.*** ((*c*.1438) *MKempe* B (Add 61823) 234/16,26, MED s.v. *labouren*, sense 3.b 'to journey painfully or with difficulty')
'She was not able to endure as long journeys as the man could [endure], and therefore she toiled [i.e. travelled painfully] as long as she could, until she fell ill [. . .] Our Lord sent lightning, thunder, and rain almost all the time, that they dared not toil [i.e. journey painfully] outward.'

b. Iacob tolde him for-quat he **swanc *So fer.*** (*a*.1325(*c*.1250) *Gen.& Ex.*(Corp-C 444) 1656–7, MED s.v. *swinken*, sense 2.c 'to toil in travelling, travel with effort') (= example (36))
'Jacob told him why he had toiled [i.e. travelled] so far.'

c. I rede þe **trauaile *to grece*** noþing; Dwelleþ at hoom & mery makeþ (*a*.1400(?*a*.1350) *Siege Troy*(1) (Eg 2862) 37/444, MED s.v. *travailen*, sense 5.a.a 'to make a journey, travel, go')
'I advise you, do not travel to Greece at all; stay at home and be merry.'

d. May he ***til a woman* wynne**, In prieuyte, he wyl not blynne Er he a childe put hir with-Inne. (*a*.1400 *Preste ne monke* (Cleo B.2) 89, MED s.v. *winnen*, sense 11.a.a 'to come or go, proceed, make one's way, manage to get somewhere, etc.')
'If he may secretely win [i.e. manage to get] to a woman, he will not cease until he has put a child into her.'

Also here, there is a metonymical relation between the verb meaning and the construction meaning—the FIGURE moves by MEANS of labouring/toiling. It is likely that the motion use of ME *threpen* belongs to the same group. According to the *MED*, *threpen* can be used to mean the following:

a. to scold, chide, rebuke; also, to threaten (sb.), menace, etc.
b. to argue, quarrel, debate
c. to compete, contend; also, to fight, battle, etc.
d. of thunder, rain, the winds: to rage
e. to assert (that sth. is so), maintain despite opposition, insist, etc.
f. ~ *awei* to hurry or rush away

The frame that is common to all these contextualized meanings of *threpen* could be seen as one of OPPOSITION. The motion meaning (f) is only attested once, in (183):

(183) He dessyrit full depely..þat bright for to braȝ in his big armes; But hym thught þat ho þroly **þrappit *awaγ***, And o fer fro the freike foundit to kepe. (*c*.1540(?*a*.1400) *Destr.Troy* (Htrn 388) 13811, MED s.v. *threpen*, sense f ~ *awei* 'to hurry or rush away')
'He deeply wanted to embrace the fair one in his big arms, but it seemed to him that she rushed away impetuously, and went to remain at a distance from the man.'

The FIGURE in (183), 'the fair one', is a lady that appears in Ulysses' dream—he wants to embrace her, but she *threpes* away out of his *big armes* impetuously. In this context, it seems probable that the 'opposition' meaning of the verb, as in 'to fight, to battle' (see the *MED*'s sense (c)), 'to rage' (d), interacts with the construction meaning of MOTION and is construed as the co-event, most likely the MEANS, of MOTION. In this interpretation, there is definitely a similarity to the 'toil' verbs in that both the 'toil' verbs and *threpen* may be used for MOTION by means of struggling.

Another possibility is that *threpen* contributes the meaning of 'scold, chide', so that it denotes not the MEANS OF MOTION, but a co-occurring event that is not in causal relation to the MOTION EVENT (Talmy's relation of CONCOMITANCE, see 2.1.2). In this reading, the lady is chiding as she moves away. Yet, since no other verbs in the Middle English motion verb inventory seem to express a concomitant, not causally related co-event, or more specifically, denote any kind of 'speaking', this possibility seems less convincing.[125]

6.2.5.4 'Modality'

Like in Old English, (pre-)modal verbs can also be used in the intransitive motion construction in Middle English. The Middle English (pre-)modal verbs that were identified as motion verbs via the *MED* definition search are *monen* 'shall, will', *moten* 'must', *mouen* 'may', and *wilnen* 'to wish, desire', as in (184a–d).

(184) a. She **mone** ***To the prynce off Aragon*** (*a*.1500(?*a*.1400) *Torrent*, MED s.v. *monen* (v.2), sense 1.b 'to be compelled, must, ought; must go, etc.')
'She must [i.e. must go] to the Prince of Aragon.'

b. Criseyde **moste** ***out of the town***. (*a*.1425(*c*.1385) Chaucer *TC*, MED s.v. *moten* (v.2), sense 5.b.c 'be under compulsion to go (come), etc.')
'Criseyde must [i.e. must go] out of the town.'

c. For it was nyght, and ***ferther*** **myghte** they noght. ((*c*.1390) Chaucer *CT. Rv.*, MED s.v. *mouen* (v.3), sense 6.a 'with implied *gon*: to be able to go')
'For it was night, and they could not [i.e. could not go] further.'

d. ***Whedire*** **wilnez** thowe, wye, walkande thyn one? (*c*.1440(?*a*.1400) *Morte Arth.(1)*, MED s.v. *wilnen*, sense 3. 'to intend sth., purpose, intend to go, etc.')
'Whither do you intend [i.e. intend to go], good sir, walking all by yourself?'

Of course, the other pre-modals, which were not captured by the *MED* definition search, can be used in the IMC as well (cf., e.g., (59c) with *shulen out* 'shall out' in 4.1.2.4).[126] As we have seen above (2.2.4), this use of modal verbs in intransitive motion constructions is typical for the Germanic languages. Berthele calls this kind of r-relation between verb meaning and constructional meaning MODAL PRECONDITION ('modale Rahmenbedingungen' 2007: 247), i.e. the verb expresses a precondition of the MOTION EVENT, such as being obliged to go somewhere or wanting to go somewhere.

Yet, this relation does not only license the pre-modals but also other Middle English verbs that express modal precondition: The native (or Scandinavian) verbs *atlen* 'to intend, plan', *menen* 'to mean, intend', *thinken* 'to think', and the

[125] In Present-Day English, the use in an IMC of a verb which expresses a merely concomitant co-event is not possible for most speakers (see 2.1.2); these verbs would be used in a *way*-construction instead (as in *to scold/rant one's way through/to. . .*).

[126] Other examples, with ME *willen* and *durren* are: *But afterward þe fuyre was i-come, he* ***wolde out*** *in no manere wise; but þere he was i-brent for love of martirdom* ((*c*.1400) *Higd.(1)*, CME) 'But after the fire had come, he did not want [i.e. did not want to go] out by any means, but was burnt there for the love of martyrdom.'; ***Ferrer*** *ne* ***draste*** *þay noȝt for fere.* (*c*.1380 *Firumb.(1)* (Ashm 33) 3726, MED s.v. *durren*, sense 1a.b 'with other constructions') 'They did not dare [i.e. dare to go] further for fear.'

impersonal verb *listen* 'to want', as well as the French loans *amen* 'to aim; also, to plan, intend' (< OF *aesmer*), *auntren* 'to venture, dare' (< OF *aventurer*), and *purposen* 'to plan, intend' (< OF *proposer*) are attested in IMCs as well. In most of these uses, like in those of the (pre-)modal verbs in (184a–d), the interaction of the verb meaning with the meaning of the IMC yields an interpretation of 'modally broken motion', such as 'intend to go to X':

(185) a. I wil **auntre** ***to þe dore***, þat i hadde mete. (*c.*1410(*c.*1350) *Gamelyn*, MED s.v. *auntren*, sense 3.a 'to venture, take a chance, attempt (sth.), to venture to go (to a place, to sb.), etc.')
'I will venture [i.e. venture to go] to the door, so that I should get food.'

b. With shippes xii ***to Italy*** had they **mente.** (1543(1464) Hardyng *Chron.B*, MED s.v. *menen* (v.1), sense 2.d 'to go toward a place')
'They had meant [i.e. meant to go] to Italy with twelve ships.'

c. Sche was a pilgryme, **purposyng** be þe grace of God ***to Seynt-Iamys-warde***. ((*a.*1438) *MKempe A*, MED s.v. *purposen*, sense 2 'to plan or intend to go, etc.')
'She was a pilgrim, intending [i.e. intending to go] to Saint James, by the grace of God.'

d. For be þe first gestenyng al brouȝt togidir & al þe sete fully besette, þe ȝates schul be loken for ones & euere, ne þar noon ***þidir*** **þenke** þat out is bilafte. (*a.*1450 *PNoster R.Hermit*, MED s.v. *thinken*, sense 9 'to intend to go (to a place)')
'For when the first guests are all brought together and all the seats fully occupied, the gates shall be locked once and for ever, that nobody who is left out intends [i.e. intends to go] thither.'

e. ***To þe holy land*** him **list**, & þider gan him spede. (?*a.*1400(*a.*1338) Mannyng *Chron.Pt.2*, MED s.v. *listen* (v.1), sense 1.a.a *him list* 'he wishes, desires, etc., also, with implied inf. of motion: *him list to* he desired to go (somewhere)')
'He desired [i.e. desired to go] to the holy land, and he began to travel swiftly thither.'

For *amen* 'to aim' and *atlen* 'to intend', by contrast, their use in the IMC does not necessarily have to be interpreted as modally broken, as suggested in the meanings given by the *MED* for (186a–b). These may be seen as metonymical extensions, from 'intending/daring to go to X' to 'going to X'.

(186) a. Antenor **Amyt**..***To the palis***. (*c.*1540(?*a.*1400) *Destr.Troy* (Htrn 388) 2023, MED s.v. *amen*, sense 5.a 'to direct one's course, go')
'Antenor went to the palace.'

b. Alixandre wiþ his ost **atlede** ***þidire***. (*c.*1450(*c.*1350) *Alex.& D.*(Bod 264) 15, MED s.v. *atlen*, sense 3.a 'advance, go (to a place), approach, etc.')
'Alexander went thither with his army.'

Note that *amen* is a loanword from (Anglo-)French, which is not used to express motion in the donor language: ME *amen* comes from AN *asmer* (< L *ad-estimare*), meaning 'to estimate, value, suppose, etc.'—according to the *AND*, *asmer* is not attested in intransitive motion uses in Anglo-Norman (nor in other varieties of Old and Middle French, according to *TL* and *DMF*). Its Middle English uses such as in (186a) are therefore most likely not borrowed from the donor language, but analogous to the uses of synonymous verbs in Middle English. AN *purposer*, by contrast, is attested in IMCs in Anglo-Norman (187):

(187) quatre niefs issint chargés des laynes **purposauntz** ***vers l'estaple de Caleys*** [. . .] furent gettez sur les rokkes (PRO SC 8/171/8529, AND s.v. *purposer*: *purposer vers* 'to be bound for')
'Four ships going out, charged with wool, bound for the market at Calais, were cast onto the rocks.'

As this kind of use, however, is not listed in *TL* and *DMF* for continental varieties of French, it is most likely another phenomenon of English contact influence on Anglo-Norman, as discussed in section 6.1 above.

6.2.5.5 'prepare, make ready'

Furthermore, not only verbs denoting the *modal* preconditions (volition, obligation, etc.) are found in the IMC, but also verbs that express other preconditions. As has been recurrently noted in the literature on Middle English alliterative vocabulary (e.g. Turville-Petre 1977: 70), many verbs that normally mean 'to prepare' are used to express motion as well, particularly in alliterative verse. Matsunami (1985: 104) mentions the verbs *busken*, *dighten*, *greithen*, and *richen* as examples of this 'prepare' type, attested in motion uses such as (188a–c).[127]

(188) a. What wise we mow best **buske** ***of þis lond*** (*a*.1375 *WPal.*(KC 13) 1653, MED s.v. *busken*, sense 4.a 'to go (to a person or place), betake oneself; esp. go hastily, etc.')
'In what way we may best go out of this land.'

b. Quen þis angel ***away*** was **diht**, Tua men þer cam were clad in quiht. (*a*.1400 *Cursor* (Göt Theol 107) 10551, MED s.v. *dighten*, sense 5 'to go; fall (to the ground), etc.')
'When this angel had gone away, two men clad in white came there.'

[127] Matsunami (1985: 104) also has *don* and *dressen* in the 'prepare'-type. ME *don*, however, is only used in a motion sense with a reflexive pronoun, as in *He dyd* **hym** *faste away* (*a*.1500 (?*c*.1400) *Triam.*

c. He **grathed** þen ***yn to grece***. (*a*.1500 *Parl.3 Ages* (Add 33994) 416, MED s.v. *greithen*, sense 5.a 'to go, hasten')
'Then he hastened into Greece.'

These uses of verbs meaning 'prepare' in an IMC, as in (188a–c), have parallels in Middle High German, as discussed by Paul ([1886] 1920: 319–20), and apparently also in Old Norse: According to Cleasby-Vigfusson (s.v. *búa*, sense III), ON *búask*, which was adopted into English as ME *busken*, is also attested in this kind of construction, as in (189):

(189) nú **býsk** hann ***út til Íslands*** (Nj. 10, CV s.v. *búa*, sense III)
'Now he "busked" out to Iceland.'

As pointed out by Paul, this kind of use is not elliptical in the sense that a MOTION verb would have to be supplemented. Rather, the 'prepare' verb is simply combined with the directional element; in other words, it is used in the intransitive motion construction. In a compositional interpretation, this combination should also give rise to a 'broken' motion reading ('prepare to go to X'), like for most of the verbs discussed in 6.2.5.4. Yet, the meanings given in the *MED* for the 'prepare' verbs suggest an additional metonymical reading 'go to X', as in the *MED* senses from which (188a–c) are taken.

In (190), it seems as if the contextualized motion reading of *richen* 'to prepare' has become a decontextualized meaning of the verb because *richchand* (if it is indeed a form of the same verb) must mean something like 'running' here. The use of *richchand* in (190) cannot be explained by the mere need to have a word starting in /r/, since a form of *rennen* would have been possible as well.

(190) Renaud com **richchande** þurȝ a roȝe greue, & alle þe rabel in a res, ryȝt at his helez. (*c*.1400(?*c*.1390) *Gawain* (Nero A.10) 1898, MED s.v. *richen* (v.1), sense 3.b 'to move, run, *refl.* make one's way, proceed')
'Renaud [i.e. the fox] came running through a rough thicket, and all the pack at once right at his heels.'

In addition to the verbs above, which are also listed by Matsunami (1985: 104), the *MED* definition search further identified the 'prepare' verbs *bounen*, *shapen*, and *trussen* in intransitive motion constructions, such as in (191).

(191) a. Þe moste party of hys princez..**Bound** [Ashm: Hastis þam] ***to ascolon***. (*a*.1500 *Wars Alex.*(Dub 213) 1116, MED s.v. *bounen*, sense 4.a 'to go, proceed, etc.')

(Cmb. Ff.2.38) 495, MED s.v. *don*, sense 5.d.a. *refl.* 'to betake oneself, go') 'He quickly betook himself away'. Since no motion use of *don* without reflexive is given in the *MED*, it seems that this is a CAUSED MOTION construction and not an INTRANSITIVE one with a pleonastic reflexive. ME *dressen* can be used in the meaning 'prepare' (cf. *MED*, sense 5), but since it also has many caused motion uses (cf. *MED*, sense 1, 7, 8, 9), I classified it as 'NEUTRAL MOTION verb'. Both the caused motion uses and the 'prepare' use may be involved in its intransitive motion use; see the discussion below.

'The majority of his princes went to Ascolon.'

b. Sum of þam hent him by þe lap þat he suld noght *oway* **schap**. (*a*.1425 *NHom.(3) Pass.*(Hrl 4196) 63/628, MED s.v. *shapen*, sense 6.a 'to go, etc.')[128]
'Some of them caught him by the shirt, so that he should not go away.'

c. Men..come with camellis and with assis, and charge hem with that golde, and thenne they **trusse** [BodeMus: hieth hem] faste *awey*. (*a*.1500 *Mandev. (3)* (Rwl D.99) 471, MED s.v. *trussen*, sense 2.a 'to go, depart, etc.')
'Men come with camels and with asses, and charge them with that gold, and then they go away fast.'

ME *trussen* is borrowed from F *trousser* 'to pack'. The only dictionary of medieval French that lists a motion use for this verb is the *DMF* (sense 3.a 'avec une idée de mouvement') with the earliest attestation from the second half of the fourteenth century (*Mir. femme*, 1368). This is considerably later than the first motion attestation given by the *MED* for ME *trussen* (sense 2a 'to go, depart, get moving') from the *Ancrene Wisse c*.1230(?*a*.1200). Therefore, the motion use of ME *trussen* might perhaps be an English development again. As always, though, we cannot be sure whether a motion use of OF *trousser* existed on the continent and just happens to not be attested or to not have found its way into the historical dictionaries.

6.2.5.6 Conclusion: r-relations in the Middle English IMC

In order to describe which types of verbs can occur in an argument-structure construction, construction grammar often uses the notion of r-relations (cf. section 2.2.2). R-relations are the relations that hold between the meaning of the construction (in our case, INTRANSITIVE MOTION) and the meaning of the verb that is used in this construction. If the verb is a MOTION verb (i.e. a MANNER, PATH, or NEUTRAL MOTION verb), it gives access to a MOTION frame itself, so the r-relation between verbal and constructional meaning would be one of INSTANCE (e.g. RUNNING as an instance of MOTION). In some MANNER verbs, where the notion of TRANSLOCATIONAL MOTION is less prominent (e.g. ME *hippen* and *hoppen* 'to leap, bounce'), the r-relation will be more adequately be described as one of MEANS (e.g. 'bouncing' as the MEANS of MOTION). The boundaries between MEANS and INSTANCE are fuzzy: RUNNING, for example,

[128] It is possible that, different from what the *MED* suggests, (191b) is not an attestation of *shapen*, but actually one of *scapen*, the apheretic form of *escapen*. Note that a manuscript variant of the same line is given as a quotation in the *MED* entry of *scapen*: *Som of þam hent him be þe lapp þat he suld noght o-way schapp.* (?*a*.1425(?*c*.1350) *NHom.(3) Pass.*(RwlPoet 175) 808). The only other *MED* attestation of *shapen* in an IMC is *Hu might it schope [Vsp: sco; Trin-C: shape; Frf: falle] into þi hert to do suilk a dede?* (*a*.1400 *Cursor* (Göt Theol 107) 15483) 'How may it come into your heart to do such a deed?' The others are all with reflexive pronouns, which might be pleonastic reflexives, but could also be reflexive objects.

can be seen as a MEANS of MOTION but, since you usually run in order to change location, also as an INSTANCE of MOTION (see the discussion in 2.2.3).

In the preceding subsections, we have seen that, next to the r-relations of INSTANCE and of MEANS (MANNER OF MOTION) which license verbs linked to a MOTION frame on their own, the Middle English IMC also combines with verbs which do not necessarily give access to a MOTION frame themselves: The non-MOTION verbs *breken* and *bresten* 'break' (6.2.5.2), as well as the verbs meaning 'labour, toil' (6.2.5.3), can also be construed as denoting the MEANS of MOTION (MOTION by MEANS of breaking something, or by MEANS of labouring, toiling). A relation of PRECONDITION holds between MOTION and verbs expressing modality (modal verbs and verbs meaning 'intend, plan', etc., see 6.2.5.4) as well as the 'prepare' verbs (6.2.5.5). The combination of these PRECONDITION verbs with the IMC usually gives rise to a meaning of 'modally broken motion' (see also Berthele 2007: 242), such as 'want to go', 'have to go', 'prepare to go'. Sometimes, however, the combination yields a non-broken motion meaning nevertheless, such as in *they trusse faste awey* (191c) 'they go away quickly' (lit. 'they prepare away quickly'). This is most likely a metonymical extension with 'preparation' as part of the MOTION EVENT standing for the whole MOTION EVENT.

As we have seen, many of the non-motion verbs attested in the IMC are roughly synonymous, i.e. in paradigmatic relation to each other. Therefore, instead of capturing them by the relatively abstract r-relations of MEANS and PRECONDITION, it may also be plausible to explain their occurrence via an exemplar model (cf. Bybee 2013).

6.2.5.7 Other non-MOTION verbs in the Middle English IMC

In the preceding subsections, we have seen a range of non-MOTION verbs which can receive a contextual motion reading through the interplay with the semantics of the intransitive motion construction. However, among the non-MOTION verbs found in the inventory (see Table 6.1 and Appendix B), there is also a group of verbs which seem hard to explain via a relation between verbal and constructional meaning, although they occur in IMCs. These are verbs which roughly mean 'to take'—ME *fon* and *fongen*[129] 'to grasp, seize', and ME *nimen* and *taken* 'to take'—in uses such as (192).

(192) a. He saw Tyrry ***away*** with Tyb **fang** (*a*.1475(*a*.1450) *Tourn.Tott.*, MED s.v. *fongen*, sense 9.b 'to proceed, go')
'He saw Thierry go [lit. take] away with Tib.'[130]

[129] *Fongen* is simply a new form of older *fon*, formed by analogy to the past participle form of *fon*, namely (*i*)*fangen*/(*i*)*fongen* (maybe also to the corresponding noun *fang*; cf. Brunner 1962: 245).
[130] The edition by Kooper (2006) also has this reading. An alternative interpretation may perhaps be 'He saw Thierry away [i.e. go away] with the taken/conquered Tib', i.e. with *fang* as a participle 'conquered, received, taken' used adjectivally: In the text, Tib is the Reeve's daughter, whom the winner

b. **Fo** [Otho: go] we ȝet ***to Beline***. (1275(?*a*.1200) Lay. *Brut*, MED s.v. *fon*, sense 10.b 'to proceed, go')
'Let us now go [lit. take] to Beline.'

c. ***Fro þe auter down*** he **nam** And to þe cherche porche he cam. (*a*.1400(*c*.1303) Mannyng *HS*, MED s.v. *nimen*, sense 4.a 'to go, travel, etc.')
'He went [lit. took] down from the altar and came to the church door.'

d. Y schal **take** ***out to anoþer pard*** & prykie fro hem anon. (*c*.1380 *Firumb*. (1), MED s.v. *taken*, sense 50.a 'to go, come, travel, etc.')
'I shall go [lit. take] out to another side, and immediately gallop from them.'

ME *chesen* 'to choose', somewhat similar to the 'take' verbs in its semantics, is attested in this kind of use as well (193).

(193) Anon ***of lond*** he **ches** ***Out of markes eiȝe sene***. (*c*.1330(?*a*.1300) *Tristrem*, MED s.v. *chesen*, sense 8 'to choose or take one's way, go, etc.')
'He went [lit. chose] immediately from the land, out of Mark's eyesight.'

It does not seem plausible that the 'take' verbs in the IMC are motivated through a relation of verb semantics and constructional semantics like the verbs in the preceding subsections—how should 'take' be related to INTRANSITIVE MOTION? It rather seems that other patterns in which the 'take' verbs occur are responsible for these uses: probably first and foremost the combination with *way* (similar to what we saw for OE *niman* and *aredian* in 5.1.4.4), but perhaps also CAUSED MOTION uses with reflexive pronouns (196) and the uses with GOAL in the direct object (197), as illustrated below. The fact that the verbs seem to be semantically rather light and flexible anyway may play an additional role in their motion use.[131]

The combination with *way* (or roughly synonymous nouns such as *course*, *journei*, *pas*, *stie*, or *strete*), as in (194a–d), has been proposed by Matsunami (1985) and Turville-Petre (1977: 80) as being the origin of the intransitive motion uses of the 'take' and 'choose' verbs.

(194) a. ***Straght unto Kaire his weie*** he **fongeth**. ((*a*.1393) Gower *CA*, MED s.v. *fongen*, sense 1.c ~ *wei* 'make one's way, go')
'Straight to Caire he takes his way.'

---

of the tournament at Tottenham will be allowed to marry. Although the past participle of *fongen* is usually weak, with *-ed*, *fang* could be the strong participle of *fon*, such as in *They went Florence to leman haue **fonge***. (1500(?*c*.1400) *Florence* (Cmb Ff.2.38) 1829, MED s.v. *fon*, sense 1.b ~ *to leman* 'obtain as a mistress') 'They would have liked to obtain Florence as a mistress.'

[131] Cf. the *MED* entry for *taken*, which has as many as 55 senses—plus many subsenses. This is a lot, even for the *MED*.

b. ***Toward Rome*** he **nom *his way*.** (*c*.1400(?*c*.1308) Davy *Dreams*, MED s.v. *nimen*, sense 4.b ~*wei* 'to take one's way, go, etc.')
'He took the way toward the castle.'

c. ***To Jerusalem*** I **take *the waye*.** (1591(?*a*.1425) *Chester Pl.*, MED s.v. *taken*, sense 50.c.a ~ *wei* 'travel a road or path, etc.')
'I take the way to Jerusalem.'

d. ***To a cheefe foreste*** they **chesen *theire wayes*.** (*c*.1440(?*a*.1400) *Morte Arth.* (1), MED s.v. *chesen*, sense 8 ~ *wei* 'choose or take one's way, go, etc.')
'To a great forest they choose their ways.'

In these phraseological structures, the PATH + GROUND-PP would originally be a postmodifier of *way*. Yet, as can be seen in (194a–d), the head *way* and the PATH + GROUND-PP are often discontinuous. This might facilitate the dropping of *way*. Perhaps analogy to MOTION verbs, which can be used with and without *way* (such as *riden* in (195a–b)), has also played a role.[132]

(195) a. ***The streight way toward surre*** **rideth** he. (*a*.1500(*a*.1450) *Gener.(2)*, CME)
'He rides the straight way toward Syria.'

b. Now ***toward perse*** **ridith** Generides (*a*.1500(*a*.1450) *Gener.(2)*, CME)
'Now Generides rides toward Persia.'

What is more, the 'take' verbs are also attested with a reflexive pronoun in CAUSED MOTION uses, such as in (196a–c). Thus, they could also join in the sporadic use of CAUSED MOTION verbs without reflexive pronoun in the intransitive motion construction as discussed in detail in section 6.2.4.

(196) a. Hie forwerpeð here ræd and here claðes, and **nemeeð *hem to ðare woreld*** (*a*.1225(*c*.1200) *Vices & V.*(1), MED s.v. *nimen*, sense 4.a ~ *refl.* 'betake oneself, visit')
'They throw away their counsel and their clothes and betake themselves to the world.'

b. He **took *hym to a strong towre*.** (1485(*a*.1470) Malory *Wks.*, MED s.v. *taken*, sense 50.a ~ *refl.* 'betake oneself, etc.')
'He betook himself to a strong tower.'

c. ***To þe chefe chaiare of þe qwene*** he **chese *him*** be-lyue. (1450(?*a*.1400) *Wars Alex.*, MED s.v. *chesen*, sense 8 ~ *refl.* 'betake oneself')
'To the throne of the queen he betakes himself [lit. chooses himself] quickly.'

[132] On verbs combining with *way* in Middle English, see also Israel (1996: 221).

Finally, *nimen* and *taken* have attestations in which the GOAL is expressed in a direct object (as in (197a–b))—similar to PDE *take refuge* or *take lodging*.

(197) a. ***The lond*** full prevyly hath She **nome** (*a*.1450 *Parton.*(1), MED s.v. *nimen*, sense 4.c ~ *lond* 'arrive, disembark')
'She went to [lit. took] land very secretly.'

b. The erle rase and **tuk *a toure*** (*c*.1440(*a*.1400) *Eglam.*, MED s.v. *taken*, sense 50.b.a 'go to (a place), resort to, etc.')
'The earl rose and resorted to [lit. took] a tower.'

On the basis of uses such as (197a–b), which describe a MOTION EVENT, it is also conceivable that *nimen* and *taken* adapted their complementation patterns to a PATH satellite, similar to what we saw in 6.2.5.1 for ME *sechen* 'to seek'.

To conclude, various factors may have been involved in the use of the 'take' verbs in intransitive motion constructions. The phraseological combination with *way* seems to be the most important factor because this combination is widely attested with all of the verbs in question. The caused motion construction with reflexive pronoun, by contrast, could only be found for *chesen*, *nimen*, and *taken*, and direct object GOALS are only attested for *nimen* and *taken*.

6.2.6 Combination of motion verbs and PATH satellites

Another aim of the Middle English inventory study is, as with Old English, to get a first impression of the combinability of MANNER verbs and PATH satellites (cf. 4.1.1). As shown in Table 6.1, where the motion verbs that are attested with a PATH satellite are printed in italics, the majority of the MANNER verbs have attestations in intransitive motion constructions: As many as 86 per cent of them (221 MANNER verbs) are attested with a PATH satellite. This is rather similar to the Old English results, where 76 per cent (81 verbs) of the MANNER verbs had attestations with PATH satellites (see section 5.1.5). It indicates, once again, that, in Middle English as well, MANNER verbs freely combine with PATH satellites, as is typical for a MANNER-conflating and satellite-framing language.

PATH satellites are also attested for the majority of the NEUTRAL MOTION verbs (94 per cent (84)), and the non-MOTION verbs (96 per cent (44)). Also here, there is no noteworthy change compared to the results for Old English. The only slight change can be found among the PATH verbs, of which 79 per cent (57) are attested with a PATH satellite in Middle English, compared to only 58 per cent of the PATH verbs (11) in Old English (see section 5.1.5). However, this increase seems to be rather due to the general increase of verbs primarily expressing PATH in Middle English. The Old English PATH verbs that were not found with PATH satellites are either no longer part of the motion verb inventory in Middle English (e.g. *grētan*, *innian*, *pæððan*, *wīcan*), or if they survive into Middle English in motion uses, they mostly still do not take PATH satellites in Middle English (ME *leven*, *leten*—OE *forlǣtan*; ME *grinden*, *grounden*—OE *gryndan*).

We can conclude that the vast majority of Middle English motion verbs, no matter to which group they belong, are attested with PATH satellites, i.e. in instances of the intransitive motion construction.

6.2.7 Middle English inventory: conclusion

This section briefly summarizes the findings of the Middle English inventory study. The first interest of this study (cf. 4.1.1) was to find out about the Middle English MANNER verb vocabulary. The definition search in the *MED* found 258 MANNER OF MOTION verbs, accounting for more than half of the Middle English motion verbs (55 per cent). Many of them have been newly formed or borrowed into the MANNER verb inventory (see 6.2.3), or represent new INTRANSITIVE MOTION uses of primarily CAUSED MOTION verbs (see 6.2.4 for factors that may have enabled these uses). Hence, the present study supports Fanego's (2012) diagnosis of continued MANNER salience in the Middle English period with more and different data.

What the inventory study has also shown is that the proportion of French and Latin loanwords among the PATH verbs is far greater than among the other motion verb groups (MANNER, NEUTRAL MOTION, and non-MOTION): More than half of the Middle English verbs primarily expressing PATH are of French or Latin origin. This is around twice as much as for the other motion verb groups, where the proportion of French and Latin loans ranges between 18.6 and 27 per cent (see 6.2.2). This indicates that (pure) PATH verbs can be seen as a rather new phenomenon in the Middle English period, due to the contact with French.

Another aim of the inventory study was to investigate the extent to which non-MOTION verbs can be used in the intransitive motion construction. We have found 46 non-MOTION verbs used to express motion, many of which can be categorized as having either an r-relation of MEANS to the constructional MOTION meaning (the 'break' and 'labour/toil' verbs such as ME *bresten* and *swinken*) or an r-relation of PRECONDITION (verbs expressing modality and the 'prepare' verbs, such as ME *wilnen* and *busken*), as shown in 6.2.5.2–6. This indicates that MOTION is very strongly associated with the intransitive motion construction in Middle English, considering that many verbs that do not evoke a MOTION frame on their own can receive a MOTION reading in this construction.

What has been shown throughout the different sections of the inventory study is that several French loanverbs are used to describe INTRANSITIVE (MANNER OF) MOTION in Middle English, although in the donor language (Anglo-French and continental Old and Middle French) they are not, or at least they are not attested in such use according to the relevant dictionaries (see, e.g., 6.2.4.4 for *butten, frounten, halen*; 6.2.5.4 for *amen*; 6.2.5.5 for *trussen*). Their use as motion verbs in Middle English must be analogous to other Middle English verbs of similar semantics, as shown in the respective sections. Moreover, some verbs are not used to express INTRANSITIVE MOTION in continental varieties of Old French according to the dictionaries, but they are in Anglo-Norman (cf., e.g., 6.2.3.2 for ME *frushen, labouren, travailen;* 6.2.5.4 for *purposen*). Since these belong to semantic groups that commonly 'provide' motion verbs in Middle English ('break', 'labour, toil', 'intend'), their motion use in Anglo-Norman suggests contact influence of Middle English on Anglo-Norman (see also section 6.1). These intricacies have to be taken into account when making statements such as that 18.6 per cent of the Middle English MANNER verbs, or a fifth of the Middle English non-MOTION verbs, are borrowed from French (as in section 6.2.2): The motion use of several of these French verbs is only an English innovation.

Finally, section 6.2.6 has shown that, like in Old English, the majority of Middle English MANNER verbs combine with PATH satellites (86 per cent), as do the majority of the PATH verbs, NEUTRAL MOTION verbs, and non-MOTION verbs (79, 94, and 96 per cent respectively). We can conclude that, judging from its motion verb inventory, Middle English is a rather typically satellite-framing and MANNER-conflating language. The only perhaps untypical aspect are the new pure PATH verbs from French, which will be investigated in chapter 9.

6.3 *Motion verb use*

After the study of the Middle English motion verb inventory, which has shown which verbs are *possible* for talking about MOTION, the present section will investigate which verbs (and which patterns) are *frequent* in usage (cf. 4.2). The texts that serve as the basis of the usage study are presented in sections 4.2.4–6. This section starts with a presentation of the general results with regard to the frequency of the different groups of motion verbs, as well as of the individual verbs (6.3.1). Like in the Old English usage study, the question of genre and style with respect to the frequency of MANNER verbs as opposed to that of NEUTRAL MOTION verbs (particularly *comen* and *gon*) will be discussed (6.3.2). Section 6.3.3 addresses the PATH verbs; section 6.3.4 the complementation patterns.

6.3.1 Frequency of the different semantic groups

The frequencies of the different semantic groups in the five texts, as well as their overall frequency across the texts, are shown in Table 6.3 and visualized in Figure 6.4. As in the study on motion verb usage in Old English (5.2.1), the most common NEUTRAL MOTION verbs (ME *comen* and *gon/went*) are shown separately from the other NEUTRAL MOTION verbs. Table 6.4 below gives the frequencies of the individual verbs.

The results are similar to the ones for the Old English texts (see 5.2.1) in that PATH and non-MOTION verbs are the least frequently used groups, ranging from two to ten per cent and zero to three per cent respectively (cf. zero to six per cent PATH, and three to ten per cent non-MOTION in Old English). The most frequently used group, as in Old English, is again the NEUTRAL MOTION verbs,

TABLE 6.3. Frequency of the different verb groups in motion descriptions from Middle English texts (highly frequent NEUTRAL MOTION verbs *comen* and *gon/went* represented separately).

	GREGCHRON		ST EDMUND		KNIGHT		MILLER		REEVE		ALL TEXTS	
MANNER	20%	39	16%	14	54%	81	33%	17	29%	22	31%	173
PATH	10%	20	6%	5	4%	6	2%	1	5%	4	6%	36
comen, gon/went	60%	120	64%	55	33%	49	60%	31	62%	47	54%	302
Other NEUTRAL	8%	16	13%	11	7%	11	6%	3	3%	2	8%	43
Non-MOTION	3%	5	1%	1	2%	3	—	—	1%	1	2%	10
Total motion verbs		200		86		150		52		76		564

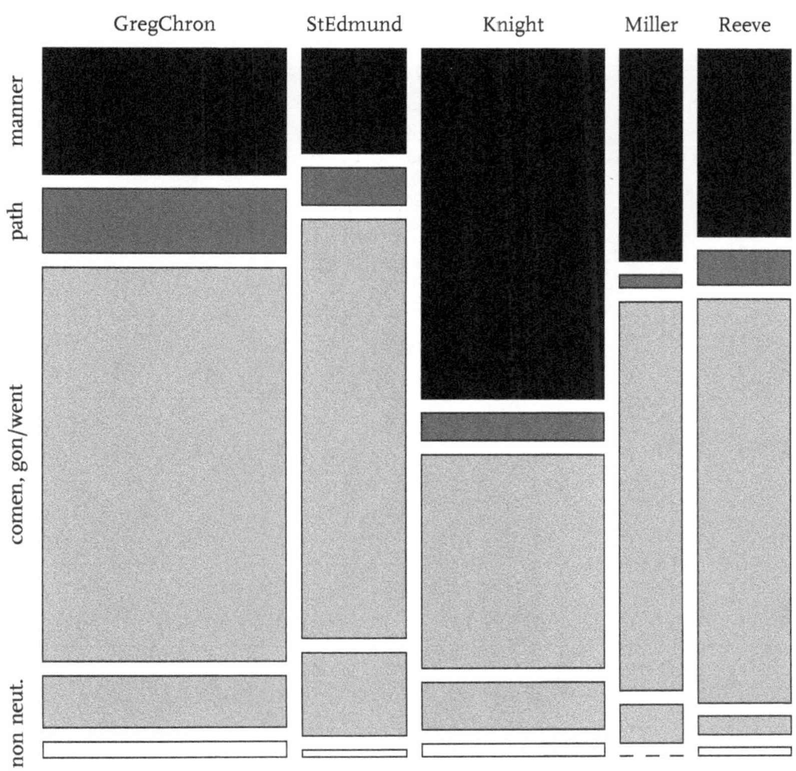

FIGURE 6.4. Frequency of the different verb groups in motion descriptions from Middle English texts (highly frequent NEUTRAL MOTION verbs *comen* and *gon/went* represented separately).

TABLE 6.4. Verbs in motion descriptions from Middle English texts.

		GREGCHRON		STEDMUND		KNIGHT		MILLER		REEVE		TOTAL		
comen	'come'	76	38%	28	33%	18	12%	7	13%	7	9%	136	24%	$Mo^{neutral}$
gon	'go'	5	3%	13	15%	20	13%	21	40%	35	46%	94	17%	$Mo^{neutral}$
went	'went' (pret.)	39	20%	14	16%	11	7%	3	6%	5	7%	72	13%	$Mo^{neutral}$
riden	'ride'	24	12%	1	1%	24	16%			1	1%	50	9%	Ma^{other}
fallen	'fall'	2	1%	5	6%	3	2%	4	8%	4	5%	18	3%	Ma^{fall}
rennen	'run'			1	1%	6	4%	1	2%	5	7%	13	2%	Ma^{speed}
londen	'land'	11	6%									11	2%	$Pa^{towards}$
sterten	'leap'					8	5%			2	3%	10	2%	Ma^{other}
crepen	'creep'			1	1%			2	4%	5	7%	8	1%	Ma^{slow}
flen	'flee'	2	1%	3	3%	3	2%					8	1%	Ma^{flee}
passen	'pass'	5	3%			2	1%	1	2%			8	1%	$Mo^{neutral}$
risen	'rise'	1	1%	2	2%	1	1%			4	5%	8	1%	Pa^{up}
romen	'roam'					7	5%	1	2%			8	1%	Ma^{other}
taken /nimen wei/journei	'take one's way'	6	3 %	1	1%	1	1%					8	1%	$Mo^{neutral}$
wenden	'wend'			5	6%	2	1%			1	1%	8	1%	$Mo^{neutral}$
sailen	'sail'	7	4%									7	1%	$Ma^{air/fluid}$
walken	'walk'					5	3%	1	2%			6	1%	Ma^{other}
drauen	'draw'			4	5%					1	1%	5	1%	$Mo^{neutral}$
alighten	'alight'			2	2%	2	1%					4	1%	Pa^{down}
shippen	'take ship'	4	2%									4	1%	Pa^{in}
departen	'depart'	1	1%			2	1%					3	1%	Pa^{away}
escapen	'escape'					1	1%	2	4%			3	1%	Ma^{flee}
faren	'go, travel'					3	2%					3	1%	$Mo^{neutral}$
lepen	'leap'					2	1%			1	1%	3	1%	Ma^{jump}

maken wei/ viage/ journei	'make one's way'	3	2 %									3	1%	$\mathrm{Mo}^{\mathrm{neutral}}$
priken	'gallop'					3	2%					3	1%	$\mathrm{Ma}^{\mathrm{speed}}$
scapen	'escape'	1	1%			1	1%			1	1%	3	1%	$\mathrm{Ma}^{\mathrm{flee}}$
springen	'spring'					2	1%	1	2%			3	1%	$\mathrm{Ma}^{\mathrm{jump}}$
climben	'climb'							2	4%			2		$\mathrm{Ma}^{\mathrm{other}}$
entren	'enter'					1	1%	1	2%			2		$\mathrm{Pa}^{\mathrm{in}}$
fleten	'float'					2	1%					2		$\mathrm{Ma}^{\mathrm{air/fluid}}$
flien	'fly'			2	2%							2		$\mathrm{Ma}^{\mathrm{air/fluid}}$
folwen	'follow'	1	1%			1	1%					2		$\mathrm{Mo}^{\mathrm{neutral}}$
gliden	'glide'			1	1%	1	1%					2		$\mathrm{Ma}^{\mathrm{glide}}$
holden wei	'hold one's way'					2	1%					2		$\mathrm{Mo}^{\mathrm{neutral}}$
pressen	'press, hasten'					2	1%					2		$\mathrm{Ma}^{\mathrm{speed}}$
shulen	'shall'	1	1%			1	1%					2		non
sinken	'sink'	1	1%			1	1%					2		$\mathrm{Ma}^{\mathrm{fall}}$
speden	'speed'					1	1%	1	2%			2		$\mathrm{Ma}^{\mathrm{speed}}$
stalken	'stalk'					1	1%	1	2%			2		$\mathrm{Ma}^{\mathrm{other}}$
turnen	'turn'					2	1%					2		$\mathrm{Ma}^{\mathrm{roll}}$
waggen	'move, etc.'	1	1%							1	1%	2		$\mathrm{Ma}^{\mathrm{other}}$
ben	'be'	1	1%									1		non
bresten	'burst'					1	1%					1		non
breken	'break'	1	1%									1		non
dressen	'go, move'							1	2%			1		$\mathrm{Mo}^{\mathrm{neutral}}$
hasten	'hasten'					1	1%					1		$\mathrm{Ma}^{\mathrm{speed}}$
hien	'hurry, hasten'					1	1%					1		$\mathrm{Ma}^{\mathrm{speed}}$

(continued)

TABLE 6.4. Continued

		GREGCHRON		STEDMUND		KNIGHT		MILLER		REEVE		TOTAL	
hosteien	'wage war'	1	1%									1	non
(bi)leven	'leave'			1	1%							1	Pa^{away}
longen	imp. 'desire'			1	1%							1	non
mouen	'may'									1	1%	1	non
parten	'part'	1	1%									1	Pa^{away}
purposen	'plan to, intend to'	1	1%									1	non
remeven	'go away'	1	1%									1	Pa^{away}
returnen	'return'	1	1%									1	Pa^{back}
rollen	'roll'					1	1%					1	Ma^{roll}
rouen	'row'	1	1%									1	$Ma^{air/fluid}$
stelen	'steal'							1	2%			1	Ma^{other}
steppen	'step'									1	1%	1	Ma^{other}
stiren	'stir'							1	2%			1	$Mo^{neutral}$
stomblen	'stumble'					1	1%					1	$Ma^{slow/impeded}$
taken REFL	'take oneself'	1	1%									1	$Mo^{neutral}$
thresten	'force one's way'					1	1%					1	Ma^{other}
treden	'tread'					1	1%					1	Ma^{other}
waden	'go'			1	1%							1	$Mo^{neutral}$
walwen	'wallow'									1	1%	1	Ma^{other}
will	'will'					1	1%					1	non
		200		86		150		52		76		564	

mainly due to the very high frequency of *comen*, *gon* and the preterite *went*,[133] which overall account for more than half of the verbs in the entire motion descriptions (see Table 6.4 below, and note how OE *faran/fēran*, which had accounted for more than a third of all motion descriptions in the Old English usage study, is now ousted by ME *gon/went*). Overall, MANNER verbs are the second most frequent group, with 31 per cent. Only in the *Knight's Tale* do they surpass the group of NEUTRAL MOTION verbs.

Before we move on to a more detailed discussion of the results, I want to briefly mention the verbs that were found in the usage study, but not in the inventory. First of all, there are the phraseologisms *maken* DET *wei/viage/journei* 'make DET way/voyage/journey', *nimen/taken* DET *wei/viage/journei* 'take DET way/voyage/journey', and *holden* DET *wei* 'hold DET way' as in (198a–c).[134] These mostly occur in the excerpt from *Gregory's Chronicle*, in which they constitute 4.5 per cent of the motion expressions. These phraseologisms were grouped with the NEUTRAL MOTION verbs (see Table 6.4).

(198) a. And on the morowe he **toke hys waye *unto Calys*** (GregChron 112.379)
'And in the morning he took his way to Calais.'

b. Ande that same yere the Erle of Northehomerlonde **made a viage *in-to Scotlonde*** (GregChron 179.1194)
'And that same year, the Earl of Northumberland made a voyage into Scotland.'

c. And that same day the Erle of Saulysbury **toke hys jornaye *towarde Fraunce*** the secunde tyme (GregChron 162.807)
'And that same day the Earl of Salesbury took his journey toward France the second time.'

Furthermore, there were a few non-MOTION verbs not found in the inventory study: The modal verb *shulen* 'shall' occurred twice in combination with a directional adverbial (*into Fraunce* and *thider* in (199a–b)), thus yielding a reading of modally broken motion (cf. the discussion in 6.2.5.4).

(199) a. And at the Parlyment be-fore hyt was ordaynyde that the Duke of Yorke **shulde *in to Fraunce*** (GregChron 178.1165)
'And at the Parliament before, it was ordained that the Duke of York should [i.e. should go] to France.'

b. But ***thider*** he **shal** by force, and there abyde (Knight 2554)
'But thither he shall [i.e. shall go] by force, and stay there.'

[133] Since *went*, the preterite of *wenden*, is gradually replacing *eode* as the suppletive preterite of *gon* in Middle English, the preterite *went* is listed separately from the verbs *wenden* and *gan* in Table 6.4. Among the five texts analysed here, forms of *eode* only occur five times in the *Life of St Edmund* (the oldest of the five texts) and once in *Gregory's Chronicle*.

[134] DET is for 'determiner', since this is not necessarily always a possessive determiner, but can also be *a* or *the*.

A similar modally broken motion reading is the result of the combination of the impersonal verb *longen* 'to desire, long' with the directional *þuder* 'thither' in (200) (cf. also *MED* s.v. *longen*, sense 1c '~ *to (into)*, long to go to (a place)'). These verbs, just like *ben* 'to be' in combination with the *toward*-PP in (201), clearly receive their motion reading only through the construction in which they are used and are thus classified as non-MOTION verbs.

(200) Þis Maide wende softe aȝen : hire rug smert ful sore. he bi-ȝat so luyte þo: þat **hire** ne **longuede** ***þudere*** nam-more (St Edmund 434)
'This girl went back quietly, her back hurt very badly. Thus she (had) gained so little that she did not desire [i.e. desire to go] thither anymore.'

(201) and toke iiij c Dolfynnys mayne that **were** ***towarde Orlyaunce*** for to have brokyn the sege (GregChron 163.840)
'And [Fastolf and companions] took four hundred of the Dauphin's troops that were [i.e. were bound/going] toward Orléans to break the siege.'

Finally, there are the verbs *hosteien* 'to wage war, make a military expedition, carry out plundering raids' (202) and *thresten* 'to crowd, throng; press forward, force one's way, push' (203).

(202) And thenne sone aftyr the kynge and hys mayne **ostyde** ***from thens*** xxj dayes ***thorowe the realme of Fraunce towarde Caleys.*** And the Franynysche men hyrde telle of his comyng that way (GregChron 111.360)
'And then soon afterwards, the king and his troops *hosteied* from thence [Harfleur] twenty-one days through the realm of France toward Calais. And the Frenchmen heard tell of his coming that way.'

(203) He ***thurgh the thikkeste of the throng*** gan **threste** (Knight 2612)
'He pushed through the thickest of the crowd.'

ME *thresten* was classified as MANNER verb (Ma^{other}), and ME *hosteien* as a non-MOTION verb. Note that *hosteien* is a loan from Anglo-Norman which is attested in an IMC in Middle English, but which is not used to express MOTION in the donor language (cf. *AND* s.v. *osteier* 'to fight, wage war', *TL* s.v. *ostoiier* 'Krieg führen', and *DMF* s.v. *ostoyer* 'guerroyer, faire la guerre, se mettre en campagne').[135] It is thus yet another example of how Middle English speakers borrow non-MOTION verbs from French and use them in a motion reading (see, e.g., 6.2.3.2, as well as the French loans among the non-MOTION verbs discussed in 6.2.5).

After this brief presentation of the additional verbs found in the usage study—phraseologisms like *taken way* or *maken voyage*, the MANNER verb

[135] Tobler–Lommatzsch also notes the figurative sense 'sich fort, sich davon machen' ('to make off'), but the only two attestations given are in combination with *(s'en) aller* 'to go', e.g. *Puis s'en est alé ostoier* 'And then he made off', literally 'And then he went waging war'. In this sense, the motion meaning therefore hinges on the combination with *aller* 'to go'.

thresten 'to force one's way' and the non-MOTION verbs *ben* 'to be', *shulen* 'shall', *longen* 'to desire', and *hosteien* 'to wage war'—the next sections will discuss the relative frequency of MANNER verbs as opposed to neutral MOTION verbs with regard to stylistic issues (6.3.2) and take a closer look at the kinds of PATH verbs found in the usage study (6.3.3).

6.3.2 Discussion: MANNER verbs, NEUTRAL MOTION verbs, and style

In the Old English usage study (5.2), we have observed that the frequency of MANNER verbs compared to the NEUTRAL MOTION verbs *faran/fēran* and *cuman* may be linked to issues of genre, topic, and style: The rather repetitive, formulaic *Anglo-Saxon Chronicles* only have MANNER verbs in 5 to 13 per cent of the motion descriptions and mainly use *faran/fēran* and *cuman* instead. The *Saints' Lives* and the *Voyages of Ohthere and Wulfstan*, by contrast, have more MANNER verbs (17 and 50 per cent, see 5.2.1). Moreover, in the more narrative and detailed passages of the *Anglo-Saxon Chronicles*, the proportion of MANNER verbs increases at the expense of the NEUTRAL MOTION verbs (see 5.2.2). Also for Present-Day English texts, Cappelle (2012: 187) shows that the frequency of MANNER verbs depends on genre and style: MANNER verbs are more frequent in fictional texts as opposed to biographical ones (see 2.1.2 above).

In the Middle English texts, the use of motion verbs is also far from homogeneous: The texts differ particularly in their respective proportion of *comen* and *gon/went* and of MANNER verbs: *Comen* and *gon/went*, which usually account for between 60 and 64 per cent of all motion descriptions, drop to only 33 per cent in the *Knight's Tale*. MANNER verbs, on the other hand, range between 54 per cent of the motion descriptions in the *Knight's Tale* and 16 per cent in *St Edmund*.[136] In the *Miller's* and the *Reeve's Tale* they account for around a third and in *Gregory's Chronicle* for 20 per cent of the motion descriptions.

Let us first briefly compare the three *Canterbury Tales* with each other. As mentioned in 4.2.2.6, the *Knight's Tale* is rhetorically more elaborate (as befits its narrator, the Knight) than the 'cherles tales' of the Miller and the Reeve, whose style has been described as 'low' (Cooper 1996: 105). The *Miller's* and the *Reeve's Tale* pattern with each other in that both have MANNER verbs in around a third of the motion descriptions, and around 60 per cent *gon* and *comen*. The *Knight's Tale*, by contrast, has 54 per cent MANNER verbs and only 33 per cent *gon* and *comen*. This higher proportion of MANNER verbs in the *Knight's Tale* may be an effect of its more elaborate, varied style, while the simpler fabliaux, despite their large variety of MANNER verbs (see below), rely more frequently on *comen* and *gon*.

---

[136] The proportion (16 per cent) of MANNER verbs in *St Edmund* is very similar to the one found in the Old English study in Ælfric's *Lives of Saints* (17 per cent). The fact that *St Edmund* is by far the earliest of the Middle English texts studied here may play a role in this.

In *Gregory's Chronicle*, the incidence of MANNER verbs is even lower than in the two fabliaux, with 20 per cent, while the proportion of *comen* and *gon/went* stays the same. Thus, like we have seen in the study on Old English, the genre of chronicles seems to discourage MANNER verbs. Furthermore, note that only eight different MANNER verb types account for the 39 MANNER verb tokens in *Gregory's Chronicle*, and that the most frequent types among them are *riden* (24 occurrences) and *sailen* (7 occurrences).[137] Thus, the MANNER verbs that are used in the *Chronicle* are also not very expressive, and are even rather objective. This is different in the three *Canterbury Tales* and in *St Edmund*: First of all, these texts feature MANNER verbs that would probably be inadequate for a chronicle, because they convey an immediacy and a sense of graphic detail that would be uncommon for this genre (such as *crepen* 'to creep', *rennen* 'run', *priken* 'gallop', *thresten* 'force one's way', or *walwen* 'wallow'). Second, the MANNER verbs in the tales and in *St Edmund* are much more varied: In the *Miller's Tale*, 11 different types account for the 17 MANNER verb tokens, in the *Reeve's Tale* nine types for 22 tokens, in the *Knight's Tale* 25 types for the 81 tokens, and in *St Edmund* seven types for 14 tokens. Thus, although a real comparison of the type–token ratios for the MANNER verbs is not sensible due to the unequal number of tokens, it is nevertheless worth noting that *St Edmund* and each of the three *Canterbury Tales* has a higher type–token ratio than the *Chronicle*, no matter whether their token number is higher or lower. We can conclude that the kinds of verbs that are chosen to talk about motion seem to strongly depend on the text type: Chronicles use fewer and less varied MANNER verbs and more NEUTRAL MOTION verbs,[138] while in the more narrative texts, the incidence and variety of MANNER verbs tends to increase.

If we compare the proportions of MANNER verbs in the Middle English texts with those reported for Present-Day English texts by Slobin (2004a: 231) and Cappelle (2012: 185), the picture is rather similar. Slobin finds MANNER verbs in 40 per cent of the motion descriptions in novels, Cappelle in 34 per cent in fictional and 20 per cent in biographical texts.[139] Among the Middle English texts analysed in the present study, the *Canterbury Tales* probably come closest to the modern genre of novel and have a comparable proportion of MANNER verbs (43 per cent on average, depending on the respective narrator's degree of sophistication). Since it relates a saint's life, *St Edmund* can be perhaps be compared to modern biographical texts; its 16 per cent MANNER verbs are similar to

[137] The other MANNER verbs occurring in the *Chronicle* are *fallen*, *flen*, *scapen*, *sinken*, *rouen*, and *waggen*, each only used once or twice, see Table 6.4.
[138] Also note that almost all of the periphrastic expressions—*taken/maken* DET *wei/voiage/journei*—occur in the *Chronicle*.
[139] Though note the different method used in Cappelle (2012) (searching for specific verbs, rather than analysing the texts for motion descriptions, see footnote 6), which entails that the proportion of MANNER verbs probably seems lower than it actually is, since several verbs will not have been searched for.

the 20 per cent in this text type found by Cappelle. The frequency of MANNER verb use in Middle English thus appears to be relatively the same as in Present-Day English.

6.3.3 Discussion: PATH verbs

As shown in Table 6.3 above, PATH verbs as a group are, once more, not highly frequent, accounting for only six per cent (36) of the motion descriptions in the Middle English texts analysed for the usage study. This is a little more than in the Old English texts, where only four per cent of the motion descriptions employed a PATH verb (5.2.3).

(204) PATH verbs used in the Middle English texts:
londen (11), *risen* (8), *alighten* (4), *shippen* (4), *departen* (3), *entren* (2), *(bi)leven* (1), *parten* (1), *remeven* (1), *returnen* (1)

Most of the PATH verbs used in the Old English texts had turned out to be not very prototypical ones (OE *forlǣtan, lendan, gewītan*). Such non-prototypical ones, which convey more information than pure PATH, are also among the Middle English PATH verbs, especially among the more frequently used ones—*londen, risen, shippen,* and also *alighten,* cf. (204)—as shall be briefly shown here. Particularly *londen* and *shippen* would perhaps better be termed 'PATH+GOAL verbs': In the texts analysed for the usage study, ME *londen* is used only where the GOAL is 'land' and the FIGURE is coming via water, as in (205) (also note that the verb's form changes from OE *lendan* (< **land-jan*) to ME *londen*, i.e. the GOAL (ME *lond*) stays transparent in the verb).

(205) he **schyppyd** at Sondewyche, and he **londyd** in Normandy (GregChron 162.807)
'He took ship at Sandwich, and he landed in Normandy.'

ME *shippen* 'to enter ship, take ship', as in (205) above, is only found to describe motion events into somewhere in which the GOAL is a ship; it is therefore not a pure PATH verb for motion 'into'. ME *alighten* was only used in the texts to describe situations of alighting from horses, a use that was already dominant in Old English (cf. 5.1.3.2), and *risen* is used for situations where change of posture is involved (like in Old English; see 5.1.3.5):

(206) Thenne **rose** he *owte of hys chayre* (GregChron 166.916)
'Then he rose out of his chair.'

Yet, compared to Old English, there are more 'good' PATH verbs in the Middle English texts, and, crucially, these are all borrowed from French. The $\text{PATH}^{\text{away}}$

verbs *departen, parten,* and *remeven,* such as in (207a–d), can be considered to be pure PATH verbs.

(207) a. And er that we **departen** ***from this place*** (Knight 3070)
'And before we depart from this place.'

b. And whanne that they hadde etyn they **departyd** (GregChron 110.352)
'And when they had eaten, they departed.'

c. And soo they **partyde** (GregChron 176.1106)
'And so they parted.'

d. And in Syn Johnys day in the Crystysmasse weke the kynge **remevyd** ***towarde Roone*** (GregChron 173.105)
'And on Saint John's day, during the Christmas week, the king removed toward Rouen.'

Also the PATH^{in} verb *entren* (208a–b) and the PATH^{back} verb *returnen* (209) can be considered pure PATH verbs:

(208) a. on that ilke nyght / That we been **entred** ***into shippes bord*** (Miller 3070)
'In this very night, when we will have entered into the board of the ships.'

b. And westward, ***thurgh the gates under marte,*** / Arcite, and eek the hondred of his parte, / With baner reed is **entred** right anon (Knight 2581–3)
'And on the west side, through the gates under Mars, Arcite, together with the hundred of his party, has entered right soon with a red banner.'

(209) And thenne the kynge **retornyd** ***ayenne into Inglonde*** (GregChron 114.448)
'And then the king returned again/back into England.'

Chapter 9 will discuss a selection of French-borrowed PATH verbs in more detail.

6.3.4 Complementation patterns

The inventory study has already shown that most Middle English motion verbs *can* be used with PATH satellites: 86 per cent of the MANNER verbs, 79 per cent of the PATH verbs, 94 per cent of the neutral and 96 per cent of the non-MOTION verbs (6.2.6). The usage study demonstrates that they also *are* used with PATH satellites most of the time, as shown in Table 6.5.

Motion verbs overall are used with satellites in 76 per cent of the cases; complementation with ground NPs is rather marginal (2 per cent). MANNER verbs are combined with satellites in 72 per cent of the cases, such as in (210a–b). This meets the expectations for a satellite-framing, manner-conflating language (cf. section 2.1).

TABLE 6.5. Complementation of the different motion verb groups in Middle English texts.

	AT LEAST 1 SATELLITE		NPGROUND		BARE VERB		TOTAL
MANNER	72%	(125)	1%	(2)	27%	(46)	(173)
PATH	81%	(29)	3%	(1)	17%	(6)	(36)
come/go	76%	(231)	3%	(8)	21%	(63)	(302)
Other neutral	81%	(35)	5%	(2)	14%	(6)	(43)
Non-MOTION	100%	(10)	—	—	—	—	(10)
Total	76%	(430)	2%	(13)	21%	(121)	(564)

(210) a. But faire and wel she **creep** ***in to the clerk*** (Reeve 4226)
'But, fair and well, she crept in to the student.'

b. Ful soft ***out*** at the dore he gan to **stele** (Miller 3786)
'Very quietly, he stole out at the door.'[140]

Note that (210a–b) are instances of MANNER verbs used for motion events that are represented as boundary-crossing (*in, out*). This kind of use is equally expected for MANNER-conflating, satellite-framing languages (see section 2.1.4.1). Furthermore, among the attestations with PATH satellites, there are several in which a whole series of PATH expressions is stacked unto one and the same verb, another characteristic of this type of language (see 2.1.4.2). An example is (211), with the GOAL *unto London*, the SOURCE *fro the Downys*, and two windows on MEDIUM PATH *thoroughe Kent* and *ovyr Scheters Hylle*; see also footnote 155 for similar examples from the *Knight's Tale*.

(211) wherefore the vyntage **come** by londe ynne cartys ***unto London fro the Downys, thoroughe Kent*** and ***ovyr Scheters Hylle*** (GregChron 178.1151)
'[Due to the frost, transport by ship was not possible], which is why the wine came to London by land, in carts, from the Downs through Kent and over Shooters Hill.'

It should be noted that in case of doubt whether a prepositional phrase was locative or directional, such as in (212a–b), I classified it as locative, rather than as a PATH satellite.

(212) a. I am, thow woost, yet of thy [Diana's] compaignye, / A mayde, and love huntynge and venerye, / And for to **walken** ***in the wodes wilde***, / And noght to ben a wyf and be with childe. (Knight 2307–10)

[140] Note that ME *ginnen (for) to*, literally 'begin to', is not always used to mark the beginning of an action, but frequently also 'as a weak auxiliary [. . .] denoting actions or events as occurring (rather than as beginning to occur)' (MED s.v. *ginnen*, sense 3.b). This is why I sometimes left it untranslated, particularly in contexts such as (210b), where the event 'stealing out at the door' is of such a short duration that it would be odd to talk about its 'beginning'.

'I am still one of your [Diana's] company, you know, a maid, and I love hunting and shooting, and walking in the wild forests, and not being a wife and being pregnant.'

b. He **rolleth *under foot*** as dooth a bal (Knight 2614)
'[of knights being tossed off their horses in the tournament:] He rolls under foot as a ball does.'

This means that I interpreted *in the wodes wilde* in (212a) as the area where the FIGURE walks, not as the GOAL to which she walks, and the same for rolling *under foot* in (212b). These are therefore counted together with the bare verb uses, such as (213a–b).

(213) a. His officers with swifte feet they **renne** / And **ryde** anon at his comandement (Knight 2010)
'His officers, at his command, they run with swift feet and ride immediately.'

b. And by-twyne ix and x of the belle þer **come** certayne men of the Byschoppys of Wynchester (GregChron 159.729)
'And between 9 and 10 o'clock, there came certain men of the Bishop of Winchester's.'

With the higher frequency of descriptions such as (212) with local, not directional adverbials, or with no explicit PATH information present at all, as in (213), the bare verb use increases from 7 per cent in the Old English texts (section 5.2.4) to 21 per cent in the Middle English texts.

In the realization of satellites, there is no significant change: In the clauses with satellites in Middle English, 71 per cent contain at least one prepositional phrase and 38 per cent at least one adverb (Table 6.6), as compared to 74 per cent and 43 per cent in Old English (Table 5.5 in 5.2.4) (again, the percentages add up to more than 100 because often, more than one PATH satellite is present). Note that PATH verbs are much less frequently combined with PATH adverbs than the other verb groups—only in 17 per cent of the cases (all instances of the verb *(a)risen* combined with *up*), as opposed to around 40 per cent on average for the other verb groups. Since different from prepositions (as in *departen from this place* in (207a)), adverbs are not needed to introduce a GROUND NP, this can be interpreted as a certain reluctance to 'double' one and the same PATH information in a motion description (there was no *departen away* in the texts, for instance).

To sum up the results, Middle English just like Old English meets the expectations for a MANNER-conflating and satellite-framing language, not only in the frequent combination of MANNER verbs with PATH satellites (72 per cent), but also in the possibility of combining multiple PATH expressions with a single verb (see (211)), and in the use of MANNER verbs for boundary crossing situations, such as (210a–b).

TABLE 6.6. Types of satellites in the Middle English motion descriptions with PATH satellite according to verb type.

	AT LEAST 1 PP		AT LEAST 1 ADVERB		TOTAL WITH AT LEAST 1 PATH SATELLITE
MANNER	67%	(84)	47%	(59)	125
PATH	83%	(24)	17%	(5)	29
come/go	71%	(163)	39%	(89)	231
Other neutral	83%	(29)	23%	(8)	35
Non-MOTION	60%	(6)	40%	(4)	10
Total	71%	(306)	38%	(165)	430

6.3.5 Middle English usage: conclusion

To conclude, this study of how MOTION is typically expressed in Middle English has essentially shown the following:

(a) PATH verbs are not very frequently used (6 per cent overall); pure PATH verbs are borrowed and account for only one per cent of the 564 motion descriptions.

(b) MANNER verbs are more frequent than PATH verbs (31 per cent overall), but depend on style (ranging between 16 and 54 per cent in the individual texts): They appear to be comparatively less frequent than NEUTRAL MOTION verbs such as *gon* and *comen* in less narrative, more objective texts such as *Gregory's Chronicle*, while their frequency (both type and token) rises in fictional narrative such as the *Canterbury Tales*. In these, their frequency is comparable to the one in Present-Day English narrative fiction as found by Slobin (2004a: 231) and Cappelle (2012: 185).

(c) All motion verbs mostly combine with PATH satellites (76 per cent); the most frequent realization of PATH satellite is a prepositional phrase (71 per cent), but adverbs also occur in 38 per cent of the motion descriptions with satellite.

After having thus established that Middle English can indeed be considered a MANNER-salient, satellite-framing language, in which the borrowed PATH verbs constitute a semantic novelty, we will now turn to the donor languages.

CHAPTER 7

Latin and medieval French in the motion verb typology

SO FAR, WE HAVE seen that Old as well as Middle English have only a very limited number of (native) pure PATH verbs (see the inventory studies in sections 5.1 and 6.2), and, more importantly, that these PATH verbs are used rather infrequently (see the usage studies in 5.2 and 6.3). To talk about MOTION events, Old and Middle English typically use either a NEUTRAL MOTION verb (such as OE *cuman* 'come', *faran* 'go, travel', ME *comen* 'come', *gon* 'go') or a MANNER verb (such as OE *rīdan* 'ride', *siglan* 'sail'), and express PATH in satellites (prepositional phrases, adverbs, prefixes). This means that the borrowed PATH verbs that come into Middle English are indeed semantically unusual. Their early use will be the topic of chapter 9. Before that, we will have a short look at which system they come from, i.e. how motion is talked about in medieval French and Latin. The present section will summarize the findings of previous studies on motion expression in Latin and Old French (7.1), and complement them with a case study on motion expression in the Old French *Aucassin et Nicolette* (7.2), i.e. another usage study.

7.1 *Previous studies*

Diachronic studies in the framework of the motion verb typology are still scarce (but see Xu 2006, on Chinese; Fanego 2012, on English). The only area in which some substantial diachronic research has been done is the development from Latin to the Romance languages. This is hardly surprising since there is a lot of historical evidence to work with and since the typological difference between satellite-framed Latin and verb-framed Romance[141] had been pointed out by Talmy as early as 1985 (62, 68–72, 75, etc.).

[141] See, however, Kopecka (2004) and Kopecka (2006) on other, minor patterns in French, and Schwarze (1985), Koch (2000: 108–9), Iacobini & Masini (2006), Iacobini (2009), and the

The diachronic studies on Latin and Romance that I am aware of are Schøsler (2008) and Stolova (2008), both on Romance languages in general, Kopecka (2004: 200–12; 2009; 2013) on the development of French (as well as Iacobini & Masini 2006, Iacobini & Fagard 2011, and Iacobini 2012 on Italian). The major findings of these studies relevant to the present one will be summarized and discussed in the following.

7.1.1 From Latin to Romance

Schøsler (2008) focuses on verbs and traces the development of the typical expression of the semantic elements of PATH and MANNER from Classical Latin to Modern French by way of analysing motion expressions in one short text per period. In Classical Latin it is possible for PATH to be expressed in prefixes of verbs, in prepositional phrases, or both, such as in (214).

(214)	**e**	**castris**	**Helvetiorum**	egressi
	'out-of	the castles	of the Helvetians	outstepped'

(*Bellum Gallicum* I, Schøsler 2008: 116)

The typical case for a Latin motion verb is to have a prefix (the satellite): In Schøsler's case study of an extract of *Bellum Gallicum*, 75 per cent of the 73 occurrences of motion verbs have a PATH prefix (such as ***transire***), and only a quarter are simple, non-prefixed verbs such as *ire*. Hence, structurally, Classical Latin can be said to follow the satellite-framed pattern. Schøsler also finds that the verb roots themselves carry MANNER information, such as in ***accurrere*** 'to to-run', in around a fifth of the cases (21 per cent) (ibid: 127).

For Late Latin (studied on the basis of *Peregrinatio Aetheriae* (*c*.400) and Gregory of Tours (*c*.600)), Schøsler discovers a similar situation with respect to satellite-framing: Most of the verbs carry PATH prefixes (52 out of 67 and 59 out of 77, respectively). MANNER-conflation, by contrast, has decreased greatly in these later texts, where she reports a token frequency of 6.8 per cent for MANNER verbs (ibid: 127). However, all of these seem to be in the form of participles (ibid: 117), such as *ambulantes* in (215), where the main verb is *pervenimus* 'we to-came'. Hence, these MANNER verbs are adverbials and not the main verb in the clause.

(215) Interea **ambulantes** pervenimus ad quendam locum (Schøsler 2008: 117)
'In the meantime, **walking**, we to-came to a certain place.'

Thus, summing up the diagnosis for Latin suggested by Schøsler, it seems that syntactically we are dealing with a satellite-framed language, since the typical case is for PATH to be expressed in a satellite, usually a prefix, but sometimes

contributions in Cini (2008) on PATH-satellites in Italian. See also Hijazo-Gascón & Ibarretxe-Antuñano (2013) for French, Italian, and Spanish.

a prepositional phrase (2008: 117). With respect to conflation though, i.e. considering what is expressed by the verb root, the language seems to develop away from the MANNER-conflating type, in that it makes increasingly less use of MANNER verbs as the main verb, a tendency which, in turn, is characteristic of verb-framed languages.

Schøsler (2008) does not comment on the frequency of verbs carrying PATH prefixes in the stages after Late Latin, but Stolova (2008: 259–60) suggests that very few of these prefixed forms survive in the Romance languages. She points out that those that do survive (*intrare, ascendere, descendere, exire*) are the ones that have become non-transparent, i.e. their prefix is no longer segmentable as a prefix but has become fused with the root, resulting in what appears as a monomorphemic verb, such as Spanish *entrar, ascender, descender,*[142] or Italian *uscire* (see also Iacobini 2012: 364).

The loss of the other prefixed verbs, according to Stolova (2008: 260), was compensated by newly formed verbs that mostly follow a specific word formation pattern: They are created from nouns referring to landscape, such as *mons* 'mountain' (e.g. F *monter* 'to mount'), *podium* 'elevated place' (e.g. Catalan *pujar* 'to mount'), *collis* 'hill' (e.g. Old Sardinian *koddai* 'to mount'), *vallis* 'valley' (e.g. F *dévaler, avaler* 'to descend'), and *ripa* 'shore' (e.g. F *arriver* 'to arrive'), from adjectives like *bassus* 'low' (e.g. Sp. *bajar* 'to descend'), *altus* 'high' (e.g. F *hausser* 'to mount'), or from directional prepositions like *abante* 'forward, ahead' (e.g. F *avancer*, Sp. *avanzar* 'to advance'). Stolova notes that the Romance tendency to derive motion verbs from 'landscape nouns' is not restricted to Latin etyma, but is also found with Slavic roots for 'valley' and 'mountain' in Romanian.

In terms of Talmy's typology, these new verbs can thus be said to, at least etymologically, express the GROUND, i.e. the place with respect to which the FIGURE moves. However, as they are also used with GROUNDS different from the ones expressed in the verb root, such as in (216) and (217), where the GROUND is not a mountain, but a horse and the heavens respectively, one can assume that the semantic element expressed by them is the PATH (in this case: 'up') rather than the GROUND.

(216) Lui quens **mounta** un destrer abrivé (*Boeve* 138, from AND s.v. *munter*)
'The count got on a swift horse.'

(217) quant il **mounta** en cyel (*Cron Lond* 38.12, from AND s.v. *munter*)
'When he ascended to heaven.'

Iacobini (2012: 364) demonstrates that some former MANNER verbs also developed into PATH verbs, such as Latin *salire* 'to jump', which becomes Port. *sair*,

[142] Although the existence of both the reflexes of Latin *ascendere* and *descendere* would suggest a certain transparency, the non-prefixed form, Latin *scandere/scendere* 'to climb, to mount up', does not survive in any Romance language apart from Italian *scendere*, which, however, has the same meaning as *discendere* 'descend' (Stolova 2008: 259). On the lack of transparency of *ascendre* and *descendre*, see also Foulet (1946: 3–4).

Sp. *salir* 'to exit', It. *salire* 'to go/come up'. The development of all these new PATH verbs, together with the retention of the non-transparent prefixed verbs, which have developed into PATH verbs as well, surely is an important factor in the overall change of lexicalization patterns, from MANNER being expressed in the verb root to PATH being expressed there.

Another factor according to Iacobini (2012: 379) is the loss of the Classical Latin means of distinguishing between location and direction through case (e.g. *in* + ablative: 'location' vs. *in* + accusative: 'direction') and through distinct prepositions (e.g. *apud* 'at' vs. *ad* 'to') and adverbs (e.g. *foris* 'on the outside' vs. *foras* 'to the outside'). With the loss of these means, he argues, there was no straightforward way of expressing boundary-crossing motion any longer, except by using PATH verbs.[143]

7.1.2 Medieval French

For Old French, Schøsler's (2008: 119) analysis of the *Roman de la Charrette* shows that, compared to Latin, MANNER verbs are increasingly used as participles in verbal periphrases, like *aller flottant* 'go floating', which again is a means of introducing MANNER information that is typical for verb-framed languages (MANNER outside the main verb). There is, however, also a strong increase in the use of adverbial PATH particles such as *aval* 'down' or *fors* 'out' (Schøsler 2008: 120). These PATH satellites combine with PATH verbs (a 'double-framing' pattern in Croft et al.'s terminology (2010: 208), see 2.1.3) in examples such as (218), but also with MANNER verbs and with neutral verbs (a satellite-framing pattern).

(218) Lors **descendi** li rois **aval** (*Charrette* 3319, Schøsler 2008: 119)
'Then descended the king down.'

The PATH particles have been linked to a possible Germanic contact influence, e.g. Schøsler (2008: 207) says they seem particularly frequent in Anglo-Norman, which is of course in heavy contact with Middle English (see section 6.1).

A short excursus on these PATH particles might be in order here, because it has been a long-standing matter of debate in Romance linguistics how to account for them (e.g. Jaberg 1939; Foulet 1946; Kramer 1981; Gsell 1982; Rohlfs 1983: 46–9; Mair 1984; Iacobini & Masini 2006: 164–6). They are not only found in Old French, but also in Present-Day Rhaeto-Romance and Italian (e.g. *andare giù* 'go down'), particularly in the northern and central varieties. The debate centers around the question whether they constitute a phenomenon of

[143] The Old French prepositions and particles, e.g. *jus* 'down', can also have both a locative and a directional reading (much like Present-Day English ones such as *under* or *over*), as pointed out, for instance, by Burnett & Tremblay (2009: 38), who argue that this ambiguity is always resolved by the context.

language contact—due to a Germanic adstrate, as captured in Ascoli's formula of *materia romanza e spirito tedesco*—or not, or only to some degree.

The opponents of the 'Germanic hypothesis' (e.g. Mair 1984, Iacobini & Masini 2006, Iacobini 2009) argue that the combination of verb and spatial adverb is a widespread feature of colloquial speech in general, a symptom of its tendency towards expressivity and pleonastic expression (Mair 1984: 418). In other words, the spatial adverbs are simply a diaphasic phenomenon. These scholars point out that these spatial adverbs are equally found in Vulgar Latin, i.e. colloquial Latin, as depicted in Plautus' comedies and in the works of Petronius (as in (219a), and even without a verb in (219b)). Directional particles also occur in the Vulgate, whose language is close to the popular language in some aspects as well (Mair 1984: 421–2).

(219) a. quisquis servus sine dominico iussu **foras** exierit (*Sat* 28,7, Mair 1984: 421)
'Any servant who exits out without the order of the master.'

b. Aqua[m] **foras**, vinum **intro** (*Sat* 52,7, Mair 1984: 422)
'Out with the water, in with the wine!'

These spatial particles would thus represent a constant in colloquial language from Vulgar Latin to the Romance languages. The fact that they are less frequent in Present-Day French compared to Present-Day Italian is explained by Mair (1984) with the different normative history of the two languages, with standardization and normative influence towards synthetic Latin models having a stronger effect in French than in Italian (on Latinization as a cause for subsequent loss of the spatial adverbs in French, see also Foulet 1946: 65). Still, many scholars also advocate an additional influence of Germanic contact languages on the development of spatial adverbs, acting 'as a catalyzer of processes which were already under way in Vulgar Latin' (Iacobini & Masini 2006: 165; see, e.g., also Gsell 1982, and Mair 1984 on Ladin).

No matter what the ultimate cause for the French PATH adverbs is, for the present study, the tendency of medieval French to combine PATH satellites with PATH verbs is relevant, because it entails that, syntactically, the PATH verbs very often occur in patterns that look exactly like the English ones. This makes it more likely that they would also be interpreted as being semantically like the English ones, i.e. as also carrying MANNER information in the verb, as discussed in section 8.3 below.

As regards MANNER information in the verb, Schøsler (2008: 127) notes that the token frequency of MANNER verbs increases from Latin (21.1 per cent in Classical, 6.8 per cent in Vulgar Latin) to Old French (23.6 per cent), to drop again in Modern French (10.8 per cent). However, even in Old French, their token frequency is never as high as in a prototypical satellite-framing, manner-conflating language such as Danish (40.2 per cent), which Schøsler uses for comparison. The present study confirms this: Among the Middle English texts analysed in the usage study (6.3), the *Knight's Tale* is probably closest in genre and subject matter (chivalric romance/epic) to the *Roman de la Charrette* analysed in Schøsler (2008) as exemplary for Old French. The proportion of MANNER verbs in the *Knight's Tale* is more than twice as high (54 per cent) as in the Old French *Roman de la Charrette.*

What should be borne in mind, though, is that these frequencies are given irrespective of the syntactic context in which the MANNER verbs occur. As Slobin and colleagues (e.g. Slobin 1997: 455) have pointed out, verb-framed languages tend to use MANNER verbs without directional complementation, and avoid them in boundary-crossing contexts (see 2.1.4.1–2). For a more thorough assessment of Old French, the syntactic contexts of MANNER verbs should thus be taken into account as well (see the case study in 7.2).

Kopecka (2009) sees Old French as a predominantly satellite-framing language, whose 'common strategy for conveying path information was to use verb prefixes [. . .] and verb particles' (2009: 416). With these prefixes and particles becoming less productive, the language shifted towards a verb-framing pattern (see also Kopecka 2006 and, on the productivity of Old French preverbs and particles, Dufresne et al. 2003). Kopecka (2013) presents a comparative study of 257 verb clauses referring to MOTION events from twelfth and fourteenth century fabliaux and their Present-Day French translations. She finds PATH verbs in 48.6 per cent of all motion clauses in the Old French fabliaux, followed by the neutral motion verbs *aller* 'to go' and *venir* 'to come' in 30 per cent of the cases, and pure MANNER verbs in only 10.1 per cent (2013: 167). In the Present-Day French translations of the fabliaux, the proportion of PATH verbs increases to 57.9 per cent, the proportion of MANNER verbs stays more or less the same (11.5 per cent), and *aller* and *venir* decrease to 10.3 per cent. (Kopecka 2013: 171). The use of PATH satellites[144] is shown to have been more widespread in Old French (where almost 40 per cent of all motion clauses, and a good 30 per cent of the ones with MANNER verbs, have a PATH satellite, ibid.: 169–70) than in Present-Day French (where it drops to a good 25 per cent, ibid.: 173).

What might be slightly problematic about these results, however, is the nature and frequency of Old French *en*, which developed from Latin *inde* 'from there'. Kopecka (2013: 180) points out its 'extensive use' in the fabliaux data and interprets this cautiously as 'it appears [. . .] that OF allocates a great deal of attention to the initial portion of the Path'. Yet on the basis of its original spatial meaning 'away, from there', OF *en* has developed the aspectualizing function of inchoative marker in which it is very frequent, including with motion verbs (cf., e.g., Buridant 2000: §328.2, Marchello-Nizia 2002: 215, and, on Middle French, Wunderli 1987: 202).[145] In view of the fact that it very often has aspectual rather than spatial meaning, an analysis that does not count *en* as PATH particle will

[144] Note, however, that Kopecka, like Talmy (see 2.1.3) does not classify prepositional phrases as satellites (2013: 166).

[145] The sheer frequency of *en* alone (in the eleventh century *Vie de Saint Alexis*, for instance, it combines with one out of eight verbs, according to Marchello-Nizia (2002: 214)) would suffice to cast doubt on its purely spatial semantics, because this would mean that in Old French the initial window

be likely to yield a different picture, with the proportion of motion clauses with PATH satellites probably diminished to a considerable degree.

7.2 *Case study: motion expression in* Aucassin et Nicolette

In order to obtain more information on the frequency of MANNER and PATH verbs in Old French, and also to have a look at the syntactic context of MANNER verbs, I conducted a case study of how intransitive MOTION is described in the prose parts of the *chantefable Aucassin et Nicolette* (anonymous, probably early thirteenth century, Picard) (ed. Suchier 1889). The text is a parodic narration of the lovers Aucassin and Nicolette, who have to go through various evasions from prisons, kidnapping by pirates, and journeys on ship and land until they are finally reunited again. As in 5.2 and 6.3, only the main verb in a clause was taken into account, i.e. in the (very infrequent) case of an adverbial participle such as in *li nes [. . .] ala tant par mer waucrant* 'the ship went so long over the see drifting', the main verb *aller* 'go' was counted, while the participle *waucrant* 'drifting' was not.[146]

7.2.1 Results: frequency of the different semantic groups

The verbs used in the 140 motion descriptions are shown in the following overview, in which the same criteria were applied that were used for Old and Middle English above:

Motion verbs in the prose parts of the Old French *Aucassin et Nicolette* (ed. Suchier)

1. **MANNER verbs: 17 per cent (24/140)**
   caïr 'fall' (2), *chevaucher* 'ride' (4), *courir* 'run' (2), *enbler* 'steal, steal away' (2), *fuir* 'flee' (2), *glisser* 'glide' (1), *nagier* 'row, swim' (2), *salir* 'jump' (1), *se lancer* 'throw oneself' (2), *sigler* 'sail' (1), *tourner* 'turn' (3), *voler* 'fly' (2)

2. **PATH verbs: 33 per cent (46/140)**
   arriver 'arrive' (4), *avaler* 'descend' (2), *descendre* 'descend' (3), *entrer* 'enter' (5), *issir* 'exit' (7), *lever* 'rise' (1), *monter* 'mount' (16), *(de)partir*[147] 'leave' (5), *passer* 'pass' (2), *repairier* 'return, go home' (< L *repatriare*) (1)

---

of a PATH (the SOURCE) would probably be more often expressed than the final window (the GOAL), whereas various studies have shown that it is usually the other way round (Kopecka 2004: 131, 160; Lakusta & Landau 2005; Stefanowitsch & Rohde 2004: 251; Verspoor & Dirven 2004: 85; see also Radden & Dirven 2007: 291; Regier & Zheng 2007).

[146] See, however, Schøsler (2008: 124–125) on the function of *aller* + participle as a progressive periphrasis. Since this occurred only twice in my data though, once with the MANNER-verb participle *waucrant* (see above) and once in *si ala vielant par le pais* 'and went fiddling through the country', the results would not change greatly had I included them.

[147] The only segmentable verbal prefix that was found was *de-*, occurring twice with *partir* as *departir*.

3. NEUTRAL MOTION **verbs: 50 per cent (70/140)**
aller 'go' (21), *errer* 'go, travel' (6), *traire* 'draw, go, betake oneself' (5), *venir* 'come' (38)

The result in terms of verb types is that the most neutral verbs *aller* 'go' and *venir* 'come' are most frequent, making up 42 per cent of all verbs, and together with the similarly neutral *traire*[148] and *errer*[149] accounting for half of all motion verbs. The second largest group are PATH verbs, which make up a third of all motion expressions (33 per cent). MANNER verbs are the smallest group, with 17 per cent (even if *errer* 'go, travel'[150] were put into the group of MANNER verbs, they would not exceed 22 per cent). Note that this comparatively low frequency of MANNER verbs can hardly be due to the topic of the text, on the contrary, because the story is full of adventurous detail. MANNER verbs in *Aucassin et Nicolette* might even be more frequent than they are in other texts—in the first hundred motion descriptions from Béroul's *Tristran et Iseut* (ed. Braet & Raynaud de Lage 1989), by comparison, there are only seven MANNER verbs.[151]

Compared to the texts analysed for Old and Middle English, the major difference appears to lie not so much in the frequency of MANNER verbs (but see 7.2.2 on their complementation) but mostly in the higher proportion of PATH verbs in *Aucassin et Nicolette* (33 per cent as opposed to 0–10 per cent in the English texts).

7.2.2 Results: complementation patterns

The second question concerns the syntactic contexts in which the MANNER verbs appear. Let us first have a look across all the verbs (MANNER, PATH, and NEUTRAL MOTION): 76 per cent of all verbs (106/140) are combined with at least one PATH expression. This is usually a prepositional phrase, as in (220a–b), sometimes with the rather general *a* (often amalgamated with the article: *au* < *a+le*, *as* < *a+les*), which mostly seems to serve as a link between GROUND-expression and verb (220a), sometimes with a more specific preposition such as *aval* 'down' (220b).

(220) a. et vint **au** palais (28.22)
'And came to the palace.'

b. si se laissa glacier **aval** le fossé (16.17)
'And let herself glide down the ditch.'

---

[148] Cf. the glossary in Suchier (1889: 115): *errer* 'wandern, gehen' ('to wander, go').
[149] Cf. the glossary in Suchier (1889: 130): *traire* 'sich begeben' ('to betake oneself').
[150] OF *errer*, according to *TL* (s.v. *errer* vb.) can mean 'irren, irregehen, sich verfehlen' ('to err, be mistaken'), but more frequently 'reisen, herumschweifen, fahren, gehen' ('travel, stray, move, go'). See also *AND* (s.v. *errer* 1.) 'to journey, travel; to wander'. Both Latin *errare* 'err, stray, wander' and *iterare* 'travel, journey' seem to play a role here. See also F *chevalier errant* 'knight-errant'.
[151] Schøsler's analysis of the motion verbs in the Old French *Chevalier de la Charrette*, by contrast, observed 23.6 per cent MANNER verbs (2008: 127). This higher proportion may be due to the inclusion of participles in '*aller* + MANNER-participle', as indicated in Schøsler (2008: 124–5).

Only eight per cent of all verbs (11/140) occur together with a PATH adverb as satellite, such as *contreval* 'down' (221a), *fors* 'out' (221b), or *ariere* 'back', *avant* 'forth', *deseure* 'on top', *outre* 'over'.

(221) a. si s'avala **contreval** el gardin (12.15)
'And descended down into the garden.'

b. Et Nicolete issi **fors** (38.22)
'And Nicolette exited out.'

En < L *inde* was not counted as a PATH adverb, for the reason that it very often does not have spatial meaning any longer (see the discussion at the end of 7.1.2). Had it been included, the proportion of verbs with PATH adverb would have doubled to 15 per cent.

The question of how prominent the Old French satellite-framing pattern is thus hinges crucially on the definition of satellite. In Talmy's definition, which excludes prepositional phrases (2000: 102, 222; see 2.1.3), it would be rather marginal, with eight per cent, not counting *en*. It would also be much less frequent than in satellite-framing Old and Middle English, where around a third of all motion clauses contain an adverb as satellite (see sections 5.2.4 and 6.3.4).[152] In the wider definition adopted in the present study, which includes prepositional phrases (see 2.1.3), PATH satellites are present in the vast majority of clauses describing motion—in 76 per cent, the exact same percentage as in the Middle English texts analysed in section 6.3.4. This is also true for the PATH verbs, which are combined with PATH satellites in 83 per cent (38/46) of their uses.

Let us now turn to the complementation patterns of MANNER verbs. Half of the MANNER verbs (12 out of 24) are combined with a PATH expression, usually a prepositional phrase as in (222a), more rarely (three times) with a PATH adverb, as *hors* 'out' in (222b). Hence, compared to the overall 76 per cent of verbs occurring with a PATH satellite, this combination is less frequent for MANNER verbs (50 per cent).

(222) a. et les gens du païs **cururent** ***au lagan*** si troverent Aucassin (34.11)
'And the people of the country ran to the shipwreck and found Aucassin.'

b. que l'espaulle li **vola** ***hors du liu*** (24.83)
'That his shoulder flew out of [its] place.'

However, it should be pointed out that not all of those are prototypical intransitive MANNER verbs: First of all, three are with the verbs *caïr* 'fall' and *fuir* 'flee', which of course already carry some PATH information in themselves (223a–b).

[152] 36 per cent of the motion descriptions analysed for Old English (201/557), and 29 per cent of those for Middle English (165/564) contain an adverb as satellite.

(223) a. qu'il **caï** si durement ***sor une piere*** (24.82)
'That he fell so hard upon a stone [from the horse].'

b. dient qu'ele **est fuie** *fors de la terre* (6.4)
'They say that she has fled out of the country.'

Second, among these MANNER verbs, there are also two instances of *se lancer* 'throw oneself' (both 'into the middle of the enemies'),[153] and one of *s'enbler* 'steal oneself', which all, although semantically describing INTRANSITIVE MOTION, are structurally CAUSED MOTION (224).

(224) si **se lance** ***en mi ax*** (32.7)
'And throws himself into the middle of them [the enemies].'

That French is rather free in expressing MANNER in the verb as long as the verb is used in a CAUSED MOTION structure (with reflexive pronoun) is noted by Cappelle for Present-Day French with respect to verbs such as *s'infiltrer* 'to seep', *se couler* 'to slip, slide', *se faufiler* 'to weave one's way, slip', *se glisser* 'to slip', *se jeter* 'to throw oneself, leap', *se lancer* 'to throw oneself, rush, dash', *se précipiter* 'to hurl oneself': 'it might be the case that the sheer presence of an object argument enables a verb to be used in a satellite-framed pattern' (2012: 176–7).

To conclude, of all the MOTION descriptions in *Aucassin et Nicolette*, only nine per cent (twelve tokens, including *caïr* 'fall', *fuir* 'flee' and the verbs in CAUSED MOTION structures) or four per cent (five tokens, excluding them) are combinations of MANNER verbs and PATH satellites.

7.2.3 Narrative styles

What is striking, though, is that a good fifth of all the motion verbs (30 out of 140) are found in a structure of 'durative motion verb + *tant* 'so much' + *que* 'that' + telic motion verb + GOAL', as shown in examples (225a–d). The verb construed with *tant* 'so much/so long' can be a MANNER verb (225c–d), a PATH verb (225b), or a NEUTRAL MOTION verb (225a), but always expresses an ongoing motion. The verb introduced with *que* 'that', by contrast, always profiles an endpoint; most often it is *venir* 'come' (225a–c), but also *arriver* 'arrive' (225d) or *entrer* 'enter'. With this kind of structure, of course, MANNER verbs are avoided in boundary-crossing contexts (see 2.1.4.1).

(225) a. et il monta u palais l'espee cainte, et **erra tant qu'il vint** en le canbre u li rois gissoit (28.23)
'And ascended to the palace, with the sword girded, and went so long til he came to the chamber where the king was lying.'

---

[153] See also Slobin's restriction of the boundary-crossing-constraint to 'activity-verbs'. He explicitly refers to the possibility of boundary crossing-expressions in combination with verbs for 'high energy motor patterns that are more like punctual acts than activities, such as equivalents of "throw oneself" and "plunge"' (2004: 226; see 2.1.4.1).

b. si **monta tant** a grans painnes **qu**'ele **vint** deseure (16.26)
'And ascended so long with great effort that she came to the top.'

c. Et li rois et Aucassins **cevaucierent tant qu**'il **vinrent** la u la roine estoit (30.16)
'And the king and Aucassin rode so long that they came there where the queen was.'

d. il **nagierent tant qu**'il **ariverent** desox le cité de Cartage (36.8)
'And rowed so long that they arrived under the city of Carthage.'

This way of narrating seems characteristic for the Old French romances and epics (we will see it again in the discussion of the Middle English translation of the French *Mélusine* in chapter 8). Fotitch (1950: 24), in her work on Chrétien de Troyes, describes it as a 'typical syntactic pattern [that] spans the distances from one point of the road to the other: the inevitable *tant que*'. This pattern appears like a sequence of frames in a comic strip, where each segment of the MOTION EVENT receives its own frame, or rather its own verb in this case: go up to the palace—go—come to the chamber (225a); climb upwards—come to the top (225b); ride—come to the queen (225c); sail—arrive in Carthage (225d). At least in Present-Day English, it would also be possible to 'stack' the different segments together with only one verb: *go up to the palace and into the chamber; climb up to the top; ride to the queen; sail to Carthage* (see 2.1.4.2).

Fleischman (1990: §6.72) links the *tant que*-pattern to the oral character of the texts because it reduces information density, so it may be just a genre-specific style.[154] Slobin (1997: 445–7), however, points out that it is also reminiscent of the different rhetorical styles found in verb-framed as opposed to satellite-framed languages, in that the former tend to have 'fewer ground elements per verb' (1997: 442) and to have separate predicates for the different PATH segments instead (see 2.1.4.2). He argues that this typological characteristic might be another factor influencing the style of the medieval Romance texts. He notes the absence of similar structures in Middle English, using two passages from Chaucer's *Knight's Tale* in which several PATH segments are united under one verb (Slobin 1997: 446).[155]

[154] On characteristic features of conceptual orality, see Koch & Oesterreicher 2011.

[155] Slobin's examples (and translations) are the following (1997: 446): *He on a courser, startlynge as the fyr/****Is riden into the feelds****, him to pleye,/****Out of the court****, were it a myle or tweye;/And* ***to the grove****, of which that I yow tolde,/By adventure, his wey ge han to holde* . . . 'He, on a courser, as spirited as fire,/Has ridden into the fields, to amuse himself,/Out of the court, perhaps a mile or two;/And to the grove, of which I told you,/By chance he took his way.'

And to the grove, that stood ful faste by,/In which there was an hert, as men him tolde,/Duk Theseus the streighte wey hath holde,/And ***to the launde he rydeth him*** *ful right,/For thider was the hert wont have his flight,/And* ***over a brook, and so forth on his weye****.* 'And to the grove that stood nearby,/In which there was a hart, as he was told,/Duke Theseus went straight./And to the glade he rides directly,/For thither would the hart take flight,/And over a brook, and forth on his way.' See also example (211) in section 6.3.4.

7.2.4 Summary: motion expression in *Aucassin et Nicolette*

The case-study of motion verbs in the prose parts of the Old French *Aucassin et Nicolette* has shown that in terms of frequency, MANNER verbs only account for 17 per cent, while PATH verbs make up 33 per cent, and other, neutral MOTION verbs (mostly *aller* 'go' and *venir* 'come') 50 per cent. Compared to the Old English texts analysed above (section 5.2), this means that even though *Aucassin et Nicolette* is such a lively and adventurous story, its proportion of MANNER verbs is the same as in the *Lives of Saints*. In the Middle English narrative texts (the *Knight's, Miller's,* and *Reeve's Tale*; see 6.3), by contrast, MANNER VERBS occur around twice or thrice as often (between 29 and 54 per cent). Old French MANNER verbs are thus not highly frequent, and, crucially, are much less frequent than Old French PATH verbs. In terms of syntactic context, half of the MANNER verbs occur with a PATH expression, although there is a tendency for these to be MANNER verbs which involve a PATH component themselves (*caïr* 'fall', *fuir* 'flee'), or which have a reflexive pronoun, i.e. are structurally CAUSED MOTION (*se lancer* 'throw oneself', *s'enbler* 'steal oneself'). We also noticed a characteristic narrative pattern of using separate verbs for the various PATH segments, often with a *tant que* formula. This can be linked to a genre-specific style, but it also corresponds to the narrative style that has been observed as typical in verb-framed languages.

We can conclude for Old French, as exemplified by this *chantefable*, that even though MANNER verbs *can* be used in a satellite-framed pattern, this is not at all the standard case. More importantly, and irrespective of their syntactic context, MANNER verbs are less frequently used than PATH verbs (17 per cent MANNER vs. 33 per cent PATH). Thus, even though the language has a satellite-framing pattern at its disposal, it does not typically conflate MANNER in the verb. This is very similar to what Cardini (2008) finds for Present-Day Italian.

7.3 *Conclusion*

We have seen in this section that Latin is predominantly satellite-framing, typically expressing PATH in prefixes and prepositional phrases, and, at least in colloquial Latin, also in adverbs. Despite this structure, however, the verb roots in Latin texts, particularly later ones, seem to carry MANNER information less frequently than in more prototypically satellite-framing, manner-conflating languages (Schøsler 2008).

The Latin prefixed verbs that survive into Old French have mostly become non-transparent, i.e. they are no longer analysable as a combination of PATH prefix and motion verb, but appear as pure PATH verbs (e.g. OF *issir*). Furthermore, new PATH verbs are created in Old French (e.g. *monter, avaler*), i.e. on the whole, more PATH verbs are appearing. At the same time, PATH is very often expressed in satellites as well (76 per cent of the clauses analysed in the case study), mostly the ones known from Classical Latin (prepositional phrases and prefixes), but sometimes also in PATH adverbs (in eight per cent of the clauses)—a phenomenon inherited from colloquial Latin and possibly strengthened by contact with Germanic languages. This means that we have verb-framing (*avale les degres* 'descends the stairs'), satellite-framing (*venés avant* 'come forward'), and double-framing (*issi fors* 'exited out') structures in Old French. With regard to the frequency of MANNER verbs in Old French, we have seen that they are used less frequently than PATH verbs (17 vs. 33 per cent), and also that they are less often combined with PATH satellites than other motion verbs.

Compared to Old and Middle English, the structures in which motion verbs overall appear in Old French are thus rather similar. The major differences are the following:

(a) the existence and widespread use of pure PATH verbs in Old French (33 per cent), which are rare in Old and Middle English (4–6 per cent),
(b) the less frequent use of MANNER verbs (17 per cent in Old French vs. 31 per cent in the Middle English texts), particularly in combination with PATH satellites (50 per cent of the Old French clauses with manner verbs as opposed to 72 per cent in Middle English), and
(c) the less frequent realization of path satellites as adverbs (8 per cent of the Old French clauses vs. 29–36 per cent in Old and Middle English).

Thus, if we need to stick simplifying labels on it, then Old French would be characterized as satellite-framing (in the wider definition of satellite; see 2.1.3) with respect to its structure, and as PATH-conflating with respect to typical verb semantics (apart from *aller* 'go' and *venir* 'come', which are the most frequent verbs, just like *faran* and *cuman* in Old English, and *gon* and *comen* in Middle English).

The PATH verbs that were found in the present study are *arriver* 'arrive', *avaler* 'descend', *descendre* 'descend', *entrer* 'enter', *issir* 'exit', *lever* 'rise', *monter* 'mount', *(de)partir* 'leave', *passer* 'pass', *repairier* 'return, go home'. Apart from *lever* 'rise', all of them have been borrowed into Middle English. Since the pure PATH meanings of IN, OUT, UP, and DOWN are not covered by any native PATH verbs in medieval English (see 5.1.3), I chose ME *entren, ishen/issuen, mounten/amounten, avalen*, and *descenden* for the analysis in chapter 9, together with ME *ascenden* as the counterpart of *descenden*, so that there are two verbs each for the paths UP and DOWN (*mounten/amounten, ascenden; avalen, descenden*), and one verb each for IN and OUT (*entren, ishen/issuen*). The meanings and complementation patterns that these verbs have in the donor languages will be briefly presented in chapter 9 as well.

CHAPTER 8 Borrowed PATH verbs in Middle English

Preliminary considerations

CHAPTERS 5 AND 6 HAVE shown that medieval English has hardly any native pure PATH verbs in its motion verb inventory (sections 5.1 and 6.2), and that in usage as well, PATH is typically expressed outside the verb, in satellites (sections 5.2.4 and 6.3.4). By contrast, medieval English has a wealth of MANNER verbs (sections 5.1 and 6.2). The verbs that are most often used to express MOTION are the neutral verbs OE *faran, cuman,* ME *gon, cuman,* as well as the MANNER verbs (5.2 and 6.3), with their respective frequency being influenced by genre and style (5.2.2 and 6.3.2).

On this basis, it is indeed fair to regard the borrowed pure PATH verbs that come into the language in the Middle English period (*entren, ishen/issuen, descenden, avalen, (a)mounten, ascenden*) as typological 'misfits' in the English system of talking about MOTION, since they carry purely PATH instead of MANNER meaning or MOTION only. The aim of chapter 9 is to investigate how these borrowed verbs have been integrated into the English system in the Middle English period.

Before that, the present chapter discusses some preliminary considerations to this investigation, together with the results of a pilot study on ME *entren* on the basis of *PPCME2* and *CEECS* (Huber 2012).[156]

8.1 *Differences between translated and autonomous texts*

This section will illustrate that, in studying how the new pure PATH verbs are used in Middle English, it is necessary to keep apart autonomous Middle English texts and texts that are translated from Latin or French. The choice of words in translations from Latin or French is likely to be influenced by their

[156] In this chapter, sections 8.1–3 are partly taken from the pilot study in Huber (2012).

original, and the borrowed PATH verbs may therefore be used differently in translations than they are in autonomous texts.

For Present-Day English, the influence of a typologically different original on the English translation has been shown by Cappelle (2012: 185–6): English translations from French make use of MANNER verbs only half as often as autonomous English texts do, i.e. they are clearly different in their motion expressions from autonomous texts. English translations from German, i.e. from another MANNER-conflating, satellite-framing language, by contrast, are not any different from autonomous English texts in the frequency of MANNER verbs.[157]

For Old English, we have seen in section 5.3 that the translator of the Gospels did *not* tend to render Latin PATH verbs with Old English PATH verbs. The reason for this, though, is most probably that Old English does not *have* any PATH verbs that are fully equivalent to the Latin ones (see 5.1.3)—apart from OE *genēalǣcan* 'to approach', which the translator indeed uses consistently to render L *adpropinquare* 'to approach'. This situation changes in the course of the Middle English period, when a range of pure PATH verbs are borrowed into the language. Therefore, a Middle English translator of a French text has the option to use the very same verb in the translation that he encounters in the original. Here is an illustration:

The late Middle English translation (*c.*1500) of the French *Mélusine* (ed. Donald 1895) was made on the basis of the 1478 print of Jean d'Arras' *Mélusine* (see Stouff 1974), available in the edition by Brunet (1972). A short passage from each—French original and Middle English translation—is given in (226). The translation features many lexemes that are identical, or almost identical, in the original (printed in bold in (226); additional French loans in the English translation are underlined): Among the motion verbs, there is F *parti* and ME *departed*, F *montèrent* and ME *mounted*, F *passèrent* and ME *passed*, F *entrèrent* and ME *entred*. Since these verbs had already been borrowed into English, it is likely that the translator was simply guided in the choice of words by their presence in the original.

(226)

Par **foy**, dist Alain, or est bien le pays **delivré** du **lignage** de Josselin; Dieu ait des ames **mercis**, combien qu'ilz ne nous amassent oncques.	'By my **feith**,' said þeire fader. 'Alayn, now is the land wel clene **delyuered** of the **lynage** of Josselin; god on theyre sowles haue **mercy**, how be it they loued vs neuer.
Or, beaulx enfants, je vous diray que vous ferez.	Now fayre sones I shall saye you what ye shal doo.
Tout premièrement vous irez au roy et lui **requerrez** qu'il vous donne **place** pour **édiffier** la **prieuré**; et luy dictes	First ye shal goo to the kinge & **requyre** hym that it plese hym to gyue you a **place** for to **edefye** the **Pryorye**

[157] Cappelle (2012) compares the group of MANNER verbs to the group of non-MANNER verbs. In terms of the present study, the latter group would subsume NEUTRAL MOTION verbs and PATH verbs.

la **manière** comment vostre **cousin** le vous a enjoinct, et je croy qu'il vous en dira bonne responce. Et ilz luy respondirent que cy feroient-ils. Et à tant se **partirent** de leur père,	/ and telle hym the **maner** how ye be commanded of your **Cousyn** to fownde it. and I byleue he shal gyue you a good ansuer.' And they said that thus shuld they doo. And thanne they **departed** fro theire fader,
(i) *et à tant chevauchèrent que ilz vindrent à Vannes*, et trouvèrent que le roy estoit **parti**, et estoit allé à Sussinnon pour soy ésbatre et desduire à la **chasse.**	(i) *and so long they rode that they camme to Vannes* and founde the kinge **departed** & was goon to Sassymon for to dysporte hym at **Chasse.**
Et adonc ilz (ii) ***montèrent** à chevau et vindrent au port, et **passèrent** et **entrèrent** en la **forestz**, et chevauchèrent tant qu'il vindrent au **chasteau**,* et	And they (ii) ***mounted** on horsbak and came to the gate and **passed** & **entred** the **Forest** and rode so long tyl they came to the **Castel.***
trouvèrent que le roy estoit allé au **parc chasser**; et les deux frères vont après, et trouvèrent le roy dessoubz ung grant arbre sus ung **estang**, où il attendoit le cherf que les chiens **chassoient.** Adonc les deux frères se trairent **à part**, pour ce qu'ilz ne voloient pas destourber au roy à veoir son deduit. (ed. Brunet 1972: 111)	and founde the kyng goon to the **park** to the **chasse** / and the two brethren yed after & founde the king nyghe a grete tree by a **staung** where he abode aftir the herte that houndes **chassed.** Thenne the two bretheren drew them self **aparte** bycause they wold not lette the kyng to see the dysporte (ed. Donald 1895: 98)

Note that there are not only striking lexical similarities, but also structural ones: The French text makes use of the *tant . . . que* structures which, whether a mere stylistic device or also a typological issue, are typical for motion event descriptions in medieval French romances and epics (see 7.2.3). These are also transferred over into the English translation; see the structures marked with (i) and (ii) in (226).[158]

It is always possible that a Middle English translation[159] will simply use the verbs that it finds in the original, as shown in the parallel passage in (226). This idea is supported by the results of the pilot study on ME *entren* (Huber

[158] A brief comparison of the French and English versions of *Mélusine* gives the impression that the translator renders the French V_{manner} *tant . . . que* constructions rather faithfully with V_{manner} *(so long) . . . tyl/that.* Only rarely, one of these rather long-winded segment-by-segment narrations is shortened, as in the following, where the italicized parts in the French are left untranslated:

> En ceste partie nous dist l'istoire que tant chevaucha Raimondin et sa mesnie en sa compaignie, *qu'il vint ès mons de Monjouet et les passa, et chevaucha tant par la Lombardie*, qu'il arriva *ung soir* à Romme *auprez Noiron. Et le lendemain vint à Saint-Pierre*, et là trouva le pape Benoit, *qui pour lors regnoit*, et se traist par devers luy (ed. Brunet 1972: 375)
>
> Here sayth thystorye [sic] that Raymondyn rode so long that he came to Romme and his companye with hym, where he fonde the Pope named Benedictus / & drew hym toward hym (ed. Donald 1895: 340)

[159] In the following, 'translation' will be used as a shorthand for 'translation from Latin or French'. There are also Middle English texts translated from other languages, though less numerous: *Caxton's History of Reynard the Fox* (included in *PPCME2* as CMREYNAR.M4), for instance, is a translation

TABLE 8.1. Frequency of ME *entren* in autonomous and translated texts (*PPCME2*) (Huber 2012: 321).

	WORDS (TOTAL)		*ENTREN* IN AUTONOMOUS TEXTS		*ENTREN* IN TRANSLATIONS	
SUBPERIOD	AUTONOMOUS	TRANSLATIONS	FREQUENCY PER 100,000 WORDS		FREQUENCY PER 100,000 WORDS	
ME 2 (1250–1350)	21,494	125,109	9	(2)	29	(36)
ME 3 (1350–1420)	307,485	178,503	16	(50)	40	(71)
ME 4 (1420–1500)	254,103	11,181	28	(72)	9	(1)
Total	583,082	314,793	21	(124)	34	(108)

2012: 321) on the basis of *PPCME2*: As shown in Table 8.1, ME *entren* is more likely to occur in a text translated from French or Latin than in an autonomous Middle English text.

Overall, *entren* has a normalized frequency of 34 in 100,000 words in translations, and of only 21 in 100,000 words in autonomous texts. With regard to the different subperiods, a direct comparison between autonomous texts and translations seems only possible for M3 (1350–1420), since the sample size for autonomous texts in M2 and for translations in M4 is simply too small (see the second column, 'words (total)', in Table 8.1). In M3, however, the imbalance noticed in the total is even more marked, with *entren* being used 40 times in 100,000 words in translations and only 16 times in autonomous texts.

The reason for this imbalance is probably that the use of *entren* in translations is often triggered by a PATH verb in the respective exemplar: A comparison of all uses of *entren* in the Wycliffite Bible translation from *PPCME2* with the Vulgate shows that the verb is consistently used to render Latin *intrare*, *introire*, and *ingredi* (see also Ogura 2002: 104; and also (227)). As one would expect for a loanword, its frequency in autonomous texts rises throughout the subperiods, although it never becomes as frequent as in the translations.

Therefore it is necessary to separate translated texts from more autonomous texts in the present study: If translations use the borrowed PATH verbs just like their French and Latin equivalents are used, they will not help to answer the question how these verbs are used in Middle English. Neither is it advisable, though, to exclude translated texts from the study entirely, because their use of the verbs has most certainly also influenced the general use of the verbs.

from Dutch. Translations from MANNER-conflating, satellite-framing originals, such as Dutch, are included in the category of autonomous texts in the present study, because a typologically similar original text is not likely to have a distorting influence on the motion expressions of the Middle English translation (cf. Cappelle 2012 on Present-Day English translations from German).

Furthermore, it is important to keep in mind that when talking about translated and autonomous texts we are not dealing with two neatly distinguishable groups. It is typical for medieval texts that the boundary between them is often extremely fuzzy: First, what is basically a translation may still contain extensive additions to the original text; an example—of textbook fame (cf., e.g., Baugh & Cable 2013: 144–5)—is Trevisa, adding his own comments to his translation of Higden's *Polychronicon*. Second, texts can be based upon Latin or French originals as regards their content, though without being a close translation. This would be the case, for instance, in Malory's *Morte Darthur* which is based on a large number of mostly French sources, which he 'systematically abridged, adapted, and even at times translated' (McCarthy 1996: 77).[160] We will come back to this in section 9.1.6., where it will turn out that in one part of the *Morte Darthur*, *entren* is used in ways highly untypical for Middle English but reminiscent of its use in French.

Third, otherwise autonomous texts may still contain translated quotations and allusions, such as to classical authors or to the Bible. The latter is obviously characteristic of religious tracts and sermons (see 10.1.6), but may occur in all kinds of texts. For instance, in a passage on cursing, the religious tract *An Apology for Lollard Doctrines* (attributed to Wycliff) cites Psalm 108 (227a). The wording is very close to the one in the Wycliffite Bible translation (227b), with which the writer must have been familiar, and which, in turn, may be influenced in its choice of ME *entren* by the Latin synonym *ingredi* in the Vulgate (227c).[161]

(227) a. And seiþ þe salm [. . .] [cursing] **entred** as water in to his inword þings, and as oyly in to his bonis (*c.*1475(?*c.*1400) *Wycl.Apol.* (Dub 245))

b. it **entride** as water in to hise ynnere thingis; and as oile in hise boonus. (*a.*1425(*c.*1395) *WBible(2)* (Roy 1.C.8), Ps 109:18))

c. et **ingredietur** quasi aqua in viscera eius et quasi oleum in ossa eius (Vulgata Ps 108:18),

Due to these overlaps between translated and autonomous texts, the translated/autonomous distinction will be revisited and looked at in more detail in the discussion when necessary in chapter 9.

The present section has argued why it is crucial to study the use of the borrowed PATH verbs separately in translations and autonomous texts. The next section will introduce another dichotomy that is needed for the present study: the one between LITERAL and NON-LITERAL uses.

---

[160] Malory also uses a few English sources, such as the *Alliterative Morte Arthure*, the *Stanzaic Morte Arthur*, or Hardyng's *Chronicle* (cf., e.g., Spisak & Matthews 1983: 622). On Malory's sources, see, e.g., McCarthy 1996, Vinaver 1929: 100–108, 128–54; Field 2001a, and, on the minor sources, Norris 2008.

[161] Of course, the bypass via the Wycliffite Bible is not even necessary, since the writer of the *Apology* may have rendered the verse as (227a) independently.

8.2 *Preference of* NON-LITERAL *over* LITERAL *uses*

As we have seen in chapters 5–6, the typical English way of talking about MOTION is to express PATH in satellites, and either neutral MOTION or MANNER in the verb. The borrowed pure PATH verbs *enter, ish/issue, descend,* and so forth are a novelty in this system, since their semantics (pure PATH) are unusual for an English motion verb. Yet, they could be expected to find an appropriate semantic niche for the description of motion events in which PATH is the most important information and there is no MANNER in the first place.

This would be the case, firstly, in metaphorical motion. Let us take the example of PDE *to enter into a contract/a profession.* In this kind of metaphorical expression, the source domain of MOTION is mapped onto more abstract target domains, such as CHANGE. What is important here is the mapping scope of the metaphor: The mapping scope is described by Ungerer & Schmid (2006: 119) as 'a set of constraints regulating which correspondences are eligible for mapping from a source concept to a chosen target concept'. In the case of *enter into a contract/a profession,* just like in Ungerer & Schmid's example of the LIFE IS A JOURNEY metaphor, the mapping scope 'relies on the image schema of "path" and is supported by the correlations "change < > motion" and "purposes < > goals"' (2006: 120). In other words, the mapping scope covers the event components of FIGURE, GOAL, MOTION, and PATH, but does not really cover MANNER OF MOTION. Therefore NEUTRAL MOTION verbs and PATH verbs are particularly well suited to describe 'metaphorical motion events'; PATH verbs perhaps even more so, since PATH and MOTION are at the centre of the metaphor.[162] Stretching the mapping scope further to include MANNER, and using a MANNER verb such as *crawl, jump, roll* instead of *enter,* together with *into a contract/a profession* would be very unusual, or at least quite unexpected and creative—it would strike the reader or listener as overtly metaphorical and expressive and would hence not be a case of conventionalized, or 'dead', conceptual metaphor.

Another semantic niche for pure PATH verbs in a MANNER-conflating language might be the use for situations where, although the motion is not metaphorical, it does not take place in any discernable MANNER either. An example would be the FIGURES *smell* or *light* entering somewhere. Although these are 'real', and not metaphorical, motion events, they lack any obvious MANNER OF MOTION. This is similar for the rising of the sun or moon, which also does not happen in any describable MANNER OF MOTION. Recall that these were also the typical TRANSLOCATIONAL MOTION uses of the Old English PATH verb *arīsan* (section 5.1.3.5), which was otherwise mostly used for CHANGE OF POSTURE.

[162] Another example would be the widespread metaphorical expression of *loanwords entering a language.*

The pilot study of ME *entren* on the basis of *PPCME2* (Huber 2012: 321–4) has shown that autonomous texts use *entren* significantly more often than translated texts to refer to motion events such as the ones described in the preceding paragraphs, i.e. for metaphorical motion events and motion events without visible MANNER OF MOTION. For lack of a better comprehensive term, these will be called NON-LITERAL motion events in the following. Translations and autonomous texts thus differ not only in how frequently they use *entren* (8.1) but also which kinds of motion events they use it for (Huber 2012: 321–4). The majority of attestations of *entren* in translations (more than three out of four attestations) is used for LITERAL MOTION EVENTS, i.e. events in which a visible, concrete FIGURE moves into a visible, concrete GOAL in a visible MANNER, such as in (228a–b) from *Mandeville's Travels*, a translation from French.

(228) a. And whan sche hadde thus seyd sche **entred** in to the fuyr (CMMandev. M3; 45.1133, PPCME2)
'And when she had said this, she entered into the fire.'

b. And whanne the monkes **entren** in to þat place þei don of bothe hosen & schoon or botes alweys (CMMandev.M3; 39.974–5, PPCME2)
'And whan the monks enter into that place, they always take off both trousers and shoes or boots.'

The FIGURE in (228a) is a girl about to be burnt on the stake, the GOAL is the fire: It would therefore be possible to describe the MANNER OF MOTION, for instance with a verb such as *steppen* 'step' or *sterten* 'leap', or also with a CAUSED MOTION structure such as *she was thrown*. Also in (228b), it would not be odd to describe the monks *stepping, going*, or *coming* into the place. The choice of *entren* as opposed to these MANNER or NEUTRAL MOTION alternatives is most likely due to the fact that the French exemplar will have *entrer* in these cases as well, i.e. to an etymological trigger in the original.

By contrast, *entren* is used more often for NON-LITERAL motion events in autonomous Middle English texts, in which half of the attestations are NON-LITERAL, such as (229a–d):

(229) a. And than **enters** thow ***into the thirde degre of lufe***, in the whilk thou sal have grete delyte (CMRollEp.M24, 69.125, PPCME2)
'And then you enter into the third degree of love, in which you shall have great pleasure.'

b. that ***wijsedom*** schal not **entre** ***into an yvel willid soule*** (CMPurvey.M3, I,51.2071 PPCME2)
'That wisdom shall not enter into an evil-minded soul.'

c. And this man was Adam, by whom ***synne*** **entred** into this world (CMCTPars.M3, 296.C2b.356, PPCME2)
'And this man was Adam, by whom sin entered into this world.'

d. loke þat þe brynkes of þe leded pottus mouþe & þe brynkus of þe pot with þe ryndus be wel dabbid with cleye & with hors-donge so þat ***non oþer flauor* entre** in-to þe pottus ne non reche come oute. (CMHorses. M3, 107.239, PPCME2)
'Make sure that the rims of the mouth of the lead pot and the rims of the pot with the barks are well covered with clay and with horse dung, so that no other flavour enter into the pots and no reek come out.'

In these, the FIGURE (*sin, wisdom, flavour* in (229b–d)), the GOAL (*the third degree of love, an evil-minded soul* in (229a–b)), or both, are not concrete and visually perceivable. Thus, in (229c) ([. . .] *synne entred into this world*) a NEUTRAL MOTION verb like *come* would be a possible alternative for *entren*, but most MANNER verbs would be odd, as in *synne ran/walked/stumbled into this world*. The same is true for the other examples in (229).

On the basis of the pilot study on *entren* (Huber 2012), we may hypothesize that the other borrowed pure PATH verbs in autonomous texts are also more likely to be used in NON-LITERAL contexts, where PATH is crucial and there is no MANNER OF MOTION to talk about in the first place. Conversely, (unless they acquire additional MANNER meaning; see section 8.3) they will be less likely to be used in LITERAL contexts, where there is indeed MANNER OF MOTION to talk about and where the use of a MANNER verb is more typical. In other words, if a language often encodes the MANNER OF MOTION in the verb, a verb such as *entren*, just like *come in*, lends itself very well for contexts where the MANNER in which a FIGURE moves is not visible or where the motion is metaphorical. These usages, also present in the donor language, may be exactly the 'peripheral portions' (see Durkin 2009: 5) of the meaning of a verb which are borrowed into Middle English. Since these peripheral meanings are often metaphorical or abstract, the distinction between LITERAL and NON-LITERAL MOTION, albeit a rough measure, will serve as a first guideline to the borrowed uses. Specific LITERAL and NON-LITERAL MOTION uses of the verbs will be discussed in the respective sections in chapter 9.

8.3 *Enrichment of* PATH *with* MANNER *semantics*

The findings of studies on the interpretation of unknown motion verbs in contemporary languages suggest that the borrowed pure PATH verbs might receive additional MANNER semantics after they are adopted into Middle English: Middle English speakers not familiar with French may interpret the newly borrowed French verbs as also carrying MANNER meaning, due to the conflation of MOTION and MANNER prevalent and typical in their own language, as did the test subjects with made-up verbs in Naigles & Terrazas (1998) and Cifuentes-Férez & Gentner (2006).

In Naigles & Terrazas (1998), speakers of (MANNER-conflating) English and speakers of (PATH-conflating) Spanish were shown short video clips of motion events, e.g. of a woman skipping towards a tree, and the event was described to them with a sentence containing a made-up verb (E *to krad*, Sp. *mecar*) such as in examples (230a–b).

(230) a. She's **kradding** the tree!
Está **mecando** al árbol!

b. She's **kradding** ***towards*** the tree!
Está **mecando** ***hacia*** el árbol!

Some test subjects received the verb in a syntactic context without directional prepositions, such as (230a). This pattern is more similar to sentences with PATH verbs like *He entered/circled/descended X* than to sentences with a preposition and MANNER verb like *He ran/jumped/slid into/down from/to X*, and hence should suggest a PATH reading of the made-up verb. Other test subjects received the verb in contexts with a directional preposition (such as English *towards* or Spanish *hacia* in (230b)) to create a bias towards a MANNER interpretation of the unknown verb. The test subjects were then shown two parallel video clips that presented the same scene as before, only that either the direction or the manner of motion was changed: for instance, on one screen, they saw a woman skipping away from a tree; on the other screen, a woman marching towards a tree. Asking the participants on which of the screens they thought the woman was *kradding* (or any other verb that had been used in the descriptions before) enabled the researchers to find out whether the verbs were interpreted as MANNER or as PATH verbs: English speakers who had encountered the verb in a syntactic context with a directional preposition (230b) more often opted for MANNER interpretations than for PATH interpretations. The reverse was true for Spanish speakers who got to know the verb in contexts without a directional preposition (230a). In the respective other contexts, the answers were less consistent (Naigles & Terrazas 1998: 367). These results can be taken to show that speakers generalize about the types of meanings motion verbs usually encode in their language and extend this to their interpretation of unknown motion verbs.

These results are corroborated by Cifuentes-Férez & Gentner (2006): When English and Spanish speakers were presented stories containing sentences with novel, made-up verbs, they tended to interpret them in line with their own language-specific semantic patterns: English speakers interpreted the unknown verbs significantly more often as MANNER verbs, Spanish speakers as PATH verbs. This again documents that the interpretation of unknown motion verbs seems to be guided by the prevailing semantic pattern in a language (see also Maguire et al. 2010 for similar results).

These findings may be transferred to the language contact situation in late Medieval England: Since in the French donor-language varieties, the PATH verbs are usually combined with PATH satellites (see 7.1.2, 7.2.2, and the donor-language portraits of the borrowed verbs in chapter 9), they occur in the same structures in which English motion verbs occur. Thus, there is nothing in principle in the complementation patterns of the PATH verbs that would suggest to a Middle English speaker that they are any different from the standard Middle English motion (+MANNER) verb. Furthermore, as we will see in the donor language portraits in chapter 9, in some of the French usages of the PATH verbs, MANNER information is present as well, particularly in the MILITARY uses of *entrer* and *issir*. If the verbs are borrowed into Middle English in these peripheral meanings, they would not be semantic misfits among the Middle English motion verbs anyway, since several native verbs express PATH and MANNER at the same time as well.

In the autonomous texts analysed in the pilot study (Huber 2012: 324–6), this hypothesis was supported: *entren* in LITERAL contexts was often—in more than half of the cases—used to refer to motion events in which a FIGURE enters the GOAL in a hostile way (see the burning, killing, and destruction in (231a–b)), or where it has to surmount an obstacle, such as when the entering is difficult or forbidden (232).

(231) a. kynge Royns of Northe Walis had rered a grete numbir of peple and were **entred** in the londe and ***brente and slew*** the kyngis trew lyege people (CMMalory.M4, 45.1482, PPCME2)
'King Royns of North Wales had gathered a great number of people and they had entered the land, and burnt and killed the king's true allies.'

b. the king of Scottys **entered** at Karlile, and ***wasted all the cuntre*** onto ȝork (CMCapChr.M4, 140.3269, PPCME2)
'The king of the Scots entered at Carlisle and devastated all the country down to York.'

(232) Estir bitook hirself to Goddis disposicioun, and to ***perel of her deth***, and **entride** to the king, ȝe, ***aȝens the lawe of the lond***, whanne sche was not clepid (CMPURVEY, I,36.1709, PPCME2)
'Esther put herself in God's hands and to mortal peril, and entered to the king, yea, against the law of the land, when she was not called.'

These contexts were less frequent in the translated texts, which, probably due to the influence of their French and Latin originals, use *entren* for all kinds of inward motion events. The fact that normal inward motion, by comparison, is less widespread in the autonomous attestations of *entren* indicates that *entren* in autonomous texts is specialized for a particular kind of inward motion: aggressive or difficult entering.[163]

Chapter 9 will show that this result for *entren* from the pilot study also bears the test of a much larger corpus. We will also investigate whether the other

[163] One may speculate whether in the Middle English period the verb *entren* was perhaps often encountered in military commands, which may have influenced this specialization.

PATH verbs develop (or are borrowed in) specific MANNER uses as well. The next section will present the material and method for the study in chapter 9.

8.4 *Material and method*

This section presents the material on which the study in chapter 9 is based and describes the method that was used to obtain and classify the attestations.

8.4.1 Corpus

For the PATH verbs that shall be investigated in chapter 9, it is necessary to use a very large text base, because some of them are rather rare. The corpus used in the pilot study, *PPCME2*, for instance, does not contain a single attestation of the verbs *ishen* and *avalen*. The study in chapter 9 is therefore based on the large *Corpus of Middle English Prose and Verse* (*CME*).[164]

Most corpora are composed with the objective of achieving a certain balance with regard to genre/text type, sociolinguistic variables, and—in the case of diachronic corpora—time periods (cf., e.g., Meyer 2002: chapter 2). This is often rather difficult to achieve for corpora representing older stages of a language, due to the unequal distribution of the extant texts (see 3.1; see also the low word count for autonomous texts in *PPCME2* for the period 1250–1350 in Table 8.1). Yet, it is still an ideal that is aspired to. The *CME* is quite different in this respect and is comparable to the *DOEC* for Old English. It does not include mere samples of texts, annotation concerning genre, sociolinguistic variables, or the like. Rather, it consists of electronic versions of entire editions, many of them from the Early English Text Society series. The long-term goal for the *CME* is to include 'all editions of Middle English texts used in the *MED*'.[165] With all the caveats thus entailed by the nature of this corpus—there is no balance for aspects like subperiod, genre, length, and so forth—the *CME* has the enormous advantage of covering an extensive amount of Middle English texts.

For a list of the 146 editions included in the *CME* (as of 2015), see the website of the *Middle English Compendium*.[166] In some of the editions, none of the PATH verbs studied in chapter 9 is attested. The texts in which they occurred are listed in Appendix C. Note also that from the nine different editions of the *Canterbury Tales* featured in the *CME* (based on different manuscripts), only the Robinson edition (1957) was taken into account, to avoid repeated attestations.[167]

---

[164] I am very grateful to Paul Schaffner from the *Middle English Compendium* for being so kind as to send me the text files.
[165] http://quod.lib.umich.edu/m/mec/about/.
[166] http://quod.lib.umich.edu/c/cme/browse.html.
[167] It would be interesting, however, to study in detail whether different manuscript versions use alternatives to the borrowed PATH verbs, or whether they use them in different complementation

8.4.2 Method

From the *CME* texts, the French-borrowed PATH verbs *entren* 'enter', *ishen/issuen* 'exit', *descenden* 'descend', *avalen* 'descend', *mounten* 'mount', *amounten* 'amount', and *ascenden* 'ascend' were extracted with the help of the concordance programme *Monoconc Pro2.2* (Barlow 2004), using various spellings. False positives (e.g. homographs or verbs that occurred in the editor's introductions and notes rather than in the body of the text) were removed. The attestations thus obtained were classified as to whether they appeared in an autonomous Middle English text or in a translation (see 8.1). This was done on the basis of the information in the volumes of the *Manual of the Writings in Middle English* (Severs et al. 1919–51) as well as the information found in Wallace (2002) and the editions themselves. An overview, together with the *MED*-stencils whenever these where identifiable, can be found in Appendix C.

I generally analysed all the attestations of the borrowed PATH verbs in the corpus, except in the case of *entren*, which had more than 2,500 attestations in translated texts: A random sample of 500 was chosen from these. The attestations were categorized as to whether they referred to LITERAL or NON-LITERAL MOTION (see 8.2). Up to this point, the procedure was the same for every verb. The ensuing categorization into subgroups then, although often roughly similar, was slightly different for the single verbs, because the subcategories only emerged from the examples themselves.

For the LITERAL uses, the subcategories are mostly based on the kind of FIGURE and GROUND and the different kinds of context in which the MOTION EVENT takes place; for the NON-LITERAL uses, the kinds of FIGURE and GROUND were categorized as well, just like the target domains when the use was metaphorical. It is in the process of this multi-level categorization that the major subcategories for the individual verbs emerge. The different subsenses in the *MED* and *OED* entries have been taken into account, but it is important to note that they do not form the basis of the subcategories used in the present study. With the very fine-grained distinctions the dictionaries draw, the resultant subgroups would in most cases have been much too small and their number too large as to lend themselves for any conclusions within the present question whether the pure PATH verbs have a tendency to be used for specific contexts.

Furthermore, different from the pilot study (Huber 2012), every attestation was also categorized as to the complementation pattern in which the verb appears—for instance, whether GROUND is expressed in a prepositional phrase introduced with *from*, with an extra adverb *out*, or with GROUND in a direct object. The complementation patterns and, particularly, the semantic contexts for which they are used, will be discussed in the respective sections in chapter 9 as well.

---

patterns. This issue will be sporadically looked into in chapter 9, but it might be worth a separate investigation.

CHAPTER 9

Borrowed PATH verbs in Middle English

THIS CHAPTER PRESENTS AND discusses the uses of the borrowed pure PATH verbs in Middle English. Each of the verbs is treated in a separate section, all of which begin with a brief 'donor language portrait' giving an overview of the meanings and complementation patterns the respective verb has in Anglo-Norman, continental Old and Middle French, and (Anglo-)Latin. In the discussion of the meanings, these portraits do not strive at a full coverage of the verbs in the donor languages (for this, see the respective dictionaries). Their aim is rather to discuss those specific, often metaphorical uses in the donor languages that were prominently adopted into Middle English. Sometimes the Middle English uses of the verbs were taken over directly from the donor languages (such as the use of *issir* and *descendre* referring to LINEAGE), sometimes they appear to have been developed only in Middle English (such as the specialization of ME *avalen* for 'taking off one's headwear'; see section 9.4). Therefore it is important to address which of the Middle English uses of the verbs are already present in the donor languages.

Middle English was in contact with various languages and varieties in which the PATH verbs occur (see 6.1). The first important contact language is Anglo-Norman, for which I will use the *Anglo-Norman Dictionary* (*AND*). Second, continental Old and Middle French are also contact languages for Middle English; for these, Tobler-Lommatzsch's *Altfranzösisches Wörterbuch* (*TL*) and the *Dictionnaire du Moyen Français* (*DMF*, for 1330–1500) are used, sometimes complemented by Godefroy. Third, particularly for *descendre, ascendre,* and *entrer,* which are still very close in their form to Latin *descendere, ascendere,* and *intrare,* Latin usage must have had an influence as well. As it is probably the Latin used in the British Isles in particular that will have played a role here, the *Dictionary of Medieval Latin from British Sources* (*DMLBS*) will be used for examining the verbs in Latin.

Following the donor language portrait, a general presentation of the usage contexts in which the verb is prominently attested in Middle English autonomous texts and translations is given. For each verb, this presentation of the general results is followed by sections on the specific LITERAL and NON-LITERAL MOTION uses, also with an eye on whether the specific uses predominantly occur in certain genres and whether they are taken over from the donor languages or represent an English development of the verb. In this, where it appears informative (and where the original is extant), Middle English translations from the *CME* will be compared to their French and Latin original, to see whether they simply copy the use of the respective verb in the donor language or whether they deviate from the exemplar.

For most of the verbs, the complementation patterns are discussed in a separate subsection, except for *avalen* and *ascenden*, where this was not a fruitful issue—*avalen* is not attested very frequently, and occurs in very heterogeneous uses, and *ascenden* simply occurs bare or with PATH satellites. When a verb has considerably changed its complementation patterns after the Middle English period (particularly *entren*, whose typical use has changed from *enter into* GOAL-GROUND to *enter* GOAL-GROUND), the focus on Middle English will be widened to include later developments.

9.1 *Entren*

9.1.1 Donor language portrait: OF/MF *entrer*, L *intrare*

OF *entrer* 'enter' has developed from Latin *intrare* 'enter', which, according to the *TLL*, Walde-Hofmann (both s.v. *intro*), and Ernout-Meillet (s.v. *in*), has been derived by adding a verbal ending to the adverb/preposition *intra* (for a discussion of alternative views, see Stolova 2008: 256). Thus, different from Latin *ingredi*, for example, which is formed by the PATH prefix *in* 'in' and the (MANNER OF) MOTION verb *gradi* 'to step', Latin *intrare* has been a PATH verb from its beginning, and has never been a combination of a PATH satellite and a (MANNER OF) MOTION verb.

A) Syntax

Latin *intrare* is found with GROUND expressed both in a prepositional phrase (233a) and as an accusative object (233b):

(233) a. predicti fratres [*sc.* minores] primo **intraverunt *in Angliam*** (s1200 *Ann. Exon.* 11v, DMLBS s.v. *intrare*, 1b)
'The abovementioned friars first entered England.'

b. juratores..dicunt quod..Robertus occisus fuit set non viderunt eum **intrare *domum Rogeri*** nec exire (1221 *PlCrGlouc* 75, DMLBS s.v. *intrare*, 1a)
'The jurymen say that Robert was killed, but they did neither see him enter nor exit Roger's house.'

In Anglo-Norman, this is similar: *entrer* is used in combination with PATH satellites, i.e. with a prepositional phrase such as in (234a) or with an additional

PATH adverb such as *enz* in (234b). It is also used with GROUND as a direct object, such as in (235).

(234) a. **entroyent** ***dedenz le temple*** (*Bible* 44.1, cf. AND s.v. *entrer*)
'They entered within the temple.'

b. **Entrerent** ***enz dedenz le mur*** (*S Brend* MUP 271, cf. AND s.v. *dedenz*)
'They entered in within the wall.'

(235) En grant dolur **entrat** ***Bretaine*** (GAIMAR 20, cf. AND s.v. *entrer*)
'He entered Brittany in great sorrow.'

The complementation of *entrer* with GROUND in a direct object, such as in (235), however, might be an Anglo-Normanism, as it appears to be rather rare in continental varieties of Old French: *TL* lists it toward the very end of the three-column-long entry, and the attestations it gives for the direct object pattern are all either from the Anglo-Norman *Romance of Horn* (236a), or from translations of Latin texts, in which Latin has an 'enter' verb with a direct object GROUND, such as (236b):

(236) a. **Entrer** veut ***la cité*** (*Horn* 2972 *Var.*)
'He wants to enter the city.'

b. Dunkes **entrerent** li Franc ***l'oratoire*** (Lat.: *ingressi oratorium*) (*Dial. Gr.* 12.3)
'Then the Franks entered the oratory.'

(both from *TL* s.v. *entrer*, sense 'betreten, eintreten in')

This first impression based on the dictionary entries is confirmed by a search in the Old French *Nouveau Corpus d'Amsterdam*[168] for the lemma *entrer* in texts classified as 'Région Parisienne'. The search yields 91 attestations, none of which is construed with a direct object GROUND. More than half of them contain a prepositional phrase with *en* 'in(to)',[169] such as in (237).[170]

(237) quant il furent **entrez** ***en la chambre*** (c.1290, *HelcanusN*, NCA)
'When they had entered in the room.'

The situation seems to be similar for Middle French, which is another contact variety for Middle English. The *DMF* lists one sense in which MF *entrer* is construed with a direct object (s.v. *entrer*, I.A.1.b 'Pénétrer (dans un pays) avec

[168] This can only be taken as a rough measure, because the corpus contains a few texts in various manuscript versions.
[169] Note that this is OF *en* 'in(to)' < Latin *in*, not the homonymous adverb *en* 'from there' < Latin *inde* discussed in 7.1.2.
[170] The rest mostly has either 'in'-adverbs such as *enz/leanz/ceanz* or *y*.

des intentions belliqueuses' 'penetrate into a country with bellicose intentions'), but the one attestation that illustrates this sense is from an Anglo-Norman text ((240c) below). I looked for the form *entra* (3rd person singular, passé simple) in the *Corpus de Textes de Français Médiéval et de la Renaissance* (*7FMR*), i.e. the corpus of the *DMF*, and the more than 600 attestations that were found contained merely two instances where *entra* has a direct object GROUND:

(238) a. Et **entra** ***le chasteau*** a telle compaignie qu'il luy pleut. (André de LA VIGNE *Le Voyage de Naples*, p. 1495, 255, 7FMR)
'And entered the castle in such company as pleased him.'

b. Codrus [. . .] courageusement **entra** et penetra ***l'ost adversaire*** (anonyme, *Le Violier des histoires rommaines*, 1521, 115, 7FMR)
'Codrus [. . .] bravely entered and penetrated the enemy army.'

Note that in (238b), the direct object *l'ost adversaire* is separated from *entra* by a second verb, *penetra* 'penetrated', so that the example is actually ambiguous—*l'ost adversaire* could be the complement of both verbs or of *penetra* alone. Furthermore, the attestation is from a translation of the Latin *Gesta Romanorum* and may therefore again be influenced syntactically by a Latin structure.[171] The direct object pattern, therefore, appears to be very infrequent in Middle French as well, vastly outnumbered again by constructions with PATH satellites, mostly prepositional phrases with *en* 'in(to)' (more than 300), and adverbial and prepositional *dedans* 'in(to)' (more than 100).

In all the relevant varieties of French, as for Latin *intrare*, there are also CAUSED MOTION uses of *entrer* (i.e. with the FIGURE as direct object, 'enter someone/something into X'), and reflexive ones, such as (239).

(239) ***S'***en **entre** en la forest (*Ferg.* 6, 6, TL s.v. *entrer*, sense refl. 'eintreten')
'He enters himself/REFL into the forest.'

B) Semantics

In the donor language(s), F *entrer* and L *intrare* are of course used for all kinds of MOTION into somewhere and for all kinds of FIGURES (including spears, bread, celestial bodies, thoughts, etc.). This section will point out specific uses which we will encounter again in Middle English below. These are, first, the military/hostile uses in (240a–d).

[171] The corresponding passage in the edition by Oesterley (1872: 340) has *Quod postquam Cosdras cognovit, mutato regis habitu arma accepit et* ***exercitum penetravit***, i.e. an accusative object, but no equivalent to *entra*. There are, however, many different manuscripts of the *Gesta Romanorum*.

(240) **Military/Hostile Uses**

a. *entrer sur quelqu'un* 'to launch an attack on somebody' (AND)
une d'Escocés vient sodeignment al huse de la sale pur espier de lour affair et si temps fuist pur **entrer** ***sur eux*** (*Anon Chr* 125.33)
'One of the Scots suddenly came to the door of the hall, to spy on their business, and whether the time was right to enter upon them [i.e. to launch an attack on them].'

b. En partic. 'Pénétrer (dans un pays) avec des vues belliqueuses' (specifically: 'enter (into a country) with bellicose intentions', DMF)
. . . nouvelles vinrent en l'ost [. . .] que les gens d'armes qui estoient establi ens es garnisons de Saint Omer [. . .] **enteroient** ***en la vallee de Cassiel***, et destruiroient le pais (FROISS., *Chron.* D., p. 1400, 444)
'[. . .] news came to the army [. . .] that the soldiers who were established in the garrisons of Saint Omer would enter into the valley of Cassel and destroy the country.'

c. trans.: 'Pénétrer (dans un pays) avec des intentions belliqueuses' (transitive, 'enter (into a country) with bellicose intentions', DMF)
le Cont de Cork [. . .] **entra** ***un fort paijs*** a grand hardivesse sur ses enemys [. . .] (*Lettres agn.* L., 1394, 208).
'The Count of Cork [. . .] entered a strong country in great audaciousness against his enemies [. . .].'

d. *Entrer dans/sur* un ennemi/*entrer l'un en l'autre.* 'Attaquer (un ennemi)' ('attack (an enemy)', DMF)
[. . .] les Englois [. . .] monstrerent desfense tres bonne ; mais il ne porent durer longement, car les François **entrerent** tantos ***en euls*** et les espardirent. (FROISS., *Chron.* D., p.1400, 654)
'[. . .] the English [. . .] presented a very good defense, but they could not last long, because the French soon entered onto them [i.e. attacked them] and scattered them.'

This specific use is also found for Latin *intrare*, given as the sense 'invasion or forcible boarding' by the *DMLBS* (s.v. *intrare*, sense 2a).

Second, the sense 'become a member of a group, a religious order, a way of life' in (241a–c) will be prominent in Middle English. It is found in both insular and continental French and in Latin:

(241) **Entering a group, a religious order, etc.**

a. *entrer (en) religion* 'to enter the religious life' (AND)
E **entrerent** ***en religioun*** (*Dial Jul* 1862)
'And they entered the religious life.'

b. S'engager dans un état, une situation, un mode de vie, une activité ('begin/take up a state, a situation, a way of life, an activity', DMF)
[. . .] l'un, c'est le mary ou la femme, puet son mary revoquier qui est **entrez** ***en religion*** sans son consentement. (*Sacr. mar.*, c.1477–81, 52).

'The other, whether husband or wife, may call back their spouse who has entered into religion without their consent.'

c. to enter institution, so as to become a member: a. (monastery, or sim.), b. (guild) (DMLBS)
in gildam *intrent* (*a*.1190)
'They should enter the guild.'

In Anglo-Norman, *entrer* is also used in the legal sense 'to enter into possession' (242).

(242) **to enter into possession** (AND s.v. *entrer* v.n., sense 1)
le dediret pur ceo ke il est **entré** par memes cely auncestre (*Fet* 94)
'He would deny it because he entered [into possession] through this same ancestor.'

As this is not given as a specific sense by *TL* and *DMF*, *entrer* does not seem to have been used as a specific administrative/legal term in continental French. For Anglo-Latin, by contrast, this use of *intrare* is attested indeed: The *DMLBS* lists the meaning 'to enter (land, building, church, etc.) so as to take up residence or take possession' (*DMLBS*, s.v. *intrare*, sense 5a). This specific use, therefore, may have been a semantic development particular to England. The same appears to be the case for the caused motion sense 'to enter in document, record, register', given by the *AND* (s.v., sense 7) as in (243).

(243) **to enter in document, record, register** (AND s.v. *entrer*, sense 7 'to enter, record')
soient **entrez** les despens de l'hostel [. . .] ***en un roule par soy*** (*Rot Parl*1 ii 241)
'The spendings of the establishment shall be recorded in a separate roll.'

The *DMLBS* mentions this use for L *intrare* as well (sense 15 'to enter in document, record, register'), which might be due to Anglo-Norman influence—the *AND* entry quotes a textbook from the thirteenth century for learning Latin, where Latin *inscribere* 'to in-write', rather than *intrare*, is given as the equivalent of AN *entrer* in this sense.

9.1.2 ME *entren*

As mentioned in section 8.4.2, ME *entren* is so frequent in translations that it was necessary to limit the analysis to a random sample of 500 attestations from the translated texts. For the autonomous texts, all the 527 attestations that could be found have been analysed. Their analysis reveals that there are considerable differences in the use of ME *entren* in autonomous texts compared to translations from Latin or French, as shown in Table 9.1 and Figure 9.1.

Attestations referring to general LITERAL motion events are considerably more frequent in translations, where they account for almost half of the attestations, while in autonomous texts, they only account for a fifth (see the black

TABLE 9.1. Usage contexts of ME *entren* in CME translated and autonomous texts.

	LITERAL: GENERAL		LITERAL:HOSTILE-OBSTACLE-POSSESSION		NON-LITERAL		TOTAL	
Translated	47%	(235)	30%	(148)	23%	(117)	100%	(500)
Autonomous	20%	(106)	34%	(180)	46%	(241)	100%	(527)
Total	33%	(341)	32 %	(328)	35%	(358)	100%	(1027)

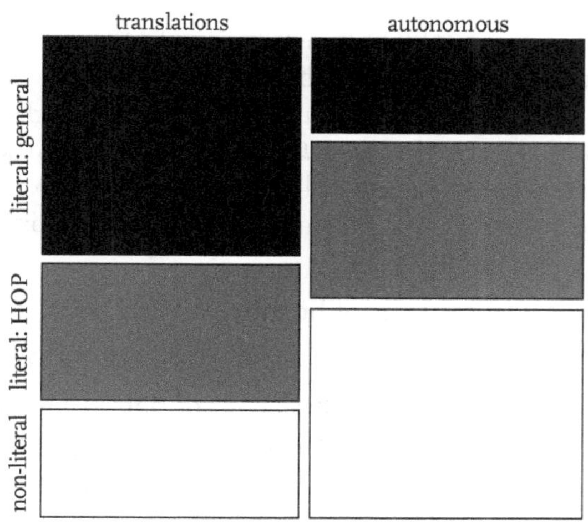

FIGURE 9.1. Usage contexts of ME *entren* in CME translated and autonomous texts.

areas in Figure 9.1 and the discussion in section 9.1.3). By contrast, in both translated and autonomous texts, *entren* is used similarly often for specific kinds of LITERAL motion, namely for contexts of hostile entering, entering in order to take possession, or entering against prohibition or obstacles (see the grey areas in Figure 9.1 and the discussion in 9.1.4). Finally, the proportion of attestations in which *entren* is used for NON-LITERAL motion is much higher in the autonomous texts than in the translated ones (the white areas in Figure 9.1, see the discussion in 9.1.5).

9.1.3 GENERAL LITERAL uses

Two typical examples of what is classified as 'general' LITERAL MOTION here are given in (244a–b).

(244) a. as blanchardyn was **entred** ***in a gardyne wythin þe kynges paleys*** alone, wythout eny feliship, for to complayne the better his hertes sorowes (*c.*1489 Caxton *Blanchardyn and Eglantine*)
'As Blanchardyn had entered in a garden within the king's palace, alone, without any company, in order to better lament the sorrows of his heart.'

b. And at the lafte he / bedene drawen and conftreyned to **entre** and dwelle with hem / fchewed hym to hem (*c*.1430(*a*.1410) Love *Mirror*)
'And at the last, [Jesus], strongly begged and pressed to enter and dwell with them [the disciples of Emmaus], revealed himself to them.'

In (244a), Blanchardyn enters a garden to be alone and lament; in (244b), the two disciples of Emmaus urge Jesus to not go any further, but to enter and stay with them. These kinds of contexts in which there is no hostility, difficulty, or taking possession involved, were classified as 'general'. They are far more frequent in the translated texts (47 per cent) than in the autonomous texts (20 per cent). This suggests that translators used ME *entren* more freely for general LITERAL MOTION events, probably because of the presence of F *entrer* or L *intrare* in the exemplar. Writers of autonomous Middle English texts, by contrast, appear to be reluctant to use *entren* for general LITERAL MOTION events. However, they often use it to describe literal motion in contexts of hostile, possessive, or difficult entering, as shown in the following section.

9.1.4 LITERAL uses in contexts of hostility, obstacles, taking possession

A closer analysis of all the LITERAL attestations of *entren* shows that the verb is used recurrently in specific contexts, which will be called HOSTILE, OBSTACLE, and POSSESSION (HOP) in the following. What they all have in common is that they highlight the fact that a boundary is traversed (or, in the case of OBSTACLE, often *not* traversed). For expository reasons, these three contexts will be first discussed separately, but since, as we will see below, they overlap in several attestations, it makes sense to treat them as a single category (as in Table 9.1 and Figure 9.1).

9.1.4.1 Hostile contexts

Let us start with the HOSTILE contexts: Many of the LITERAL uses of *entren* refer to contexts where a FIGURE enters a GOAL in a hostile, aggressive way. This is a contextual meaning that can also be found in the donor languages (as in (240a–d) in 9.1.1). Prototypical examples are given in (245–246), where the italicized parts show that the context is a HOSTILE one, such as the Babylonian carnage in Jerusalem in (245a) and the description of the large and horrible hostile army burning the towns in (245b).

(245) a. He brek þe bareres as bylyve, and þe burȝ after, And **enteres** in ***ful ernestly, in yre of his hert*** [. . .] ***Þay slowen of swettest semlych burdes, baþed barnes in blod, and her brayn spylled, prestes and prelates þay presed to deþe, wyves and wenches her wombes tocorven, Þat her boweles outborst aboute þe diches, And al watz carfully kylde þat þay catch myȝt.*** (*c*.1400(?*c*.1380) *Cleanness*)

'He [i.e. Nebuzaradan] broke the barriers immediately, and then the city [i.e. Jerusalem], and enters in in fury, with angry heart [. . .] They slew some of the sweetest pretty ladies, bathed children in blood and spilled their brain, they crushed to death priests and prelates, they cut up the bellies of women and girls so that their bowels burst into the ditches, and everybody they could catch was sadly killed.'

b. I witter þe þe Emperour es **entirde** into Fraunce, With ***ostes of enmyse, orrible and huge***; ***Brynnez*** in Burgoyne thy burghes so ryche [. . .] (*c*.1440(?*a*.1400) *Morte Arth.(1)*)
'I inform you that the emperor has entered France with hostile armies, horrible and huge; he is burning your wealthy towns in Burgundy.'

In (246a), it is the robbery and taking of prisoners after the entry, which makes the context a hostile one; in (246b) it is the characterization of the FIGURES as enemies as well as the fact that they kill their opponents.

(246) a. And whanne he hadde **entrid** the cite anon he and his men fille to ***roborie***, and robbid certayn worthi men of the cite, and ***put some of thaym in to prison*** til thay hadde paid notable summe of money to saue thair livis. And the said capteyn rood aboute the cite ***beryng a nakid swerd*** in his hand [. . .] (?*a*.1500(*a*.1471) *Brut-1461(2)*)
'And when he [i.e. the captain of the Kentish men] had entered the city, he and his men immediately started robbery, and robbed some worthy men of the city, and put some of them in prison until they had paid a substantial sum of money to save their lives. And this captain rode about the city with a naked sword in his hand [. . .]'

b. þat straunge þennes [. . .] hadden **entred** ***as enemyes*** into þe toun of Wynchelsee, & ***slowen al þat euere withstode ham and wiþseyd her comyng*** (?*a*.1425(*a*.1400) *Brut-1377*)
'That strange men [. . .] had entered as enemies into the town of Winchelsea and killed everybody who withstood them and opposed their coming.'

9.1.4.2 Obstacle contexts

Another recurrent usage context for *entren* are attestations that mention the prevention or prohibition of an entering, such as in (247a–b).

(247) a. And who-so **entryes** into ye chaumbre yer ye ale lyggys in, ***wyth-oute leue of ye officers*** he shal pay, to amendement of ye lyght, di. pound of wax. (*Doc.*in *EGilds* XXII Gild of St. Thomas of Canterbury)
'And whoever enters into the chamber in which the ale is stored, without permission of the officers, he shall pay—for the improvement of the light—half a pound of wax.'

b. I comaunde stably to freris þat þei haue not suspect companyes or conseilis of wymmen, & þat þei **entren *not*** þe abbeies of nunnes (?c.1430(c.1400) *Rule & T.St.Francis(1)* (Corp-C 296), in *Wycl. English Works*)
'I firmly advise friars not to have dubious company or meetings with women, and not to enter the abbeys of nuns.'

Other attestations mention that the entering is difficult. In (248a), for instance, Lancelot cannot find an entrance into the chapel, and in (248b), Amoryus' entering is made difficult by nettles, bushes, and thorns. These contexts were categorized together with the type above as OBSTACLE contexts.

(248) a. he ***coude fynde no place*** where he myghte **entre** (1485 Malory *Wks.*)
'He could find no place where he could enter.'

b. 'The ryuyng?' quoth he, 'qwere ys that?' and forth thru thyk and thyn He gan lepe; that ***nowdyr nettyl, busche, ner thorn Myght hym let,*** tyl he was **entryd** in. ((1449) Metham *AC*)
'"The cleft in the wall?" he said, "where is that?" and started to run on through thick and thin, and neither nettle, bush nor thorn could stop him, until he had entered in.'

9.1.4.3 Contexts of taking possession

The third specific context in which *entren* is frequently found is to describe situations where the FIGURE enters to take possession of the GOAL, such as in (249a–b). This use is probably adopted from Anglo-Norman legal language (as in (242) in 9.1.1; see also 9.1.5).

(249) a. whan his fadyr is ded the eyur schall **enter *the lond*** and pay a releeff (?a.1509 *Doc.Melton* in *Bk.Brome*)
'When his father is dead, the heir shall enter the land and pay a relief [i.e., 'a feudal payment to the overlord made by an heir upon taking possession of an estate', MED s.v. *relef(e)*].'

b. William Jenney come to Filby and **entred *in-to Holme Halle londes,*** seyng that he **entred** in the ryght and titell of his douterlawe (1475 *Paston*)
'William Jenney came to Filby and entered into the lands of Holme Hall, saying that he entered in the right and title of his daughter-in-law.'

This kind of use is most widely attested in the *Paston* and *Stonor Letters*, which often treat matters of business. Unlike those in (249a–b), however, many of these POSSESSION contexts also explicity involve HOSTILITY or OBSTACLES. This is naturally related to the contemporary legal system:

By the thirteenth century an occupier who was left in peaceful possession of land for a considerable period could develop actual rights to it. An owner had to make an entry and establish his claim in order to preserve his title [. . .]. A claim was made if the entrant came to the land in an open manner in daylight, entered it, declared he was taking control, and commanded the occupier to leave. [. . .] Because forcible entry always presented a danger to public order, rules governing an entry were altered periodically to keep pace with social realities [. . .] By the late thirteenth century assaults on persons were not permissible, and by 1411 threats by intruders were illegal. In the late fourteenth cenutry, if such attacks did occur, retaliation by the aggrieved party was considered lawful.

When the purpose of an entry was merely to assert a claim, the entrant needed followers sufficient in number to bear witness and protect him from assault. When the aim was to occupy and hold the land in dispute, entrants preferred to take as many men as they could muster to overawe the enemy. Bands of as many as 500 adherents were not unknown in the fifteenth century, though actual fights were the exception, and deaths were few (Foust 2002: 208).

Such a large band of adherents 'in war-like gear', brought by the entrant as support, is mentioned in (250), which relates an incident in the dispute between a mother and her son about the inheritance of the husband.

(250) hur sonne a forseide ***with a grete multitude of people arraied in manere of werre*** [. . .] come ***with force*** and **entrid** (into the seid) manere londeȝ; and tenementeȝ; And hit kepith with force (1453 *Chanc.E*, Petition of Thomas Fitz Harry and his Wife)
'Her abovementioned son came with force with a large number of people in war-like gear [. . .] and entered into the lands and tenements of the manor, and he is keeping it with force.'

The context of (251) is similar; it is an episode in the struggle of the Paston family about Caister, where the entrant lays siege to the manor:

(251) the said Duc ***with great force assegid*** and **entred** þe said manoire of Castre and other landes of your said suppliant, putting hym from þe lawful possession and estate þat he had in þe same, and also take from him vjc shepe and xxx nete (1475 *Paston*)
'The Duke besieged and entered the manor of Caister and other lands of your suppliant with great force, putting him from the lawful possession and estate he had in the manor, and took 94 sheep and 30 cows.'

The aggressive setting of these POSSESSION contexts is by no means an exception. Aggressive and hostile connotations of *entren* are thus supported by the POSSESSION uses, perhaps even by some of those where no army or fight is involved—after all, for a person to take possession of something, quite often another person has to be dispossessed of something, an act which may understandably enough be regarded as hostile. This idea is backed up by the fact that some of the POSSESSION attestations explicitly stress that the entry should be made peacefully, such as (252a–b).

(252) a. and to suffre hym ***pesibilly*** to **entre**, or any in his name (1496 *Paston*)
'And to allow him, or anyone in his name, to enter peacefully.'

b. Item, it is also let me wet þat my lord of Norffolk wyll send a man or two ***in pesybyll wyse*** for to **entre** the maner of Gwton (1496 *Paston*)
'Also, I was also told that my Lord of Norfolk will send a man or two to enter the manor of Gwton in a peaceful way.'

This may be interpreted as an indication that the usual way of taking over somebody else's property involves conflict or violence—therefore also some difficulty—, and if it does not, this is signalled by adding that it happened, or should happen, in a peaceful manner.

Thus the three prominent LITERAL contexts of *entren*—HOSTILE, OBSTACLE, and POSSESSION—very much shade into each other: In many cases, HOSTILE entering has the aim of taking POSSESSION of the GOAL and comes up against OBSTACLES (as in, for instance, (246b) where it is explicitly mentioned that the FIGURE encounters resistance). It makes sense, therefore, to treat these three contexts in combination, as in Table 9.1 and Figure 9.1 above, which show that in 63 per cent of the overall LITERAL attestations in autonomous texts, *entren* occurs in HOSTILE-OBSTACLE-POSSESSION (HOP) contexts. In the translated texts, attestations in HOSTILE-OBSTACLE-POSSESSION contexts account for only 39 per cent of the overall LITERAL attestations of *entren*.[172] The different proportion may be taken as an indicator that *entren* is specialized for these contexts in Middle English. Its use is freer in the translated texts, where it occurs more often in descriptions of a FIGURE entering a GOAL in a general way. This must be due to the fact that in the translations, *entren* will often be triggered by a F *entrer* or L *intrare* in the original and thus reflect the use of these verbs in the donor languages, where they are used for all kinds of inward motion.

9.1.5 NON-LITERAL uses

Many NON-LITERAL uses of ME *entren* pertain to the discourse domains of ADMINISTRATION and RELIGION, two domains which were strongly influenced by French and Latin, whose use of *entrer* and *intrare* is reflected in many of the Middle English attestations.

[172] Among the 500 random attestations of *entren* in translated texts, 227 come from Bible translations, so one could suspect the comparative lack of HOP-contexts to be due to the biblical texts. If the Bible translations are not taken into account, the proportion of HOP-contexts among the LITERAL attestations of *entren* indeed rises a bit, but still does not quite reach half of the LITERAL attestations, i.e. it is still lower than in the autonomous texts.

9.1.5.1 ADMINISTRATION

Several recurrent usage contexts of *entren* found in the *CME* can be seen as belonging to the domain of ADMINISTRATION. These include *entren* in the senses of

(a) 'to enter (an item) in a record or list; to record' (see *MED* s.v. *entren*, sense 12)
(b) 'to assume possession of property or title', and, in combination with *upon (on)*, 'to proceed against (a holder of property) so as to dispossess him' (see MED s.v. *entren*, sense 3.b–c)
(c) 'enter into a membership or group' (see *MED* s.v. *entren*, sense 5.a).

These can also be found in the donor languages (9.1.1) and will briefly be illustrated in the present section.

Attestations of the caused motion use of *entren* in the sense of 'to enter (an item) in a record or list' (cf. *MED* s.v. *entren*, sense 12) are frequent in administrative documents, such as in (253a). But they also occur in other texts, as for instance in the *Pardoner's Tale* in (253b), where the pardoner enters into his roll the names of those who have bought an indulgence. This must have been taken over from Anglo-Norman administrative usage (cf. *AND* s.v. *entrer*, sense 7 'to enter, record'), as illustrated in (243) above.

(253) a. Also, þulke [. . .] þat haldeþ, byggeth Get, Shep, swyn, and a-ȝen selleþ vn-shlawe, shulde to þe kynge fyue pans by þe ȝere of þe custome of swyn. And to þe clerke of þe town, a peny for to **entre** ***here names*** (*a*.1400 *Usages Win.*)
'Also, those that hold or buy goat, sheep, and swine, and sell [them] again unslaughtered, owe five pence per year to the king, due to the swine tax, and a penny to the clerk of the town for entering their names.'

b. Cometh up, ye wyves, offreth of youre wolle! ***Youre names*** I **entre** heer ***in my rolle*** anon ((*c*.1390) Chaucer *CT.Pard.*)
'Come up, you women, offer some of your wool! I will enter your names here in my roll right away.'

Another NON-LITERAL sense of *entren* taken over from Anglo-Norman administrative usage is 'to assume possession' (cf. (242)), related to the LITERAL POSSESSION uses described in 9.1.4.3. In (254), for instance, *entren* is clearly used NON-LITERALLY, since 'moveable goods' (such as cattle) are not a GOAL into which a FIGURE can LITERALLY enter.

(254) Also, I will that myn executours **entre** ***vpon all my goodes moveable*** (*Lincoln Diocese Documents* 1529)
'I also want my executors to enter upon all my moveable goods.'

This is similar in attestations where the metaphorical GROUND is the person who was in possession before, i.e. the person that is *entered upon*, as in (255).

(255) some sey that ye tendryd lytyll my lordys dethe in as myche as ye wold so sone **entre** ***vp-on hym*** aftyr hys dyssease (1476 *Paston*)
'Some say that you were little moved by my lord's death, as you would so soon enter upon him [i.e. upon his possessions] after his death.'

Actually, also in some of the POSSESSION attestations that were classified as LITERAL in the present study, since they describe concrete entering into property (cf. 9.1.4.3), the important aspect is not necessarily that the FIGURE has entered the GOAL, but rather that he or she has taken POSSESSION of it. This can be seen, for instance, in (256), where it is less important that the seller, Toppis, went back into the place, but that the place became his property again. Thus, there is a considerable overlap between NON-LITERAL POSSESSION and LITERAL POSSESSION.

(256) Pyte, a lystere, bowgth itt of Toppis, and now for defawt of payment Toppis hath **enterid** ayen ***þer-inne*** and shall selle itt in hast, as it is told me. (1451 *Paston*)
'Pyte, a dyer, bought it [i.e. "a handsome place in St. Laveransis parish"] from Toppis, and now because of default in payment, Toppis has entered in it again and shall sell it in haste, as I am told.'

The third prominent sense of *entren* in the domain of ADMINISTRATION is the one of 'becoming member of a group' (cf. (241) for this use in the donor languages). Attestations in the GROUP sense are mostly from ordinances of gilds, in the context of becoming a member of a gild, as in (257a–b).

(257) a. And quo-so **enter** ***into yis gyld***, he schal makyn feythe to ye alderman for holdyn of yeyse statutis ((*c.*1329) *English Gilds*, Gild of St Peter)
'And whoever enters into this guild shall swear to the warden to observe these rules.'

b. And who-so **entres** ***in to yis fraternite***, he schal pay for is entre ij. s. ((14th/15th cent.) *English Gilds*, Gild of St Mary)
'And whoever enters into this fraternity, he shall pay two shillings for his entry.'

In general, for the ADMINISTRATIVE uses, the fact that French was an important language of administration in the first centuries of the Middle English period will have been the decisive factor. When English became the language of administration again, much of the native vocabulary for this domain that had existed in the Old English period was long gone for lack of use, and the French terminology was adopted (see Durkin 2014: 235–6).

9.1.5.2 RELIGION

Numerous other NON-LITERAL uses of *entren* belong to the domain of religion. These include descriptions of metaphysical FIGURES such as the devil or the holy ghost entering into someone or something as in (258a–c), and the sense 'to enter religion, take up monastic life', discussed below.

(258) a. the dedely enemy of mankynde [. . .] **entryd** the ymages Withyn the temples (*c.*1500(?*a.*1475) *Ass.Gods*)
'The deadly enemy of mankind entered the images within the temples.'

b. Satan: [. . .] I **entered *in Judas*** (*a.*1450 *Yk.Pl.*)
'Satan: "[. . .] I entered in Judas."'

c. he [. . .] openes a place in his saule to þe haly gast: þe whilke, whene he es **entred *in***, he lightenes þe saule & thurgth his blyssed presence comfortes (*c.*1475(*a.*1400) *Boasting & P.*)
'He opens a place in his soul to the Holy Ghost, which, when it has entered in, lightens the soul and comforts through its blessed presence.'

Illustrations for *entren* in the sense of 'to enter religion, take up monastic life' are supplied in (259). As we have seen in 9.1.1, it can equally be found in contemporary insular and continental French as well as Latin (see (241)), from where it will have been taken over.

(259) a. folk that been **Entred *into ordre***, as subdekne, or dekne, or Preest, or hospitaliers. ((*c.*1390) Chaucer *CT.Pars.*)
'People that have entered a religious order, as subdeacon, or deacon, or priest, or hospital warden.'

b. aftir þei ben **entrid *in to þe kny*ȝ*thod of Crist*** (?*a.*1425(*a.*1415) *Wycl. Lantern*)
'After they have entered into the knighthood of Christ.'

Overall, the fact that Latin was the dominant language for all things religious throughout the Middle English period (e.g. Käsmann 1961: 6) will have played a role in the adoption of ME *entren* in religious contexts in the vernacular.

9.1.6 Autonomous texts revisited

At a closer look at the attestations of *entren* in autonomous texts, it turns out that, although not occurring in translations strictly speaking, many of them are actually strongly influenced by French and Latin originals. This concerns (a) quotes from or references to the Bible, as well as (b) attestations of *entren* from the *Grail Book* in Malory's *Morte Darthur*. The following subsections will discuss these in turn; section 9.1.6.3 will then present the results of a revised categorization, in which these attestations are grouped with the ones from translated texts.

9.1.6.1 Quotes or paraphrases from the Bible

At least 48 attestations of *entren* in the autonomous texts represent quotes or paraphrases from the Bible, where the corresponding passage in the Vulgate contains a form of *intrare, introire,* or *ingredi.* In some of them, the link to scripture is already apparent in their being introduced by phrases such as *Luk seiþ þat* [. . .]; *Þe story telliþ hou* [. . .]; *And þus meneþ þe gospel þat* [. . .]; *And þus seiþ Crist þat* [. . .]. But there are many other, less explicitly quoting, examples as well, such as the ones in (260), where the respective Latin verse is given in brackets.

(260) a. And whanne þes kingis sawen þis sterre, þei hadden myche joie wiþal. And þei **entriden** ***in to þe hous***, and foundun þe child, wiþ Marie his modir (*a.*1425 *Wycl.Serm.*) (Matt 2:10–11 *videntes autem stellam gavisi sunt gaudio magno valde et* ***intrantes domum*** *invenerunt puerum cum Maria matre eius*)

b. And while þei wenten to bie oile, þe spouse cam. And þes virgyns þat weren redi **entriden** ***in*** wiþ þe spouse (*a.*1425 *Wycl.Serm.*) (Matt 25:10 *dum autem irent emere venit sponsus et quae paratae erant* ***intraverunt*** *cum eo*)

c. Y am not worþi þat þou **entre** ***under my roof*** (*a.*1425 *Wycl.Serm.*) (Matt 8:8 *non sum dignus ut* ***intres sub tectum meum***)

d. Jesus **entride** ***in to a castel***, and a womman þat hiȝte Martha toke Crist in hir hous (*a.*1425–*a.*1500 *Wycl.Serm.*) (Luke 10:38 ***intravit in quoddam castellum*** *et mulier quaedam Martha nomine excepit illum in domum suam.*)

e. Jesus **entride** ***in to a Pharisees hous*** on a Satirday to ete wiþ him, and þei aspieden to take him in defaute (*a.*1425–*a.*1500 *Wycl.Serm.*) (Luke 14:1 *cum* ***intraret in domum cuiusdam principis Pharisaeorum*** *sabbato manducare panem et ipsi observabant eum.*)

f. biddeþ his servantis go out into weyes and hegges and constreynen men to **entre** þat my hous be fillid (*a.*1425–*a.*1500 *Wycl.Serm.*) (Luke 14:23 *et ait dominus servo exi in vias et sepes et conpelle* ***intrare*** *ut impleatur domus mea.*)

g. And whanne þe douȝter of þis womman was **entrid** ***in to þe halle***, and pleside to Heroude and his gestis bi tumbleris lepyng (*a.*1425–*a.*1500 *Wycl.Serm.*) (Mark 6:22 *cumque* ***introisset*** *filia ipsius Herodiadis et saltasset et placuisset Herodi*)

h. Þou schal **enter þis ark** wyth þyn aþel barnez And þy wedded wyf (*c.*1400(?*c.*1380) *Cleanness*) (Gen 7:1 ***ingredere*** *tu et omnis domus tua* ***arcam***.)

i. ȝete uus out þose ȝong men þat ȝore-whyle here **entred**, Þat we may lere hym of lof, as oure lyst biddez (*c.*1400(?*c.*1380) *Cleanness*) (Gen 9:15 *ubi sunt viri qui* ***introierunt*** *ad te nocte educ illos huc ut cognoscamus eos*)

j. And whan þei ***into Egipt*** **entred** were, Þe Egipciens fast behelden here (*a.*1450(1412) Hoccl.*RP*) (Gen 12:14 *cum itaque* ***ingressus*** *esset Abram* ***Aegyptum*** *viderunt Aegyptii mulierem quod esset pulchra nimis.*)

k. For-þi þan **entrid** barsabe (*a.*1400 *Cursor*) (Kings 1:28 *Bethsabee quae cum fuisset* ***ingressa*** *coram rege.*)

l. Thay were talde but sexty and ten Whan thei **enterd** ***into this lande***. (*a.*1450 *Yk.Pl.*) (Gen 64:27: *omnis anima domus Iacob quae* ***ingressa*** *est* ***Aegyptum*** *fuere septuaginta.*)

m. but ȝif man be born of water and þe Holy Goost he may not **entre** ***in to Goddis rewme***. (*a.*1425–*a.*1500 Wycl.*Serm.*) (John 3:5 *nisi quis renatus fuerit ex aqua et Spiritu non potest* ***introire in regnum Dei***)

n. It is liȝter a camel to passe throuȝ a needelis iȝe, than a riche man to **entre** ***into the kingdom of heuenes*** ((*c.*1449) Pecock *Repr.*) (Mark 10:25 *facilius est camelum per foramen acus transire quam divitem* ***intrare in regnum Dei***)

o. ***In to what hous*** ȝe **entren**, ȝe shal first seie, Pees be to þis hous (*a.*1425–*a.*1500 *Wycl.Serm.*)
And ***Into what hous*** that ȝe **entren**, first seie ȝe, Pees to this hous ((*c.*1449) Pecock *Repr.*)
Into what euere hous þei schullen **entre** seie þei first, pees be to þis hous (?*c.*1430(*c.*1400) *Rule & T.St.Francis(1)*)
(Luke 10:5 ***in quamcumque domum*** **intraveritis** primum dicite pax huic domui)

These attestations all reflect the impact of the Bible, or of Middle English Bible translations, which very often employ *entren* to render Latin *intrare, ingredi, introire* (cf. 8.1), quite different from Old English Bible translations, which usually have *gān in*, including for 'going to heaven' (see 5.3). In spite of occurring in the autonomous texts, these attestations are therefore actually not 'autonomous' at all, since they all quote or paraphrase verses from the Bible. They have been recategorized as translated attestations in 9.1.6.3.

9.1.6.2 *The Grail Book* in Malory's *Morte Darthur*

Besides the biblical quotes and references, there is one other prominent donor-language influenced group which emerges in the attestations from autonomous texts: Attestations coming from Books XIII–XVII in Caxton's edition of Malory's *Morte Darthur*, the so-called *Grail Book*. The *Morte Darthur* is based on French sources (cf., e.g., McCarthy 1996; Vinaver 1929: 100–8, 128–54; Field 2001a), but cannot be regarded as a translation, which is why it was classified together with the autonomous texts in this study: The editors of the 1983 edition of Caxton's Malory, for instance, write that '[e]ven when he most closely followed his source, as he did with the *Queste del Saint Graal* [i.e, the *Grail Book*], Malory was not simply translating but also selecting, compressing, and rearranging as he saw fit' (Spisak & Matthews 1983: 606).[173]

Some passages in the *Grail Book*, however, according to McCarthy, 'can be traced back to the French text word for word' (1996: 81). To illustrate this claim, McCarthy compares a short extract from the French *Queste del Saint Graal* (261a) with the corresponding passage in Malory's text.[174]Also Caxton's edition, which is the version included in the *CME*, still remains very close, as shown in the respective passage in (261b).[175]

(261) a. Quant li rois Varlans vit l'espee si trenchant, si pensa qu'il retorneroit por prendre le fuerre. Et lors revint a la nef et **entra** dedenz et remist l'espee ou fuerre; et si tost come il ot ce fet, si chai morz devant cest lit. Einsi fu esprovee ceste espee, que nus ne la treroit qui ne fust morz ou mehaigniez. Si remest li cors le roi devant cest lit, tant que une pucele l'en gita fors; car il n'avoit ilec home si hardi qui dedenz ceste nef osast **entrer** [. . .] (*La Queste del Saint Graal*, pp.204–5, cited from McCarthy 1996: 82)

b. And when kynge Hurlame sawe this suerd soo keruyng / he torned ageyne to fetche the scaubard / And soo came in to this shyp and **entred** and putt vp the suerd in the shethe / And as soone as he had done it / he felle doune dede afore the bedde / Thus was the swerd preued that none ne drewe it but he were dede or maymed / So laye he ther tyl a mayden **cam in to** the shyp / and cast hym oute / for there was no man so hardy of the world to **entre** in to that shyp [. . .] (1485 Malory *Wks.* (Caxton:Sommer) from Book XVII "Syre Galahad")

[173] This can be seen, for instance, in the collation by Vinaver (1929: 155–88) for the 'Syre Launcelot'-part of the *Grail Book*, where the French part of the page is always considerably larger than the English one, which sometimes consists of only a single line.

[174] For a long time, the *Morte Darthur* was only known in the printed edition by Caxton in 1485 (this is the text that is included in the *CME* (ed. Sommer 1889–91)), until in 1934 a manuscript was found in Winchester College Library (edited by Vinaver 1947). The two texts are not identical, which probably shows that Caxton not only printed, but also revised Malory's text (which is most probably not exactly the one that is extant in the Winchester manuscript, see, e.g., Archibald & Edwards 1996; Field 2001b; see also Spisak & Matthews 1983: 604–5, who doubt that Caxton did any revising).

[175] That there should not be so many changes in the Caxton version is not surprising, given Caxton's revision policies, which Blake describes as follows: 'In Caxton's edition of Malory there are two

There are two motion events of 'entering the ship' mentioned in the French version, each described with the verb *entrer*. The respective counterparts in the English version each involve ME *entren* (although the first *entred* is coordinated with the synonymous *came into*). Note that the English version describes a third 'entering the ship', which is not mentioned in the French version: *a mayden cam in to the shyp*. Here, the English version does not use *entren*, but the native *came in to*—this may have to do with the lack of a triggering *entrer* in the French version.

It seems that with respect to its use of *entren*, the *Grail Book* is strongly influenced by its source. This is already evident in the sheer number of attestations of *entren* in these chapters: The *Grail Book* (corresponding to books XIII–XVII in Caxton's edition) contains as many as 58 occurrences of *entren*. Another three occur in the titles referring to chapters from the *Grail Book*. In stark contrast to this, the rest of the *Morthe Darthur*—although more than six times larger than the *Grail Book*[176]—has a mere 32 attestations of the verb. *Entren* is thus overwhelmingly more frequent in the *Grail Book* than in the other ones. This fits with Spisak & Matthews's assessment that 'Malory was undoubtedly influenced in writing particular tales by the style of whatever sources he was using at the time [. . .] in his reduction of the *Queste* we see a heavy use of Gallic vocabulary' (1983: 623).

It is not only its sheer frequency, however, that distinguishes the use of *enter* in the *Grail Book* from the rest of the *Morte Darthur*. The contexts in which the verb occurs are also different: In the *Grail Book*, *entren* is far more often used for general kinds of LITERAL entering than in the rest of the *Morte Darthur*, as shown in Table 9.2.

Among the 51 LITERAL attestations of *entren* in the *Grail Book* we find two thirds in GENERAL contexts (34) and only one third in HOSTILE-OBSTACLE-POSSESSION

TABLE 9.2. Usage contexts of ME *entren* in different parts of the *Morte Darthur*.

	LITERAL: GENERAL		LITERAL: HOSTILE-OBSTACLE-POSSESSION		NON-LITERAL		TOTAL	
Grail Book	56%	(34)	28%	(17)	16%	(10)	100%	(61)
Rest of *Morte Darthur*	34%	(11)	66%	(21)	—		100%	(32)
Total	48%	(45)	41 %	(38)	11%	(10)	100%	(93)

types of changes: the elimination of common words which had become obsolete and the substitution of words in the alliterative style by those from the fashionable, French-influenced style' (Blake 1969: 185). This is particularly noticeable in the *Roman War Book* from the *Morte Darthur*, where Malory is very much influenced by both the alliterative style and vocabulary of the English *Alliterative Morte Arthure* that he used as the main source for this book, and where Caxton's edition differs considerably. It also shows up in the respective numbers of *entren* in Malory's vs. Caxton's text of the *Roman War Book*: According to the Norton edition by Shepherd (2004), Malory uses *entren* three times, with a fourth one being an emendation from the Caxton text. Caxton's version of the same book has seven attestations of *entren*. Since Malory's *Grail Book*, by contrast, is very 'French-influenced' anyway, Caxton will not have felt the need to change it very much.

176 Over 330,000 words, as compared to around 50,000 in the *Grail Book*.

contexts (17, though actually in OBSTACLE contexts only). In the rest of the *Morte Darthur*, by contrast, only a third of the attestations of *entren* (11) occur in GENERAL LITERAL contexts, while two thirds are in HOSTILE-OBSTACLE-POSSESSION contexts (21, with 11 clearly HOSTILE and 10 OBSTACLE). It might be a coincidence, or owing to the plot of the *Grail Book*, but these proportions of the HOP-contexts of *entren* in the different books of the *Morte Darthur* pattern with the ones in translations and autonomous texts respectively (Table 9.1 and Figure 9.1). For the reasons explained here, in the following section the attestations from the *Grail Book* are categorized with the translated texts.

9.1.6.3 Revised results

In 9.1.6.1–2, we have seen that there are two types of attestations of *entren* from autonomous texts which are actually due to imitation of Latin and French sources: 48 quote or refer to a verse from the Bible containing L *intrare/introire/ingredi*, and 61 are from the strongly French-influenced *Grail Book* part of Malory's *Morte Darthur*. If we take this into account and group all those into the category of translations, the differences noted in Table 9.1 and Figure 9.1 above become even more pronounced, as shown in Table 9.3 and Figure 9.2.

TABLE 9.3. Usage contexts of ME *entren* with revised corpus (*Grail Book* and Bible quotes grouped with translations).

	LITERAL: GENERAL		LITERAL:HOSTILE-OBSTACLE-POSSESSION		NON-LITERAL		TOTAL
Translated	48%	(292)	27%	(166)	25%	(151)	609
Autonomous	12%	(49)	39%	(162)	50%	(207)	418
Total	33%	(341)	32%	(328)	35%	(358)	1027

FIGURE 9.2. Usage contexts of ME *entren* in CME translated and autonomous texts with revised classification (*Grail Book* and bible quotes grouped with translations).

With this revision, the percentage of attestations in which *entren* is used for GENERAL LITERAL entering drops to 12 per cent in autonomous texts (20 per cent in Table 9.1), while it stays roughly the same in translated texts (48 per cent; 47 per cent in Table 9.1). This indicates that *entren* is preferred for specific contexts: those in which the entering is aggressive or difficult, i.e. where the crossing of a boundary is highly salient, and those where the verb is used NON-LITERALLY. In the autonomous texts, these special contexts together account for 88 per cent of all attestations of *entren*.

9.1.7 Complementation patterns

In 9.1.1 we have seen the complementation patterns of F *entrer* and L *intrare* in the donor languages: They combine with PATH satellites (prepositional phrases and adverbs), and in Anglo-Norman and Latin, they can also combine with GROUND in a direct object (instead of a preposition complement). The present section will discuss the complementation patterns of *entren* in Middle English (9.1.7.1), while the following section will deal with their subsequent changes (9.1.7.2). We will see that in Middle English the verb is typically used with PATH satellites, while the complementation with GROUND in a direct object (e.g. *enter the house*) is still very rare. Direct object GROUND is more common, however, in POSSESSION and HOSTILE contexts than in others. Later on, the direct object pattern becomes more frequent and supersedes the prepositional one with *in(to)*. In Present-Day English, *enter into* is largely restricted to a few NON-LITERAL uses of the verb.

9.1.7.1 Complementation patterns in Middle English

Over all of its attestations, disregarding their type and provenance, ME *entren* combines most often with PATH satellites. In more than half of its attestations (637, i.e. 62 per cent), its PATH satellite contains the information PATH^{in}—which is of course already expressed by the verb—as well, such as the prepositional phrases with *in* and *into* in (258b) and (259) above, the adverb *therin* in (256), or the adverb *in* in (258c). This is also the case in the earliest attestation of ME *entren*, the one in (262), dating from *c*.1275.

(262) We..bieþ i **entred** ***in to cristes seruise***. (*c*.1275 *Ken.Serm.*, MED s.v. *entren*)
'We have entered into the service of Christ.'

The bare use of *entren* is also not uncommon (191 attestations, i.e. 19 per cent), such as in (263a–b). The respective context makes clear that in (263a) the GROUND is heaven, and in (263b) it is an estate (i.e. a POSSESSION context).

(263) a. and thou3 withoute such grace it is ouer hard to him being riche for to **entre**. ((*c*.1449) Pecock *Repr.*)

'And yet, without such grace, it is exceedingly hard for him, being rich, to enter [i.e. into heaven].'

b. for and he **enter** by his awne auctorite ye may put him owte and take ye the profetes of the land. And gif he **enter** by auctorite of the Sheref by a writ, is called *habere facias sesinam*, ye most suffer the Sherefs Officers to **enter** by vertu of the writ or waraunt, and as sone as thai ben gone **enter** ye ageyn and take the profetez. (1482 *Stonor*)
'For when he should enter by his own authority, you may put him out again and take the profits of the land yourself. And if he enter by authority of the Sheriff, by a writ called *habere facias sesinam*, you must let the Sheriff's Officers enter because of the writ or warrant, and as soon as they are gone, you enter again and take the profits.'

By contrast, it is comparatively rare for GROUND to be expressed in a direct object. This is the case in only ten per cent of the attestations (103 attestations; see, e.g., (251), (252b), and (258a)). This complementation pattern also appears later in Middle English: The first attestation of ME *entren* with direct object GROUND given by the *MED* is (264) from *Cleanness*, i.e. more than a hundred years after the earliest attestation with an *into*-PP in (262).

(264) Þou schal **enter þis ark** wyth þyn aþel barnez And þy wedded wyf (*c*.1400(?*c*.1380) *Cleanness*, MED s.v. *entren*)
'You shall enter this ark with your noble sons and your wedded wife.'

This attestation is likely to have been modelled on the corresponding passage in the Vulgate *ingredieris arcam tu et filii tui, uxor tua* (Gen 6:18), which also has a direct object. However, looking at all the attestations from the *CME* sample in which the rare direct object pattern does occur, it becomes obvious that this pattern is definitely more frequent in contexts of HOSTILE entering and of TAKING POSSESSION (LITERAL and NON-LITERAL) than in other contexts. Table 9.4 compares the attestations with GOAL-GROUND in the direct object with those in which GOAL-GROUND is introduced by a preposition.[177]

In a good quarter of the attestations (mentioning GOAL-GROUND) in which ME *entren* is used in contexts of HOSTILITY and POSSESSION, GROUND is expressed in the direct object (for examples, see sections 9.1.4.1 and 9.1.4.3). In other contexts, only around a tenth of the attestations have GROUND in a direct object.

[177] We can only compare two structures that really are alternatives to each other. This is why, obviously, attestations such as *and enteres in ful ernestly* are not included in the numbers in Table 9.4, because they make no mention of GROUND and are thus no alternative to the direct object pattern. For the same reason, bare verb uses of *entren* are not taken into account either, and neither are combinations with prepositional phrases which do not introduce GOAL-GROUND, but introduce other GROUND information, such as *enter under my roof* (which is not an alternative to *enter my roof*). Attestations with adverbial *there*, such as *they enteryd there*, are also not included in Table 9.4, because we cannot be sure whether *there* 'replaces' a noun or a prepositional phrase.

TABLE 9.4. ME *entren* + direct object GROUND vs. *entren* + preposition + GROUND in CME.

	DIRECT OBJECT GROUND		PREPOSITION + GROUND		TOTAL	
HOSTILE/POSSESSION	27%	(47)	73%	(126)	100%	(173)
Other context	11%	(56)	89%	(453)	100%	(509)

(χ^2=26.2008, df=1, $p<0.001$)

If the distribution of the direct object versus the prepositional complement GROUND pattern was random, we should have only 26 direct object attestations in the HOSTILE and POSSESSION uses, and 77 in the other contexts. Thus, even if *enter something* is rare in Middle English, it has a clear preference for the HOSTILE and POSSESSION contexts. Why should this be the case?

That the pattern should have been taken over from French is unlikely, because, as we have seen in 9.1.1, *entrer* is very rarely used with a direct object in continental Old and Middle French. In Latin, it is of course possible to construe *intrare, introire, ingredi* with a direct object, but it is unlikely that this should have been responsible for Middle English *enter something* in the HOSTILE and POSSESSION contexts: These contexts are far more frequent in non-religious texts, such as romances, chronicles, and administrative documents, i.e. exactly in those domains where the influence of Latin was less strongly felt.[178]

For Anglo-Norman, however, the direct object pattern is commonly attested with *entrer*, including in the contexts in question here (cf. *AND* s.v. *entrer*; see also section 9.1.1). So, *enter something* in HOSTILE and POSSESSION contexts might be a 'homegrown' pattern.[179] A probable explanation for it might lie in constructional semantics: The prepositional pattern *enter into something* is an instance of the INTRANSITIVE MOTION construction, i.e. it is associated with MOTION. MOTION can be seen as the crucial meaning component in the attestations where *entren* is used for general LITERAL MOTION, as well as in the OBSTACLE contexts. By contrast, MOTION seems to be less important in the HOSTILE and POSSESSION examples: What is more important here is the notion that something is happening to the GROUND—it is taken by the FIGURE, or it is attacked by the FIGURE. This notion would be more typically expressed in a direct object construction, which is also the typical complementation pattern for native verbs that denote 'taking possession' and 'entering in an aggressive way' in Middle English, such as *nimen, taken, winnen, breken* as well as the French-borrowed *seisen* in examples (265a–d).

178 It may, however, have played a role in the direct object pattern in religious attestations such as *Yhe sal noght entre..Hevenryke* (*a*.1425(*a*.1400) *PConsc.*) 'You shall not enter the kingdom of heaven'.
179 Note also that the *DMF* entry for *entrer* has a subsense 'enter with bellicose intentions' for *entrer* with direct object, illustrated with an Anglo-Norman attestation and no continental Middle French ones (see section 9.1.1).

(265) a. Þrouȝ heir strengþ and myȝt þai **nomen** and **token al þe landes** (*c.*1400 *Brut-1333*)
'Through their strength and power they seized and took all the lands.'

b. Þis quene was þo glad inou; aboute he[o] wende anon Forto **saisi al þe lond and þe maners echon.** (*a.*1325 *SLeg.Kenelm*)
'Then this woman was quite glad and went immediately to seize all the land and each of the manors.'

c. wendon þæt hi sceoldon **winnon eall þæt land** (*a.*1121 *Peterb.Chron.*)
'They thought that they should conquer all that land.'

d. Thy men cam and **breeke our parke** by nyght, and tooke out of hyt two of our horses. ((1400) *Let.Hen.IV* in *RS 18.1* (Cleo F.3))
'Your men came and broke our park by night, and took out of it two of our horses.'

It seems plausible that *entren* should develop this slight preference for the direct object pattern in the meanings 'take possession' and 'enter in an aggressive way' due to the fact that the GOAL in these meanings is somewhat more of a PATIENT, i.e. a semantic role for which a direct object is more typical than a prepositional phrase.

9.1.7.2 Complementation patterns in Early Modern and Modern English

Before we conclude this section, let us take a brief look at what happens to the complementation patterns of *enter* after the Middle English period. Stefanowitsch (2013) has investigated the distribution of *enter*+direct object and *enter*+PP*into* from 1450 until 1700, using the *Corpus of Early English Correspondence Sampler* and the *Lampeter Corpus of Modern English Tracts*. He finds that in literal uses of *enter*, the direct object pattern becomes more frequent than the *into*-pattern in the sixteenth century, while in metaphorical uses, the *into*-pattern stays dominant.

For the present study, I conducted a similar analysis with material from a different corpus: For Early Modern English (*c.*1500–*c.*1700), I used the *Penn–Helsinki Parsed Corpus of Early Modern English* (*PPCEME*), and for Present-Day English the *BNCweb*. Figure 9.3 gives the frequency of the patterns *enter something* (the black areas) and *enter in(to) something* (the white areas) in Middle English (see above), Early Modern English (on the basis of the PPCEME, without the Bible—273 attestations),[180] and Present-Day English (on the basis of 216 random attestations from BNCweb).[181]

[180] *Enter* is more often attested in the material in *PPCEME*, but, like in Table 9.4 above, only those attestations were taken into account that are technical alternatives to direct object ground: those with preposition *in* or *into*.
[181] I extracted 300 random attestations from *BNCweb*, 216 of which were in complementation patterns with GROUND expressed either in a direct object or in a prepositional phrase with *in(to)*.

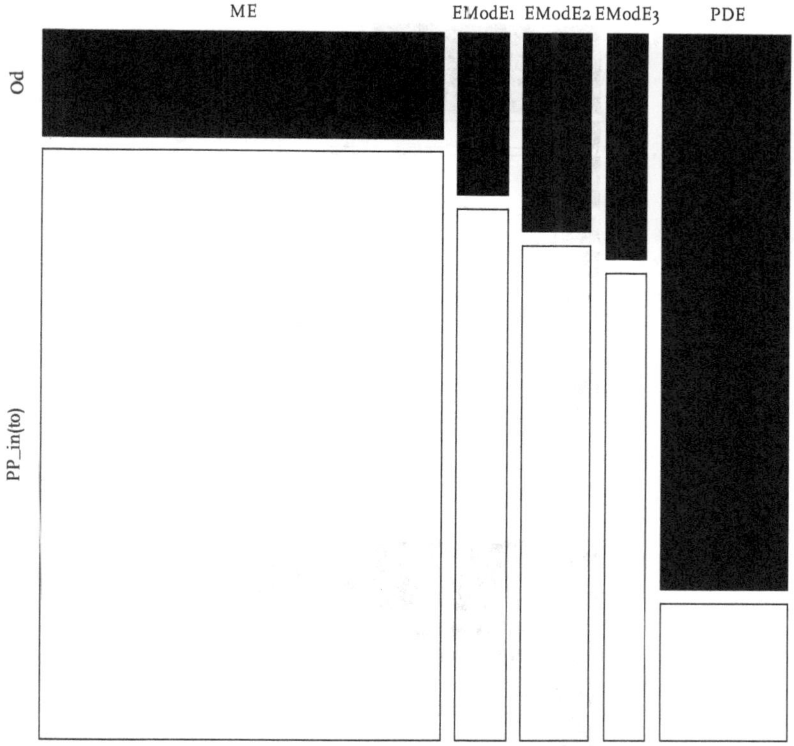

FIGURE 9.3. Complementation of *enter* in Middle English (CME), Early Modern English (PPCEME without Bible), and PDE (273 random attestations from BNCweb).

Figure 9.3 shows that over the periods, the direct object pattern is gaining ground over the prepositional one: *enter something* becomes more and more frequent, while *enter into something* recedes. Figure 9.4 shows the results for the same attestations separately, with HOSTILE/POSSESSIVE contexts on the left, GENERAL LITERAL MOTION contexts in the middle, and NON-LITERAL MOTION contexts on the right. The percentages and numbers of the attestations in the different contexts and periods are given in Table 9.5.

In Middle English, the direct object pattern already accounted for a good quarter (27 per cent) of the hostile and possession contexts, while it was still less frequent in the other contexts (12 and 10 per cent). In Early Modern English, the predilection for the direct object pattern in HOSTILE and POSSESSION contexts continues: it now accounts for more than half of these attestations, such as (266a–b) (see Figure 9.4a):

(266) a. she... **entered** and ***possessed Edenburgh*** (Hayward, E2, PPCEME)

b. the duke of Burgoyn, ***with a companye of .vi.M. knyghtys*** **entryd** ***Brytayn***, there ***by strength*** toke from hyr, hyr .iii. sonnys (Fabyan, E1, PPCEME)

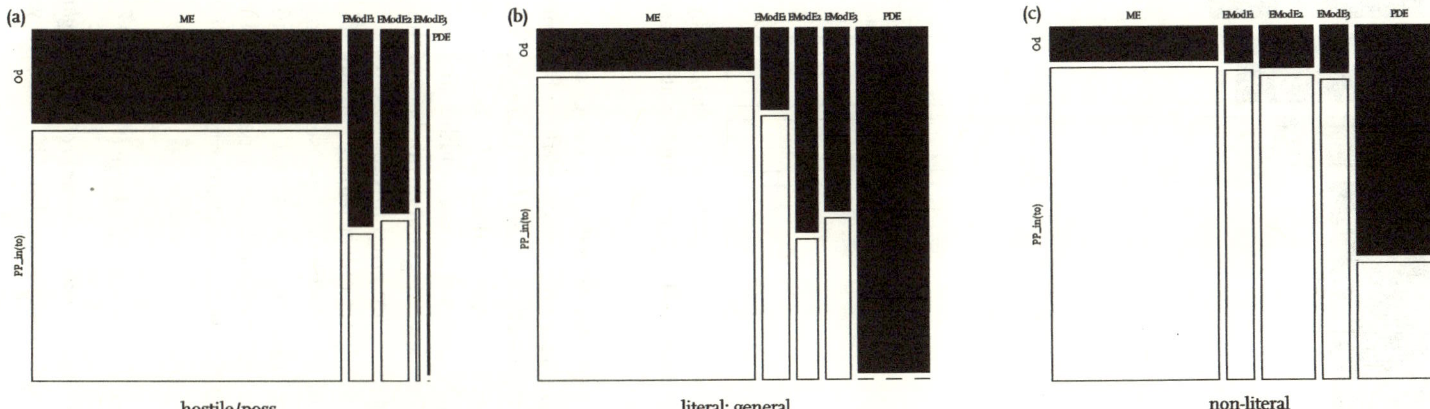

FIGURE 9.4. Complementation of *enter* in different usage contexts (ME, EModE, PDE). Black: *enter something*; white: *enter in(to) something*.

TABLE 9.5. Complementation of *enter* in different usage contexts (ME, EModE, PDE).

	ME		EMODE1		EMODE2		EMODE3		PDE	
	OD	PP	OD	PP	OD	PP	OD	PP	OD	PP
HOSTILE/POSS.	27% (47)	73% (126)	57% (8)	43% (6)	53% (8)	47% (7)	50% (1)	50% (1)	100% (1)	—
LITERAL:general	12% (33)	88% (241)	24% (8)	76% (26)	59% (16)	41% (11)	53% (17)	47% (15)	99% (89)	1% (1)
NON-LITERAL	10% (23)	90% (212)	11% (4)	89% (34)	12% (9)	88% (65)	14% (5)	86% (32)	66% (83)	33% (42)
Total	(103)	(579)	(20)	(66)	(33)	(83)	(23)	(48)	(173)	(43)

In general contexts of LITERAL MOTION, it is becoming slightly more widespread to use *enter something* rather than *enter into something* in the course of the Early Modern English period (see Figure 9.4b). The pattern seems to have spread from its origin in the HOSTILE and POSSESSION contexts, probably aided by the fact that in Latin *intrare* + direct object is possible as well. For NON-LITERAL MOTION contexts, *enter into something* is still preferred in Early Modern English (see Figure 9.4c).

In Present-Day English, *enter something* has become the standard pattern for LITERAL MOTION (see Figure 9.4b).[182] Among the NON-LITERAL MOTION contexts in Present-Day English, *enter into something* is still used particularly in the collocation *enter into a contract* (nine times in the *BNC* sample) and in the roughly synonymous *enter into an agreement, alliance, arrangement, association, covenant, transaction,* and *treaty* (another 15 times), all denoting some bond to which the figure binds itself (cf. *OED* sense I.8.c). Another collocation is with words denoting 'talk': *(enter into) correspondence, discussion, negotiation, return,* and *talk* account for another seven examples. Also *to enter into force* (for *law*) is a common expression. Why should it be exactly those uses for which speakers stick to the *enter into* pattern? I think one reason, apart from their being more or less fixed expressions, is that all these Present-Day English collocations of *enter into* can be characterized as dead metaphors, in which the source domain of MOTION is no longer evoked (on the life cycles of metaphorical expressions, see Croft & Cruse 2004: 204–6). Their deadness is illustrated by how odd the effect is if *enter* is replaced with another motion verb in *enter into a contract*: *?pop into a contract*, *?slide into a contract*, etc. Of course, the dead

182 See also the *OED* entry for *enter: enter into*: 'now largely superseded by the trans[itive] use 10'. There was only one single description of literal motion with *enter into* in the *BNC* sample: *As I* ***entered*** *through an archway* ***into the cobbled farmyard*** *the shelling had increased* (*BNCweb*). In this, however, the GOAL-complement is conspicuously distant from the verb itself, separated by a piece of additional GROUND-information (*through an archway*). Here, a direct object would not be possible if the chronological order *through an archway – into the farmyard* is to be maintained.

metaphor can be revived by expressions such as these, but the fact that other motion verbs are not frequent at all with *into a contract* can be taken as an indicator that the metaphorical mapping that was at the origin of *enter into a contract* is no longer productive. This is clearly different with other NON-LITERAL attestations of *enter*, such as (267).

(267) But the thought of returning to publishing had not seriously **entered *his head*** until the spring of 1981 (BNCweb)

Thoughts can not only *enter* a head, but, according to the *BNC*, also *come, pop, creep, slip, slide, swarm, flash, flood, burst,* or even *be invited* or *implanted into* it. Once they are in a head, they can *race, jumble and whirl, settle,* and *grow in* it and *be driven, vanish,* or *fly from* it again.[183] For the target domain of THOUGHTS, the metaphorical mapping from the source domain of MOTION is extremely productive, which indicates that the MOTION metaphor is very much alive here.

It is exactly these kinds of NON-LITERAL expressions, the ones arising from productive MOTION metaphors, that account for the bulk of the Present-Day English NON-LITERAL attestations of *enter something* (the black area in Figure 9.4c above). As the MOTION metaphor is alive in these expressions, speakers employ the complementation pattern that they use for *enter* in LITERAL MOTION contexts: the direct object. By contrast, for the dead metaphors in the fixed expressions *entering into contracts, negotiations, treaties,* and so forth the complementation pattern has not been updated.

9.1.8 Summary

What the study of ME *entren* has shown is that, above all, it is not adopted into Middle English as a motion verb as such, i.e. not for GENERAL LITERAL MOTION: GENERAL LITERAL MOTION contexts only account for between 12 and 20 per cent in the autonomous texts (Tables 9.1 and 9.3). Rather, the verb is borrowed in specific usages, many of them in NON-LITERAL MOTION contexts (around half of the attestations in autonomous texts, see Tables 9.1 and 9.3, often in the domain of RELIGION or ADMINISTRATION, e.g. *enter into moveable goods*). Another specific use in which the verb occurs frequently in autonomous texts are contexts where the entering is aggressive, difficult/prohibited, or aimed at taking possession of the GOAL (HOSTILE-OBSTACLE-POSSESSION contexts). These contexts account for around 40 per cent of the non-translated attestations of *entren* (see Table 9.3). Particularly in comparison to the translated texts, in which the GENERAL LITERAL MOTION contexts account for around half of the attestations, this shows that ME *entren* is rather restricted in its use: Only when it is triggered by one of its etyma in the French and Latin text that is translated into Middle English is it used more freely for GENERAL LITERAL MOTION as well.

---

183 Query '({thought}|{idea}) * _{VERB} (into|in|from|out of) ** ({mind}|{head})'.

The HOSTILE and POSSESSION contexts are also the ones in which ME *entren* is most frequently combined with a direct object rather than in its usual pattern with a PATH satellite. This must be due to analogy to the direct object pattern of synonyms such as *nimen, taken, seisen, winnen,* and *breken,* and, more generally, to the constructional semantics of the direct object construction (with the direct object as PATIENT), which fit these contexts better than the ones of the intransitive motion construction. Had *entren* been adopted only in these uses—one might even speculate—it could have turned into a synonym of *seisen, breken,* or the like, but of course, the contact situation would have needed to be a different one (for instance, without the continuing influence of Latin).

In a little excursus on the diachronic development of the complementation patterns of *enter* (9.1.7.2) we have seen that the direct object pattern spreads to other LITERAL MOTION uses in the Early Modern English period. In this spread, the fact that Latin *intrare* can take a GROUND object without a preposition may have played a role as well. In Present-Day English *enter,* direct object has become by far the most common way to express GROUND. The only uses that have resisted this change from *enter into something* to *enter something* are idiomatic instances of dead metaphors such as *enter into a contract,* in which the original motion metaphor is no longer active.

9.2 *Ishen/Issuen*

9.2.1 Donor language portrait: OF/MF *issir,* L *exire*

OF *issir* has developed from Latin *ex-ire,* literally 'to out-go'. At the origin of OF *issir* is thus a Latin form with a PATH satellite (the prefix *ex-*). As discussed above (7.1), this former prefix formation has become fused with respect to both meaning and pronunciation. In Old French, it is no longer analysable into separate morphemes; the PATH meaning 'out' that used to be conveyed by L *ex-* is carried by the whole verb OF *issir.*

A) Syntax

Similar to what we saw for *intrare/entrer* in 9.1.1, also *exire* and *issir* are used with GROUND expressed in prepositional phrases in the donor languages (cf. the Latin, continental French and Anglo-French examples in (268a–c)), sometimes with additional PATH adverbs (cf. *hors* in (268b–c)). Again, GROUND in the direct object seems to be common only in Latin and Anglo-Norman, as exemplified in (269a–c).

(268) a. ***de domo*** illa **exivit** (1458, *SelPlCrown* 83, DMLBS s.v. *exire,* sense 1c)
'She exited from the house.'

b. Ha! mauvaise et fausse personne, **Is, *hors,* is *de celle chaudiére.*** (*Mir. st Panth.,* 1364, 357, DMF s.v. *issir*)

'Ha, you bad and evil person, exit, out, exit from this cauldron.'

c. **istrai *hors del reaume de Engleterre*** [. . .] et jamés ne retourneray (BRITT i 64, AND s.v. *issir, issir fors de* 'to leave, depart from')
'I will exit out of the realm of England [. . .] and never will return.'

(269) a. **ecclesiam exivit** (ALEX. CANT. *Mir.* 29 p.217, DMLBS s.v. *exire,* sense 1e)
'He left the church.'

b. **issist la chambre** (*Rot Parl*i iii 13, AND s.v. *issir,* v.a. 1 'leave, depart')
'He left the room.'

c. le sire Bertreme od toutz ces gentz **ischa la ville** (*Anon Chr* 72, AND s.v. *issir,* v.a. 1 'leave, depart')
'Sir Bertram left the city with all these people.'

For continental Old and Middle French, *TL* has no attestations of *issir* with a direct object GROUND, and even the very long entry in *DMF* includes only two instances (s.v. *issir,* sense A.4), as opposed to a vast number of attestations with PATH satellites.[184] The direct object complementation pattern as in (269b–c) might thus be another Anglo-Normanism, perhaps modelled on the complementation of ME *leven* 'leave' and maybe supported by the fact that Latin *exire* can also combine with a direct object.

B) Semantics

The verb is used for all kinds of motion out of somewhere, as seen in (268–269), as well as with all kinds of FIGURES (including arrows, tears, sweat, etc.). I will only mention those specific uses in the donor language which are also prominent in Middle English (see 9.2.2–5). The first one is the figurative use in the sense of LINEAGE, listed in *AND, TL,* and *DMF* for the relevant varieties of French (see (270a–c)), and also present in Anglo-Latin (cf. *DMLBS* s.v. *exire*).

(270) LINEAGE: **'be descended from/born of'**

a. to be born, descend from (as heir) (AND s.v. *issir, issir de*)
Et ***de cele Beatrice* issit** William de Vescy le primer (*Percy Ch* 457)
'And from this Beatrice descended [lit. exited] William de Vescy the first.'

b. *Naître de, descendre de* ('be born from, descend from', DMF s.v. *issir,* sense C.2)
. . . [elle] estoit née et **yssue *de grand maison et noble parentage.*** (*C.N.N.,* c.1456-1467, 549)
'[She] was born and descended [lit. exited] from a great house and a noble family.'

[184] Moreover, they are slightly different from the Latin and Anglo-Norman ones in (269a–c), in which the direct object really denotes the SOURCE-GROUND. In the continental Middle French attestations, the direct object denotes the *boundary* of the SOURCE-GROUND, rather than the SOURCE-GROUND itself

c. *hervorgehen, abstammen* ('be born from, descend from', TL s.v. *issir*)
Li prince e li rei, Tuit ***en*** sunt **eissu** (*Reimpr* 27)
'The princes and the kings are all descended [lit. exited] from there [i.e. from Noah and his sons].'

In Anglo-Norman, the uses in the sense of LINEAGE are further supported by the fact that the past participle, *issu(e)*, is used nominally to denote 'offspring', 'heir, descendant', as in (271a–b).[185] The alternative form of ME *ishen*, namely ME *issuen*, is derived from this participle.

(271) *issue*, n. 'offspring, progeny'; 'heir, descendant' (AND s.v. *issue*, sense 4)

a. [. . .] issint qe par defaute de **issue** de dit Monsieur William les ditz manoirs ové les appurtenances remayndrent au dit conte (*Private Indentures* 131)
'So that due to lack of offspring of the said Sir William, the said manors with their appurtenances will remain with the said count.'

b. le seconde, terce, quart **issue** en la line avera mesme l'avauntage com le primer (*YBB* 18–19 Ed III 203)
'The second, third, fourth descendant of the line will have the same benefit as the first.'

The second specific use that we will find in Middle English (9.2.2) is the MILITARY use of *issir* in the meaning 'sally forth, make a sortie, make a charge against', as in (272a–f). This kind of use is not listed in *DMLBS* for the Latin verb, but present in all varieties of Medieval French:

(272) **MILITARY: 'sally forth, make a sortie, make a charge against'**

a. *'Sortir d'un lieu assiégé, d'une forteresse, faire une sortie'* (DMF s.v. *issir*, sense B.1 'exit from a besieged place, from a fortress, make a sortie')
...ilz **issirent *contre leurs ennemis***, lesquelz tantost ilz vainquirent (COLART MANS., *Dial. créat.* R., 1482, 139)
'They made a sortie [lit. exited] against their enemies, which they soon defeated.'

b. *issir a* '(mil.) to sally forth, make a charge against' (AND)
par l'enchesoun si ascunes des Engleis ***a eaux*** (=the Scottish, who are besieging Carlisle) vouldrent **yssir** ou od eaux combatre ou bataille doner (*Anon Chr* 11.28)
'In case that any of the English wanted to sally forth [lit. exit] on them, or fight with them, or give battle.'

('avec un compl. d'obj. interne marquant une limite'): *Seinne [. . .] issait* ***ses termes*** 'The river Seine exited its limits', and *issir* ***les portes de Paris*** 'exit the gates of Paris'.

[185] The *DMF* similarly has 'Part. passé en empl. adj. *Issu de* "Descendant de" ' (past participle used as adjective: 'descended from', sense C.2), but in the respective attestations, *issu* is nevertheless more verbal than in the clearly nominal attestations from Anglo-Norman in (271a–b). See, e.g., the coordination with *nez* 'born' in *Car les filz, pour ce que ilz sont* ***nez et issus*** *des parenz, [. . .]* (ORESME, *E.A.*, c.1370, 442, *DMF*) 'Because the children, since they are born and descended [lit. exited] from the parents [. . .]').

c. *issir a, en bataille* '(mil.) to go out and fight' (AND)
Agolaunt vout meut **isseer *a batayle*** ke rendre la cité e morer vil-leynement dedens (*Turpin* 453)
'Agolaunt prefers to go out and fight [lit. exit to battle] rather than to render the city and die in it like a coward.'

d. *issir contre* '(mil.) to sally forth, make a charge against' (AND)
Ebrahim e Laumaçor **isserunt *countre ly*** ou lour oosz (*Turpin* 855)
'Ebrahim and Laumacor sallied forth [lit. exited] against him with their army.'

e. *issir sur* '(mil.) to sally forth against' (AND)
Et quaunt [. . .] ceaux del chastelle ***sur eaux* yssent**, mentenaunt se tournent en fuyte (*Anon Chr* 6.40)
'And when those from the castle sally forth against them, they flee instantly.'

f. *s'issir fors encuntre* '(mil.) to make a sortie, sally out' (AND)[186]
Cest chastel n'est mie fort [. . .]! **Issum** nus ***fors encuntre***, si lur frum asaillie! (*Fant* OUP 187)
'This castle is not strong at all [. . .]! Let's sally REFL out against, and make an attack on them!'

Judging from the respective prominence of the MILITARY uses in the dictionary entries, they seem to be more widespread in Anglo-Norman than in Medieval French in general: The *DMF* only has one subsense with two attestations for them (see (272a)), while the *AND* has a whole range of them (272b–f). This impression, however, may also be due to different editorial practices—the *AND* lists all kinds of combinations with prepositions and adverbs separately.

One last use of *issir* in insular and continental French that needs to be pointed out here involves tears or other liquid FIGURES:

(273) **LIQUID FIGURE: 'flow out, pour out'**

a. '(of tears), flow, pour out' (AND s.v. *issir*, sense v.n.2)
lermes **isserent** par grantz russeals (*Sz Med* 137)
'Tears flowed out in large streams.'

b. *d'un liquide*: issir de *'couler, jaillir de'* ('flow, gush from', DMF s.v. *issir*, sense B.2)
Qui fu norry du saint lait excellent Qui du jardin **issi** en habondance. (*Mir. Berthe*, *c.*1373, 252)
'Who was nourished with the excellent holy milk which flowed abundantly from the garden.'

186 Cf. similar military senses for *issir defors* 'to make a charge, sortie' (lit. 'exit out'); *issir encontre* 'to make a charge against, sortie against' (lit. 'exit (out) against'); *issir fors (de)* 'to make a sortie, sally out' (lit. 'exit out (of)'); *s'en issir* 'to make a sortie, sally out'; all *AND* s.v. *issir*.

9.2.2 ME *ishen/issuen*

The search in the *CME* for *ishen* 'exit' and its variant forms yielded 192 attestations of the verb.[187] Around a fifth of these (38 attestations) have no <u> or similar letter, and thus represent the verb *ishen* (from French *issir*). The rest all represent the verb *issuen*, derived either from the noun *issue* or from *issu*, the past participle of French *issir* (cf. 9.2.1). The meanings of the two verbs are very much the same, therefore both forms are treated together in this section.[188] In what follows, unless indicated otherwise, *ishen* refers to both of the verbs.

As expected, *ishen* has far more attestations in texts that are translations from Latin or French than it has in autonomous Middle English texts: Of the 192 occurrences, only around a sixth (30) come from autonomous Middle English texts. Conspicuously though, these autonomous texts are but four; the verb thus is not well dispersed: Most examples were found in the non-translated parts of the *Brut Chronicle* and in Caxton's edition of Malory's *Morte Darthur* (though not in the *Grail Book*). Two attestations come from the *Paston Letters*, and one example (274) from the *Alliterative Morte Arthure.*

(274) Thay **iche** on þe *e*nmyse and *e*gerly strykkys (c.1440(?a.1400) *Morte Arth.(1)*)
'They issue on the enemies and strike eagerly.'

In (274) though, the use of *ishen* may have been triggered additionally by the alliterative verse (italicized in the example) in which it occurs.

The mosaic plot in Figure 9.5 gives an overview of the contexts in which *ishen* is used in translated (left column) and in autonomous Middle English texts (right column): The darker areas on the top represent the LITERAL uses (GENERAL LITERAL motion in black, MILITARY contexts in dark grey), the white areas on the bottom the NON-LITERAL uses. The numbers are given in Table 9.6 below.

Similar to what we saw for *entren* (9.1.2), GENERAL LITERAL motion contexts (discussed in 9.2.3) are rarer in the autonomous texts than in the translated ones. Most of the attestations from autonomous texts are in MILITARY contexts (presented in 9.2.4). The rare NON-LITERAL uses are discussed in section 9.2.5.

9.2.3 GENERAL LITERAL uses

General literal uses of *ishen* mostly refer to people normally 'exiting' from somewhere, such as from one's inn (275a) or room (275b).

---

[187] Search terms *issh**, *isse**, *ish**, *issu**, *issw**, *issv**, *yss**, *vss**, *iche*, *icen*, *iced*, and *ushen*.

[188] There is also variation between the forms with and without <u> in different manuscripts of the same text: In Lydgate's translation *The pilgrimage of the life of man*, there is one *ysswe* (i.e. a form of *issuen*) and six *ysse** (i.e. forms of *issen*), for two of which the variant manuscript readings also give forms with <w> or <u>, hence forms of *issuen*.

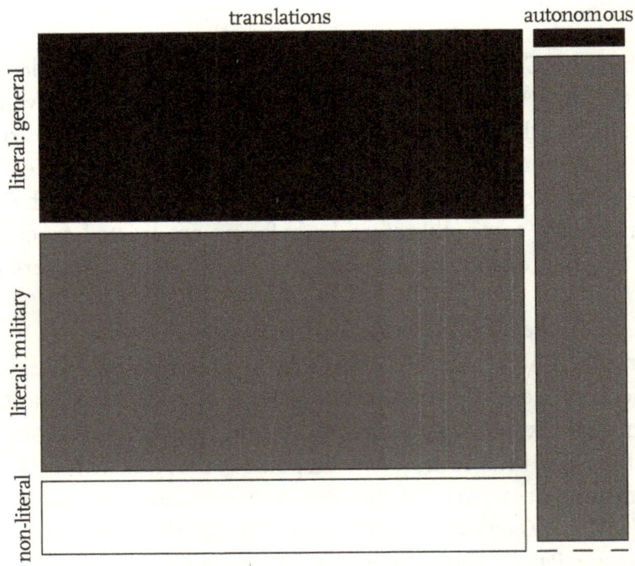

FIGURE 9.5. Usage contexts of ME *ishen* in CME translated and autonomous texts.

TABLE 9.6. Usage contexts of ME *ishen* in CME translated and autonomous texts.

	LITERAL: GENERAL		LITERAL: MILITARY		NON-LITERAL		TOTAL	
Translated	38%	(61)	48%	(77)	15%	(24)	100%	(162)
Autonomous	3%	(1)	97%	(29)	0		100%	(30)
Total	32%	(62)	55%	(106)	13%	(24)	100%	(192)

(275) a. Bot do crie þorgh þe toun, þat non in strete walk vp & doun bot to þer innes go. Whan þe crie was cried, walkand was non sene, Bot to innes hied [. . .] þei durst not **isshen** ***oute*** (?*a*.1400(*a*.1338) Mannyng *Chron. Pt.2*)
'But announce in the town that nobody walk up and down in the street, but go to their inns. When the announcement was made, nobody was seen walking, but hastened to their inns [. . .] they did not dare to exit out.'

b. barefote, and wythout ony wepyn, [. . .] **yssued** ***out of his chambre*** / and went oute of ye paleis / and came to the gate of the towne (1498 *Four Sons of Aymon*)
'Barefoot and without any weapon [. . .] he exited out of his room, went out of the palace, and came to the gate of the town.'

In the autonomous texts, there is only one such attestation; in the translated texts there are 43 of them. The 18 GENERAL LITERAL attestations of *ishen* in which the FIGURE is not human mostly describe liquid FIGURES, including fountains (276a), streams of tears (276b), and blood (276c), but also bowels issuing from a wound in the stomach (276d) (cf. the bold italicized parts). This is also attested in the donor languages, as in (273).

(276) a. Saynt Domynyk sawe a meruaylous grete gyaunt [. . .] / of whome **yssewed** out ***fyue fountaynes.*** (*Katherine of Rome*, in *Companion to the English prose works of Richard Rolle*)
'St Dominic saw an awe-inspiring great giant [. . .], of whom five fountains issued out.'

b. But with strokys manyfold / I made hyr tendre, (yt ys no doute,) / That the ***terys*** **yssede** oute, / Out off hyr brest, so gret plente, / That men myghte the lycour se / By hyr chekys renne a-down / Off verray sorwe, so gret foysown [. . .] (*a*.1475(?*a*.1430) Lydg.*Pilgr.*)
'But with many a stroke I made her tender, so that the tears ished out, out of her breast, in such large quantities that you could see the liquid of true sorrow run down on her cheeks, in such abundance [. . .].'

c. For olyuers swerde entred in to his thye an halfe foot depe, and of the ***blood*** that **yssued** oute alle the grasse was reed. (1485 Caxton, *Charles the Great*)
'For Oliver's sword entered into his thigh half a foot deep, and all the grass was red from the blood that issued out.'

d. I shall bynde my syde and wounde soo that my ***bowelles*** maye not **yssue** oute of my bely (1490 Caxton, *Four Sonnes of Aymon*)
'I shall dress my side and my wound, so that my bowels may not issue out of my belly.'

It is perhaps not a coincidence that the verb *ishen* should be used for the kinds of FIGURES in (276a–d): Due to its prominent fricative, the verb seems highly imitative of the sound produced by gushing water, such as the fountains gushing out from the giant in (276a). Soundwise, it also fits perfectly into the series of MANNER verbs *lashen*, *gushen*, *rushen*, which all have the same postalveolar fricative and are all used for liquid FIGURES (tears, blood, and bowels) gushing out (see Appendix B, Ma^{speed}). Though the motion events described in (276b–d), with tears, blood, and bowels *ishing* out, probably do not really involve such a sound, an onomatopoetically motivated verb may still be chosen in them for expressive reasons—note that, particularly in (276b–c), the quantities of liquid are stressed: the *gret plente* and *gret foysown* of the tears in (276b), and the fact that all the grass is red due to the amount of blood in (276c).

Onomatopoeia may also be involved in the few attestations in which the FIGURE is neither human nor liquid: (277a) describes dreadful flames issuing out of mouth and nose of the metal bulls guarding the golden fleece, like from a flame-thrower. (277b) describes a hail of arrows *ishing* up into the air, with the choice of verb probably motivated by the hissing sound of flying arrows.

(277) a. With huge boolys of metal, /With ***flavme dredful and mortal***, / Which **yssed** out at nasse and mouthe / [. . .] Brent[en] al that kam be-syde (*c*.1450(?*c*.1408) Lydg.*RS*)
'[The golden fleece was guarded] with huge metal bulls, with dreadful and mortal flame which issued out at nose and mouth [. . .]; they burnt everyone that came there.'

b. Speiris into sprottes spronge ouer hede / ***Arowes*** vp in the aire **ysshit** full þicke. (*c*.1540(?*a*.1400) *Destr.Troy*)
(MED s.v. *ishen* '~ up, of arrows: to shoot up', sense 2.d)
'Spears sprung into splinters over heads, arrowes ished up in the air in great numbers.'

The attestation involving the arrows in (277b) is particular in that *ishen* 'exit' is used with *in*. The meaning of this preposition is directly opposite to the usual meaning of the verb *ishen*. This is not necessarily suspect to begin with: Of course, a motion description often contains multiple PATH windowing (cf. Talmy 2000), i.e. expressions that refer to the initial, medial, or final elements of the PATH.[189] With (277b), however, the situation is different, because there is no salient GROUND element that would be the SOURCE for the PATH 'out' that *ishen* describes (at least etymologically). *Arowes vp in the aire ysshit full þicke* can hardly be meant to highlight the fact that arrows exit from somewhere (even though a line of archers might perhaps be conceptualized as a 'container' from which arrows can exit). In the onomatopoetic use of *ishen* in (277b), the meaning PATH 'out' that the verb had originally (OF *issir* < L *exire*, cf. 9.2.1) appears to have been very much backgrounded against MANNER in this onomatopoetic use.[190]

9.2.4 MILITARY uses

More than half of the attestations of *ishen* refer to a very specific 'exiting', namely to people moving out of a—typically besieged—city or a castle in order to attack, as the prototypical attestations in (278a–b) show (see also the alliterative example in (274)):

(278) a. And thanne there **Issewid *out of the Cite*** many men of armys, of Frensshe men on hors bakke, and countrid with oure Engelisshe men. (*c*.1450(*c*.1430) *Brut-1430*)

[189] Sometimes, the specific GROUND element to which the PATH verb refers does not even have to be mentioned but can be inferred from the context, such as in *Hill said he has felt 'a sense of ease and calmness' when he **enters out of the bullpen*** (from the bostonherald.com website). In baseball, the bullpen is the area where the pitchers warm up before they enter the playing field. With this background knowledge, the GOAL of the PATH expressed in *enter*, i.e. the playing field, need not be made explicit, and it is sufficient to mention the bullpen as the PATH's initial element.
[190] At least if it is indeed one and the same verb. We can, of course, never rule out the possibility that we are dealing with an independent formation. Incidentally, next to our verb *ish* borrowed from French, the *OED* has a homonymous lemma *ish* v.2, with only one attestation (from the late

'And then there issued out of the city many men of arms, French men on horseback, and fought with our Englishmen.'

b. And soo Kynge Marke by the aduyse of his Knyghtes **yssued** ***oute of the Castel of Tyntagyl*** vpon his enemyes (1485 Malory *Wks.*)
'And so, on the advice of his knights, King Mark issued out of the castle of Tintagle, upon his enemies.'

This use is also found in the donor language, particularly in Anglo-Norman (see 9.2.1), and can be paraphrased as 'make a sortie, sally forth'.[191] It is also related to similar specific MILITARY meanings of the noun ME *issue* (see meaning 1b. in *MED*: 'a military sally or sortie', s.v. *issue* (n.)).[192]

As shown in Figure 9.5 (right column), this MILITARY sense is particularly prominent in the autonomous texts: In all but one of the attestations, *ishen* can be paraphrased as 'exit in order to attack' (29 out of 30; see the dark grey areas in Figure 9.5). The verb therefore seems to be very strongly tied to these MILITARY contexts of 'making a sortie'. (279) is an attestation from a petition by John Paston, complaining about the marauding gang of Robert Ledham:

(279) [. . .], Cristofir Bradlé, Jon Cokkow, assemblyng and gadderyng to hem gret multitude of mysrewled people, kepe a frunture and a forslet at the hows of the seid Robert Ledeham and **issu** ***ought*** at her pleser, sumtyme vj, sumtyme xij, sumtyme xxxti and mo, armed, jakked, and salettyd, wyth bowis, arwys, speris, and bylles, and over-ride the contré and oppresse the people and do many orible and abhomynable dedis (1452 *Paston*)

'[. . .] Christopher Bradley and John Cokkow, assembling and gathering a great multitude of misbehaved people, keep a stronghold and a fortification at the house of the abovementioned Robert Ledham, and issue out at their pleasure, sometimes six, sometimes twelve, sometimes thirty and more, armed, dressed with defensive garments, and helmeted, with bows, arrows, spears, and pikes, and over-ride the country and oppress the people, and do many horrible and abominable deeds.'

But the high proportion of MILITARY attestations of *ishen* is interesting in the translated texts (left column in Figure 9.5) as well. Since *issir* is the usual Medieval French verb for all kinds of 'exiting, leaving' (cf., e.g., *AND* and *DMF* s.v. *issir*; see also 9.2.1)—just like *entrer* for the reverse motion—the French originals must feature this verb in a larger number of different contexts. For

nineteenth century), which is, however, extremely similar to the one in (277b). The *OED* entry is the following: '*ish*, v.²—Etymology: echoic. *intr.* 'to make the sound *ish!* or *sh!* as in striking the air forcibly' *Bullets went "ishing" just over it with curious monotony*' (1898 G. S. Robertson *Chitral* xxi. 201).

[191] It corresponds to one of the meanings given in the *MED*, sense 1.a in the entries for both *issuen* and *ishen*.

[192] The relation to the noun may be another reason for the development of the form ME *issuen* (with <u>), next to ME *ishen*.

entren, the comparison of translated and non-translated Middle English texts in 9.1 (see also Huber 2012) has shown that translations use the verb in a much broader spectrum of neutral contexts without special connotations. A similar broad spectrum of contexts may therefore have been expected for its converse, *ishen*, in translated texts. This however, as we have seen, is not the case. How can this be explained? One possibility is that the French originals also use *issir* for MILITARY contexts most of the time and that this is just reflected in their Middle English translations. The other is that the French originals use *issir* for all kinds of motion with PATH^{out}, but that it tends to be translated with a different Middle English verb if the context is not a military one. If this were the case, it would be another strong indicator that ME *ishen* used LITERALLY does not only mean 'to move out of somewhere', but that it is a technical MILITARY term, evoking a frame of 'moving out in order to attack'.

The correct explanation seems to be the latter: The Middle English translation of the French *Mélusine*, for instance, tends to render French *issir* with ME *ishen* when it is used in the LINEAGE sense (see 9.2.5) and in MILITARY contexts such as in (280b), the translation of (280a):

(280) a. A tant **issirent** Sarrazins ***de leurs tentes*** à desroy, qui les rechassèrent jusques aux barrières, et là eut grant escarmouche et fiere, et maint homme mort et navré d'ung costé et d'autre (*Mélusine* ed. Brunet (1972: 141))

b. And þen **yssued** paynemys ***out of theire tentes*** without eny aray, that rechaced them vnto the barrers, where they had grete scarmusshyng & fyers, and many men slayn & wounded of bothe partes. (*c.*1500 *Melusine*)

'And then the heathens issued out of their tents without any formation, who rechased them unto the barriers, where they had great and fierce skirmish, and many men were slain and wounded on both sides.'

The instances of French *issir* that are translated with a different verb in the Middle English version, by contrast, are neither LINEAGE nor MILITARY uses, as illustrated in (281–283):[193]

(281) a. Et quant le souldan perceut sa perte, il **issist** ***de la bataille*** et tourna la targe derrière le dos et ferit le chevau des esporons, et s'en alla grant alleure vers l'ost des paiens (*Mélusine* ed. Brunet 1972: 318)

b. And whan the saudan perceyued the dyscomfyture, he **went** ***out of the batayll*** & tourned the targe behynd, and sporyd hys hors, & fled fast toward thoost of the sarasyns. (*c.*1500 *Melusine*)

[193] Nine instances in which the French original has *issir* are not translated at all, because the Middle English version shortens or rephrases the passage in which they occur.

'And when the Sultan saw his defeat, he went out of the battle, put the shield on his back, spurred his horse and fled fast towards the Sarrazene army.'

In (281), F *issir* is rendered with ME *went out* and, crucially, it does not describe a FIGURE attacking, but a FIGURE fleeing from the battlefield after he realizes he has lost. In (282), the brothers Lusignan have just won the battle against the Dukes of Freibourg and Austria who had besieged the king of Anssay. When the king, now free, comes out of his fortress to happily thank the brothers for their help, this is rendered by the translator as *comen out*. ME *ishen* with its connotation of 'attack' was probably not an option for the translation of F *issir* in this context of 'coming out to greet and thank'.

(282) a. Adonques **issist** le roy d'Anssay **dehors**, et vint aux logis où les frères estoient logez ès tentes qu'ilz avoient conquises; et le roy là arrivé les festoya moult amoureusement (*Mélusine*, ed. Brunet 1972: 398)

b. Thenne **came** the kyng of Anssay **out of the Fortres** glad & joyous of the dyscomfyture of hys enemyes, and came to the brethern tentes where he thanked them moche of theire noble socour and gretly fes-tyed them. (*c.*1500 *Melusine*)

'Then the king of Anssay came out of the fortress, glad and joyful that his enemies had been defeated, and came to the tent of the brothers, where he thanked them a lot for their noble help, and praised them greatly.'

Another instance where the Middle English translator opts for a different verb to render the French *issir* in the original is (283). Here *issir* refers to coming out of a dream, and the translator renders it with *awaken*.

(283) a. elle estoit comme une personne qui **est issue** nouvellement ***de son somne*** (*Mélusine*, ed. Brunet 1972: 169)

b. she was as a personne that **is awaked** newly ***fro her dreme*** (*c.*1500 *Melusine*)

'She was like a person who has just awakened from her dream.'

In all of these examples, the fact that the translator does not choose ME *ishen* to render *issir*, although he usually has a tendency to stick to the lexical material provided by the source (see 8.1), seems indicative of an awareness that *ishen* just does not fit in these contexts. This supports the idea that ME *ishen* when used LITERALLY is indeed a MILITARY term.

To conclude, these two sections on the LITERAL uses of *ishen* have shown that it has a marked preference for MILITARY contexts of 'making a sortie, exit to attack', particularly in autonomous texts. When it is used for the motion of non-human FIGURES, it mostly seems motivated by its onomatopoetic character.

9.2.5 NON-LITERAL uses

As has been shown in Figure 9.5 and Table 9.6 above, *ishen* is very rarely used for NON-LITERAL motion events. There are only 24 NON-LITERAL attestations (13 per cent), all of which come from translated texts. Two thirds of them (16 attestations) are instances of the specific LINEAGE sense that has been adopted from French (see 9.2.1), such as in (284a–c).[194]

(284) a. 'Ha, traytour,' said then Raymondyn / 'thou folowest wel ***the right euyl lynee of whiche*** thou **yssued**. [. . .]' (*c*.1500 *Melusine*)
'Ha, traitor,' said Raymondin then, 'you really take after the quite evil line from which you issued.'

b. 'my frende, ye seme not to be **yssued** ***of a pour hous*** / for ye are more like a kyng than a mason or laborer [. . .]' (*c*.1489 Caxton, *Four Sons of Aymon*)
'My friend, you don't seem to be issued of a poor house, because you look more like a king than like a mason or labourer.'

c. And ***of these two*** **yssued** fflorymond, whiche afterward toke moche of peyne & traueyll. (*c*.1500 *Melusine*)
'And of these two issued Florymond, who afterwards had much pain and woe.'

The LINEAGE sense of ME *issuen* is supported by the noun ME *issue* in the sense of 'offspring, lineal descendant' (note that the verb in this sense always has the <u>). The noun in this sense is attested from the end of the fourteenth century (*a*.1393, in Gower's *Confessio Amantis*, cf. *MED* s.v. *issue*), i.e., slightly earlier than the verb (*a*.1450(*c*.1410), cf. *MED* s.v. *issuen*). Judging from the large number of attestations of the noun *issue* that were found in the *CME* while searching for the verb *issuen*, the noun seems to have been more frequent than the verb as well.

Most of the attestations of *issuen* in this sense occur in one and the same text, the translation of Jean d'Arras' *Mélusine*. The clustering of LINEAGE attestations in *Melusine* may be due to the fact that this romance tells the genealogy of the house of Lusignan, which traces its ancestry back to the mythical creature that lends the story its name. However, since the translator also opts for *issuen* in LINEAGE contexts in which the original has another verb (such as *venir de* 'come from'[195]), the use of *issuen* in this sense cannot be purely due to imitation of the original.

---

194 The remaining ones refer, for instance, to words issuing out of mouths or to scents issuing from flowers.

195 Cf. (288) below, where the corresponding passage in the French version is *et* ***de toy viendra*** *moult noble lignée qui sera grande et de haulte proesse* (*Mélusine* ed. Brunet) 'And a very noble lineage will come from you, which will be grand and of great prowess.'

9.2.6 Complementation patterns

Unlike *entren* above, *ishen* is never attested with a direct object GROUND in the present corpus.[196] In a vast majority of attestations (182 out of 192), ME *ishen* combines with PATH satellites, in the rest of them, it is used bare. In most of the cases (163 attestations, i.e. 85 per cent), particularly in the LITERAL ones, the PATH satellite contains *out*, either alone, as in (285a), or in combinations such as *out of*, as in (285b).

(285) a. he wold **issue *out*** and set vpon the kynge (*c.*1500 *The Three Kings' Sons*)
'He would issue out and make an assault on the king.'

b. the good erle of Castelforde [. . .] **yssued *out of the castell*** wyth a thousaund knyghtes, that brake vpon their enmyes (*c.*1489 Caxton *Blanchardyn*)
'The good Earl of Castleford [. . .] issued out of the castle with a thousand knights, who broke upon their enemies.'

Thus, the PATH meaning mostly does not hinge on the verb alone, but is supported by PATH satellites with *out*. *Ishen* thus occurs in the very same structures as any other Middle English motion verb (see chapter 6), and probably this is also what enables the verb to receive the additional MANNER meaning of 'make a sortie' (see 8.3).

By contrast, the attestations where the verb refers to LINEAGE usually only have an *of*-PP, and no *out*, such as in (286) and (284a–c).

(286) the noble lynee whiche **yssued *of the said woman*** (*c.*1500 *Melusine*)
'The noble family which issued from the said woman.'

In this lexicalized metaphorical use, PATH^{out} is less important, and this is probably the reason why there is no additional *out* in (286) and (284a–c). There is a parallel to this in the complementation patterns of the Germanic equivalents of *issuen* in the sense of LINEAGE: When ME *(a)springen* 'spring' is used in this sense, such as in (287a–b), it is also usually combined with *of* and *from* only, and not with *out* as in other senses (cf. (287c), see also the quotations in the *MED* s.vv. *aspringen* and *springen*).

(287) a. Heber, ***of þan* asprang** þat eberisce folc. (*a.*1225(OE) *Vsp.A.Hom.Init. Creat*, MED s.v. *aspringen*, sense 2)
'Heber, of whom the Hebrew people is descended.'

b. My name is knowe kyng Joathan, the ixe kynge **spronge *of jesse*.** (?*a.*1475 *Ludus* C., MED s.v. *springen*, sense 4.c)
'I am known by the name of King Jonathan, the ninth king descended from Jesse.'

[196] The *MED* entry for this verb also gives only a single attestation with a direct object GROUND: *With joye* ***isshed*** *thow* ***the Citee*** *[L exivisti de urbe] of his swete birth Bethelem* (*a.*1500 *Mirror Salv.* (Beeleigh)).

c. Þe stan to-chan and fouwer walmes of watere **sprungen *ut*** þer of. (*a*.1225(?OE) *Lamb.Hom.*, MED s.v. *springen*, sense 1b.)
'The stone cracked open and four wells of water sprung out of it.'

There is only one attestation of ME *issuen* in the sense of LINEAGE which has an adverb *out*, namely (288).

(288) and ***out of thy body*** shall **yssue** a fayre lynee, whiche shalbe gret & of highe proesse (*c*.1500 *Melusine*)
'And out of your body shall issue a faire lineage, which shall be great and of high prowess.'

This, however, is slightly more LITERAL than the prototypical LINEAGE uses in (286) and (284a–c) above, since it refers to the offspring issuing out of the *body* of Melusine. The additional *out* could thus be due to the reactivation of the motion metaphor, so to speak, which is otherwise quite 'dead' in the LINEAGE uses.

To conclude, this section has shown that the complementation patterns of *ishen/issuen* depend on the meaning of the verb: In the LINEAGE uses, it mostly combines with *of*-PPs, in the LITERAL uses, it usually has an adverb *out*. It thus behaves like any other English MOTION verb, in that it combines with PATH satellites. Note that, interestingly, the Anglo-Norman complementation pattern of *issir* with direct object (e.g. *issist la chambre* 'he left the room'; see 9.2.1), which may have developed on the model of Middle English *leten* 'leave' and *leven* 'leave' (see Appendix B, Pa^{away}), has not spread (back) onto Middle English *ishen*. This probably has to do with the fact that AN *issir* with direct object is only used for GENERAL LITERAL MOTION contexts of leaving and going out (cf. *AND* s.v. *issir*). These general contexts, as we have seen in section 9.2.3, are hardly attested for autonomous Middle English *ishen*, which is usually used in the specific MILITARY sense of 'making a sortie'. In this MILITARY sense, it combines with PATH satellites also in Anglo-Norman (cf. *issir contre, a/en bataille, sur, fors encuntre*, etc. in (272b–f)).

9.2.7 Summary

This section has shown that ME *ishen* is rarely used in autonomous English texts, and that where it occurs, it is almost always as a MILITARY term, in a context of war and attack. This tendency is mirrored, but less pronounced in the translated texts. The preponderance of MILITARY uses of *ishen* draws the verb close to one of the classic domains of French loan influence, the one of military vocabulary (see, e.g., Burnley 1992: 429; Baugh & Cable 2013: 166). In the

---

'With joy you [i.e. Mary] exited the city of his sweet birth, Bethlehem'. The *CME* does not contain all the texts that provide attestations for the *MED*.

LITERAL uses that refer to motion of non-human FIGURES, such as liquids and arrows, the use of the verb is possibly motivated by its onomatopoetic character. NON-LITERAL uses are rare and mostly refer to LINEAGE. They differ from the LITERAL uses in the complementation pattern of *ishen*: When the verb refers to LINEAGE, it typically combines with an *of*-PP, when it refers to LITERAL motion, it typically combines with an additional adverb *out*.

Therefore, rather than being a misfit among Middle English motion verbs, *ishen* apparently is (or has become) a fairly normal motion verb in Middle English: Judging from its usage, it is definitely not a pure PATH verb, but has additional MANNER meanings of 'making a sortie, sally forth to attack'. In some attestations (as in (277b)), it is even doubtful whether the notion of PATH^{OUT} is present at all. Thus, *ishen* seems to have been adopted into Middle English not in all the uses it has in the donor language(s), but only in a small subset of them, particularly the LITERAL MILITARY ones. These are also the ones that fit well into the general Middle English system of talking about motion.

9.3 *Descenden*

9.3.1 Donor language portrait: OF/MF *descendre*, L *descendere*

Fr. *descendre* 'to descend' has developed from a Latin PATH-prefixed MANNER verb, originally *de-scandere* 'to off-climb'. As the simple form *scandere/scendere* in the sense of 'to climb' has dropped out of use, the prefixed form *descendere* has become largely intransparent. It can therefore be considered a pure PATH verb, definitely in Medieval French, probably even already in Latin (Stolova 2008: 259–60; see 7.1).

A) Syntax

In Latin and in continental and insular French, the GROUND with *descendere* is usually expressed in a prepositional phrase, such as in (289a–c), sometimes with additional PATH adverbs, such as F *aval* 'down' in (289d).

(289) a. vidit animam [. . .] fratris [. . .] cum agmine angelorum **descendere *de caelo*** (BEDE *HE* IV 3p.211, DMLBS s.v.)
'He saw the soul of the brother with a band of angels descending from heaven.'

b. Et Gieffroy [. . .] **descend *de la fallize*** et vint en l'abbaye (ARRAS, *c.*1392–1393, 278, DMF s.v.).
'And Geoffrey descends from the cliff and comes into the abbey.'

c. ***De la croiz* descent** ore (*S Aub* 246, AND s.v.)
'Descend from the cross now!'

d. Li grans maistres de l'Ospital **Descendi** ***dou chastel aval*** (MACH., *P. Alex.*, p.1369, 57, DMF s.v.)
'The grand master of the Knights Hospitaller descended down from the castle.'

According to *DMF* and *DMLBS*, GROUND can also be expressed in a direct object in continental French and Anglo-Latin, such as in (290), but this only seems to be the case for MEDIUM GROUND, i.e., the portion that is traversed, such as the sea, the valley, and the stairs in (290a–c), but not the SOURCE GROUND or the GOAL GROUND.

(290) a. Patricius [. . .] flatu prospero **mare Hibernicum** cum navi **descendit** (NEN HB 196, DMLBS s.v., sense 2.b trans. 'to sail the sea')
'With a favourable wind, Patricius descended the Irish sea with the ship.'

b. Li autre **descendoient le val** [. . .] (MACH., *P. Alex.*, p.1369, 165, DMF s.v., sense 1.B *'aller vers le bas (de ce lieu)'* ('go toward the bottom (of this place)'))
'The others descended the valley.'

c. . . . en **descendant les degrez** par où il estoit montez en ladite chambre (*Reg. crim. Chât.*, I, 1389–1392, 563, DMF s.v.)
'[. . .] descending the stairs by which he had mounted into the above-mentioned room.'

In Anglo-Norman, the pattern of *descendre* with GROUND in a direct object is apparently absent: According to the *AND*, *descendre* is only construed with the GROUND in prepositional phrases, i.e. within a PATH satellite (i.e. different from AN *entrer* and *issir*, where GROUND can also be expressed as direct object).

B) Semantics

Descendre is used for all sorts of downward motion. Specific uses include 'descending' from horses, other riding animals, or vehicles, as in (291). This use seems to be rather entrenched, as it occurs in contexts where the riding animal is not even mentioned, such as in (291c).

(291) **RIDING ANIMAL**

a. 'to dismount' (AND s.v.)
Marie [.. .] **descendyst** ***de son arne*** (*Bible* 25.5)
'Mary dismounted from her ass.'

b. *"Quitter une monture"* 'dismount from a mount' (DMF s.v., sense 2.A.I.B)
Tant erra Gieffroy qu'il y vint, et **descend** ***du cheval***, et monte en la sale (ARRAS, *c.*1392–3, 268).
'Geoffrey wandered so long until he came there, and dismounts from the horse, and goes up to the hall.'

c. *"Quitter une monture", Empl. abs.* 'dismount from a mount' (DMF s.v., sense 2.A.I.B)
Auprés de la peult on voir sans **descendre** L'ostel de ville (LA VIGNE, V.N., p.1495, 233)
'From there, you can see the *hôtel de ville* without descending [i.e. from your horse].'

Another specific use is the one referring to LINEAGE, as in (292), similar to the one sketched for *issir* in 9.2.1. It is interesting that both the PATH^{out} as well as the PATH^{down} metaphor are used for the target domain of LINEAGE.

(292) **LINEAGE**

a. 'descendant (in lineage)' (AND s.v., p.pr. as a.)
hoyrs, masles ou femelles, **descendens *du dit fihl*** (*Foedera* i 795)
'Heirs, male or female, descending/descendants of the said son.'

b. *[Domaine de la parenté]* Descendre de. *'Être issu de'* ('be descended, born from', DMF s.v., sense IV.B.1)
Regnaut, li quiex est **descenduz *de lignie non noble*** (*Doc. Poitou* G., t.2, 1336, 131)
'Regnaut, who has descended from a non-noble line.'

c. (w. ref. to lineage or inheritance) to descend: (DMLBS s.v., sense 6a a. (of person))
Reginaldus **descendit *de Eularia*** (1207 *CurR V* 46)
'Reginaldus descended from Eularia.'

While in the LINEAGE sense of *descenden*, the metaphorical FIGURE is a person, there is another use closely connected to it, where the FIGURE is an inheritance, moving metaphorically down the family tree like the descendants of the family. This use is also attested in both insular and continental French (cf. *TL* and (293a–b)), as well as in Latin (293c).

(293) **HERITAGE**

a. '(law) to descend (by way of inheritance)' (AND s.v., v.n. 2)
[. . .] terre, le quell aprés sa mort **dessende *a le pleintif*** (*Readings* 33.20)
'Land which after his death descends to the plaintiff.'

b. '*[D'un héritage]* Descendre (de qqn) à / en qqn' ('of an inheritance: descend from someone to someone', DMF s.v., sense IV.B.2)
. . . la coronne imperiale de Germanie **descendoit** ancyennement ***du père au filz ou à son prochain heritier par succession*** (MOLINET, *Chron.* D.J., t.1, 1474–1506, 487)
'The Germanic imperial crown traditionally descended from the father to the son, or to his next heir through succession.'

c. (w. ref. to lineage or inheritance) to descend: b. (of property or title) (DMLBS s.v., sense 6b)
nec est heres [. . .] Yonis nec aliquid **descendit** ***ad eam*** jure hereditario ***de hereditate ipsius Yonis*** (1230, *LTRMem* 11r, 6*d*)
'And she is not John's heir and nothing descended to her by law of inheritance from the heritage of the same John.'

Another use mentioned specifically by the *DMF* is the one with divine FIGURES, such as Christ descending on Mary, the Holy Ghost descending on somebody, or, as in (294a), Venus descending into a room. These senses are also explicitly mentioned for Latin *descendere* by the *DMLBS* (sense 1.b), which calls them 'theological'. (294b), for instance, describes Christ's descent to hell. For Anglo-Norman, the *AND* does not mention these kinds of uses specifically, but it also has an attestation of *descendre* used in combination with the Holy Ghost (294c).

(294) **THEOLOGICAL**

a. '*[du Christ]*', '*[du Seint Esprit]*', '*[d'un dieu, d'une déesse]*' ('of Christ, of the Holy Ghost, of a god or goddess' (DMF s.v., 1.a))
Si fu tost la ***deesse*** preste, Car tout en l'eure est **descendue**, Couverte d'une obscure nue, Plainne de manne et de fin bausme Qui la chambre encense et enbausme. (MACH., *Voir*, 1364, 356).
'Then the goddess was soon ready, because she descended immediately, covered in a dark cloud full of manna and fine balsam, which incenses and embalms the room.'

b. 'theol.' (DMLBS s.v., 1.b)
dum ***Deus ad herebi vagas*** **descenderit** ***umbras*** (ALDH. *VirgV* 455)
'When God descended to the wandering shades of hell.'

c. ***Li seint esperit*** i **decent** (*Jos* 871, AND s.v., v.n.1)
'The holy ghost descends there [into the room where Josaphat is baptized].'

Furthermore, *descendre* is used in the discourse sense of 'moving on to a topic', 'descending into the details'. This sense is specifically mentioned in the *DMF* for Middle French (295a), and *TL* gives Old French attestations of it as well (295b). It is equally found in Latin, such as in (295c). The *AND* does not refer to any such use in Anglo-Norman.

(295) **DISCOURSE**

a. *P. ext. [Idée figurée de passage à qqc., en partic. dans le domaine du discours]* ('by extension: figurative notion of moving on to something, particularly in the area of discourse', AND s.v., sense IV.C)
Ainsi vueil mettre fin à ce Prologue, pour **descendre** et retourner ***à la matière principalle*** [. . .]. (BUEIL, *I*, 1461–6, 60).

'Now I want to finish this prologue, in order to descend and return to the main subject.'

b. Toutesvoies, puis qu'elles sont dictes en general, nous **descendrons *en especial*** (*Ménag.* I 210, TL)
'However, since they [the abovementioned things] are said generally, we will descend to the particular.'

c. 'to proceed, turn, "get down" (to)' (DMLBS s.v., sense 8.b)
postquam [. . .] quesivimus an expediret mundo uni imperatori subesse [. . .] ***ad Romanum imperium* descendamus**, inquirentes primo a quo Romanum processit imperium (Ockham *Dial.* 885)
'After [. . .] we have asked whether it is conducive for the world to be subject to one emperor, [. . .] we will descend to the Roman empire, asking first from what the Roman empire emerged.'

9.3.2 ME *descenden*

The search in *CME* for *descenden* and its variant forms yielded 499 attestations.[197] Similar to the results for *entren* and *ishen* presented in 9.1–2, also for *descenden*, there is a noticeable difference in how it is used in autonomous as compared to translated texts: In the autonomous texts, *descenden* is used far more rarely to refer to GENERAL LITERAL motion (only in 16 per cent of the attestations) than it is in the translations, where almost half of the attestations refer to GENERAL LITERAL motion (see the black areas in Figure 9.6). In turn, NON-LITERAL uses, particularly in the domains of THEOLOGY, LINEAGE, and HERITAGE (the light grey areas in Figure 9.6), are more frequent in the autonomous texts. The numbers and percentages are shown in Table 9.7.

The differences in frequency of the GENERAL LITERAL uses of *descenden* suggest, like those for *entren*, that in Middle English this PATH verb tends to be used far less readily in its basic meaning of LITERAL 'moving down', but that it is mostly used for specific, often NON-LITERAL meanings. Many of them, as we will see below, are adopted from the donor languages (cf. 9.3.1). The LITERAL uses of *descenden* are discussed in 9.3.3, the NON-LITERAL ones in 9.3.4.

9.3.3 LITERAL uses

A recurrent use of *descenden* in the LITERAL attestations is 'dismounting from horses'. This accounts for a good third of the LITERAL uses of *descenden* in the autonomous texts (though all from Malory), such as in (296a). We have seen in section 9.3.1 that this use is also prominent in French, and, like in French, in Middle English it does not need explicit mention of a horse, such as in (296b).

[197] Search terms *descen**, *dessen**, *decen**, *discen**, *dissen**, and *disen**.

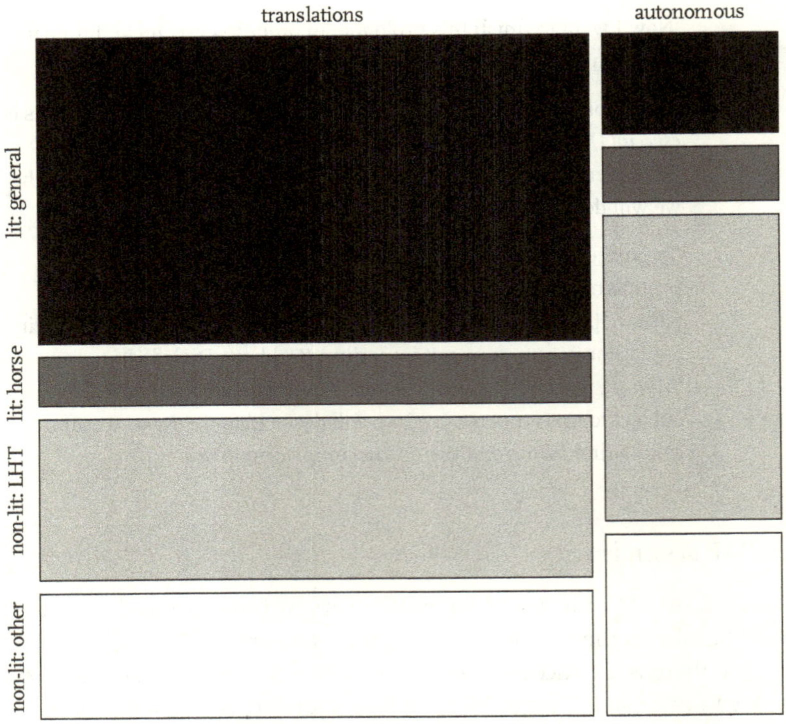

FIGURE 9.6. Usage contexts of ME *descenden* in CME translated and autonomous texts.

TABLE 9.7. Usage contexts of ME *descenden* in CME translated and autonomous texts.

	LITERAL: GENERAL		LITERAL: FROM HORSE		NON-LITERAL: LINEAGE/ HERITAGE/ THEOLOGICAL		NON-LITERAL: OTHER		TOTAL	
Translated	48%	(181)	8%	(29)	25%	(95)	20%	(74)	100%	(379)
Autonomous	16%	(19)	8%	(10)	48%	(57)	28%	(34)	100%	(120)
Total	40%	(200)	8%	(39)	30%	(152)	22%	(108)	100%	(499)

(296) a. Thenne that knyght **descended *from his hors*** lyghtly lyke a valyaunt man (1485 Malory *Wks.*)
'Then the knight descended lightly from his horse, like a brave man.'

b. By thys was sir Launcelot comen to the gates of that Castel / and there he **descended *doune*** and cryed that alle the Castel range of it (1485 Malory *Wks.*)
'With this, Sir Lancelot had come to the gates of this castle, and there he descended down, and shouted so that the entire castle was ringing with it.'

The use of *descenden* for dismounting from horses is probably also present as a connotation in (297), in the old Januarie's pompous introductory words to the wedding night with his young wife.

(297) Allas! I moot trespace / To yow, my spouse, and yow greetly offende, / Er tyme come that I wil ***doun* descende** ((*c.*1395) Chaucer *CT.Mch.*)
'Alas, I must trespass to you, my spouse, and greatly offend you, before the time comes that I will descend down [i.e. dismount from you].'

Also in the translations, *descenden* is used to describe people dismounting from their horses (cf. the dark grey areas in Figure 9.6). The crucial difference between the translations and the autonomous texts lies in the frequency of the GENERAL LITERAL attestations. In the autonomous texts, these account for only 16 per cent of all the attestations, which suggests, again, that the writers were reluctant to use *descenden* for GENERAL LITERAL motion events. In the translations, by contrast, these GENERAL LITERAL uses constitute almost half of all the attestations of *descenden*. The FIGURES are mostly people, but rain, water, blood, and tears are also sometimes said to *descend*. In many cases the occurrence of *descenden* in the Middle English translation will be due to its occurrence in the French or Latin original, such as in (298a–b), another parallel example from the French *Mélusine* and its Middle English translation. Note that the Middle English translator in (298b) slips in an additional adverbial satellite *doun* 'down' which has no equivalent in the French version in (298a).

(298) a. Raimondin, qui moult estoit énamouré de sa dame, se partit sur heure de Poetiers, tout seul, moult hastivement, et chevaucha tant qu'il vint en la haute forest de Colombiers, et **descendit *de dessus la montaigne au val*,** et vint à la fontaigne (*Mélusine* ed. Brunet 1972: 55)

b. Raymondyn, whiche was moche enamoured of his lady, departed at this ooure fro Poytiers hastly al alone, and rode tyl he camme vnto the high Forest of Coulombiers, and **descended *fro þe hylle doun in to the valey*** and camme to the fontayne (*c.*1500 *Melusine*)

'Raymondyn, who was much in love with his lady, hastily departed at this hour from Poitiers, all alone, and rode until he came to the high forest of Colombiers, and descended from the hill down into the valley and came to the fountain.'

Not only the comparison of translations to their originals is revealing, but also the comparison of different translations of the same original to each other: As many as 63 LITERAL attestations of *descenden* (and another 20 NON-LITERAL ones) come from the Earlier Version of the Wycliffite Bible, while in the Later Version there is not one single instance of this verb. An exemplary comparison of the Vulgate text of Genesis 38–46 and Exodus with the translation of the Earlier Version shows that all ten instances of Latin *descendere* in the Genesis passage, and 12 out of the 15 in Exodus are rendered with ME *descenden*, like the ones in (299b and 300b).[198] The Later Version, in contrast, consistently renders Latin *descendere* either with ME *gon down* (cf. *ȝede doun* in (299c)) or with ME *comen down* (cf. *cam doun* in (300c)) in these two excerpts.

(299) a. ***descendit**que ad fontem et hausit aquam* (Gen. 24:45)
'And [Rebecca] descended to the well and drew water.'

b. and she **descendide** to the welle, and drowe water (*a*.1425(*a*.1382) *WBible(1)*)

c. and sche **ȝede *doun*** to the welle, and drowe watir. (*a*.1425(*c*.1395) *WBible(2)*)

(300) a. *Cumque **descenderet** Moyses de monte Sinai, tenebat duas tabulas testimonii* (Ex. 34:29)
'And when Moses descended from Mount Sinai, he held two tables of the testimony.'

b. And whanne Moyses **descendide** fro the mownt of Synay, he held two tables of testymonye (*a*.1425(*a*.1382) *WBible(1)*)

c. And whanne Moises **cam *doun*** fro the hil of Synai, he helde [helde in his honde IS] twei tablis of witnessyng (*a*.1425(*c*.1395) *WBible(2)*)

This systematic difference in wording is interesting in relation to the programmatic preface to the Later Version, which lays out some principles of how to make a good and idiomatic translation of the Scripture:

> First it is to knowe, that the best translating is out of Latyn into English, to **translate after the sentence, and not oneli after the wordis**, so that the sentence be as opin, either openere, in English as in Latyn, and go not fer fro the lettre; and **if the lettre mai not be suid in the translating, let the sentence euere be hool and open, for the wordis owen to serue to the entent and sentence**, and ellis the wordis ben superflu [sic] either false (ed. Forshall & Madden 1850: 57).

Although the direct advice that follows concentrates on the resolution of Latin non-finite structures such as ablativi absoluti, and on recommended translations for frequent Latin words such as *autem* or *vero*, the preference of *gon/cumen down* over *descenden* is fully in line with the aspiration towards greater idiomaticity of the Later Version. The difference between the two versions, with the earlier one being very close to the original in both vocabulary and structure, the later one more idiomatic, has been shown and discussed, for instance, in Lindberg (e.g. 1984, 2007) and Meichtry-Gruber (2008: 58–62).

[198] One of the remaining three is translated with *comen doun*, and the other two—apparently erroneously— with *stien up*.

Lindberg (2007) sees the two versions as two subsequent steps in the translators' work, in much the same way as today's translators often proceed: first produce a literal version, second, 'test the literal possibility and draw on idiomatic resources of the language' (2007: 199). If this is the case in the Wycliffite Bibles, it means that *descenden* was rejected in the revised version for lack of idiomaticity, and *gon/cumen down* was decided on as an idiomatic alternative instead. For the question of the present study, this is another indicator that apparently it was not idiomatic to use *descenden* for GENERAL LITERAL motion events in Middle English. More widespread, in contrast, are the NON-LITERAL uses of *descenden*, which will be discussed in the following section.

9.3.4 NON-LITERAL uses

The major sub-groups that emerge within the NON-LITERAL uses of *descenden* are all also present in the donor languages: LINEAGE, HERITAGE, and THEOLOGICAL uses (see 9.3.1). Together, these account for roughly half of the attestations of *descenden* in autonomous texts, and for a quarter of those in translated texts (see Table 9.7 and the light grey areas in Figure 9.6). We will look at them in turn, starting with the theological uses.

9.3.4.1 Theological uses

Many of the attestations of *descenden*, in both autonomous and translated texts refer to divine FIGURES, such as God in (301a) or the Holy Ghost in (301b–c), descending to earth or onto someone. Note that this is not only the case in texts that treat religious matters, but also in romances, such as Caxton's *Blanchardyn and Eglantine* in (301a).

(301) a. [. . .] þe swete voyces of the ladyes & gentyl women, that sange there, that it semed that ***god*** had **descended** among them ***from heuen*** (*c.*1489 Caxton *Blanchardyn*)
'[. . .] the sweet voices of the ladies and gentlewomen that sang there, that it seemed as if God had descended among them from heaven.'

b. besyde the awteer weren all the Aposteles on Wytsonday when ***the holy gost*** **descended *on hem*** in lykness of fuyr (?*a.*1425(*c.*1400) *Mandev.(1)*)
'Next to the altar were all the Apostles on Whitsunday when the Holy Ghost descended on them in likeness of fire.'

c. Þise byeþ þe guodes / þet þe ***holy gost*** makeþ ine herte huer he **descendeþ** be þise zeue yefþes ((1340) *Ayenb.*)
'These are the goods that the Holy Ghost makes in the heart where he descends by these seven gifts.'

Similar uses of *descenden* refer to the divine power descending on someone, such as (302a), or divine gifts and interventions descending from heaven, such as manna in (302b).

(302) a. þe haligast in þe sal lend, And ***godds might in þe* descend** (*a*.1400(*a*.1325) *Cursor*)
'The Holy Ghost shall land in you, and God's might shall descend in you.'

b. Þis es þe ***brede with-owten ende*** Þat ***fra þe hegh of heuyn*** **descend** (*a*.1425 *NHom.Corp.Chr.*)
'This is the bread without end that descended down from the height of heaven.'

All these uses of *descenden* are classified as THEOLOGICAL (see the THEOLOGICAL senses of French *descendre* and Latin *descendere* in section 9.3.1). They make up a good quarter of all attestations of *descenden* in autonomous Middle English texts, i.e. they are even a bit more frequent than all the LITERAL attestations in autonomous texts taken together (see Table 9.7 above).

In all these theological attestations, whether autonomous or translated, the Latin use of *descendere* will have played a role, because all things religious are necessarily tied to the Latin language in the Middle English period. In the frequent use of *descenden* for Jesus descending to hell, i.e. the so-called 'Harrowing of Hell', as in (303a–b), for instance, the choice of *descenden* will be influenced by the Latin term *Descensus Christi ad Inferos*, which also features in the Apostles' Creed (*descendit ad inferos*).[199] Note that the treatise in (303b) explicitly refers to this article in the creed.

(303) a. And syþen his blessed body was in a ston byried, And **descended *a-doune to þe derk helle*,** And fet oute our formfaderes (*c*.1600(?*c*.1395) *PPl. Creed*)
'And then his [=Jesus'] blessed body was buried in a cave, and he descended down to the dark hell, and rescued our forefathers.'

b. In the tyme of Austyn [. . .] the comune crede hadde not withynne him this seid article—***Crist*** in his deeth of bodi **descendid *to hellis*** ((*c*.1456) Pecock *Faith*)
'In the time of Augustin [. . .], the common creed did not contain the abovementioned article—Christ in death in body descended to hell.'

Of course, the idea of heaven being above and hell being below is also present in these attestations, in keeping with the GOOD IS UP metaphor, which here is also a GOD IS UP metaphor: In the medieval worldview, God and all divine intervention necessarily comes from above, from the heavenly spheres. For these 'theological' descents, Old English mostly uses loan formations like *(of) dūne(ā)stīgan* or *niðerstīgan* (Gneuss 1955: 138–9) to render the Latin *descendere*. In the course of the Middle English period, these are replaced by *descenden*.

[199] The same wording is still retained in the Book of Common Prayer (*He descended into hell*) and in today's Common Worship (*He descended to the dead*).

Like in many of the words studied in Käsmann (1961), this is yet another case where a French word, still formally very close to its Latin equivalent, is acquired as a sort of theological terminus technicus in Middle English.[200]

9.3.4.2 LINEAGE and HERITAGE

The next prominent use of *descenden* is to refer to LINEAGE and HERITAGE, which are two further specialized senses that have been adopted from the donor language (see 9.3.1). Typical attestations for *descenden* in the LINEAGE sense are given in (304a–b).

(304) a. for their moders weren cosyns germaynez **descended *of Sir Milys Beauchamp***, knyght (*a.*1474 *Stonor Letters*)
'Because their mothers were first cousins, descended of Sir Miles Beauchamp, knight'

b. that glorious Martir Seint Thomas somtyme Archebisshop of Caunterbury ***of whos blode*** the seid Erle of wiltshire his fader and many of his Auncestres are lyneally **descended** (*Chancery*, 1454SC8/28/1358 Petition of James, Earl of Wiltshire)
'The glorious martyr St. Thomas, formerly archbishop of Canterbury, of whose blood the abovementioned Earl of Wiltshire's father and many of his ancestors are descended by lineage.'

The LINEAGE use of *descenden* is not restricted to official documents, but also appears in fictional narrative, such as in (305a–b).

(305) a. Orcays felle in questionyng with his prisoner, & askid him his name, & ***of what blode*** he was **descendid** (*c.*1500 *Three Kings' Sons*)
'Orcays started questioning his prisoner and asked for his name and of what blood he was descended'

b. And ***of this lord*** **descended** Tideus By ligne (*a.*1425(*c.*1385) Chaucer *TC*)
'And of this lord descended Tideus by lineage.'

In the LINEAGE attestations, the SOURCE is usually mentioned, because this is what these attestations are about. The use of *descenden* in the *Reeve's Tale* in (306) is a little peculiar in not mentioning the SOURCE from which *holy churche's blood* (in this case arguably the one of the Miller's wife, who is the illegitimate daughter of the parson) descends. Although the choice of *descenden* in this verse may also be due to the rhyme, Chaucer might also punningly play with the connotations of divine intervention and miracle that we have seen above.

---

[200] What exactly the reason was for the replacement of the *stien* forms by *descenden* would merit further investigation. Possible factors may include (a) a general desire to stay as close to scripture as possible; (b) that clerks wanted to imitate the morphological structure of Latin *descendere*, which had been possible with the prefixed forms of OE *stīgan* (Gneuss 1961: 137–8), but was then hindered by the increasing tendency to place PATH satellites in postverbal position (see 5.4).

(306) For hooly chirches good moot been despended On ***hooly chirches blood***, that is **descended**. ((*c*.1390) Chaucer *CT.Rv.*)
'For the goods of the Holy Church must be spent on the blood of the Holy Church that is descended.'

Very similar to the LINEAGE uses are the attestations where *descenden* refers to an inheritance descending on an heir, such as (307a–b), which are from a Chancery document (307a) and a will and testament (307b):

(307) a. [. . .] a place with certaine Rentes / with ynne the toun of Claworth [. . .] / whiche by the deces of Robert hertishil his Cousin / been **descended** ***vnto oure saide seruant by waye of heritage*** as to the Right heire of the saide place [. . .] (1420 *Chancery*, p.132)
'A place with certain rents in the town of Clayworth [. . .], which by the death of his cousin Robert Hertishil are descended by way of heritage unto your said servant as the right heir of the place.'

b. [. . .] my lande þat is tayled to him in Leicestre, Whytewell and Litel Hamildon shulde **descende *to þe said Iames*** (1425, *Earliest English Wills*, III. Codicil. Oct. 1425)
'My land that is entailed to him in Leicester, Whitwell, and Little Hambledon should descend to the said James.'

Like its LINEAGE use, also the HERITAGE use of *descenden* is not restricted to official documents, but also occurs in fictional narrative, such as in (308a–b).

(308) a. ye haue no mo heyres ***to whom*** youre londe shall **descende** after youre deth (*a*.1500(?*c*.1450) *Merlin*)
'You have no more heirs to whom your land shall descend after your death.'

b. for ye be rightfull heire vnto the Turke & his inheritaunce most **descende *vnto you*** (*c*.1500 *Three Kings' Sons*)
'Because you are the rightful heir unto the Turk, and his inheritance must descend unto you.'

Both LINEAGE and HERITAGE attestations of *descenden* involve the concept of a family tree along whose branches something descends, be it an offspring or an inheritance—in both cases the motion is metaphorical. These uses account for almost another quarter of the attestations of *descenden* in autonomous texts. Since both meanings are already present in the donor language (see 9.3.1), a specific legal and administrative use of *descenden* has been adopted here, which is not surprising in this type of contact situation (see 6.1).

9.3.4.3 Other NON-LITERAL uses

Similar to the family tree concept that underlies the LINEAGE/HERITAGE uses, other attestations of *descenden* equally invoke the concept of a system within which you can descend. In (309a–b) it is the family tree again, though without the FIGURE being a person or some inheritance—actually, there is no FIGURE involved, but *in descending* could be paraphrased as 'down the line' here.

(309) a. And so, in lyke wyse, if any of them dye, it remayne to his brother in **descendyng** [. . .] (1465 *Lin.DDoc.*)
'And so, in the same way, if any of them die, it [i.e. the heritage] shall pass to his brother in descending [i.e. the next oldest].'

b. I let the wit, quhen that the father offendis, I will cheris nane off his successioun, And off his barnis, i may weill tak amendis, ***Vnto the twentie degre* descending *doun*.** (1571(*c.*1450) Henryson, *Fables*)
'I let you know that if the father does wrong, I will not treat any of his offspring well, and I may take amends from his children, down to the twentieth generation.'

In a number of other attestations, the underlying concept seems to be the discourse as a system, within which you can descend into details, or to further, related topics such as in (310a–c). Note that this use is attested very early in Middle English, in the 1340 *Ayenbite of Inwit* (310a). It is probably adopted from the Latin rhetorical sense of *descendere* (see the DISCOURSE uses in 9.3.1).

(310) a. Ac erþan ich **decendi *to þe uirtues*** þet byeþ contraries to þe zeue zennes ich þe wylle speke ssortliche / of zeue oþre uirtues. ((1340) *Ayenb.*)
'But before I descend to the virtues that are opposed to the seven sins, I will briefly tell you about seven other virtues.'

b. But nathelees, lat us now **Descende *to the special.*** ((*c.*1390) Chaucer *CT.Mel.*)
'But still, let us now descend to the particulars.'

c. He gan first fallen of the werre in speche / Bitwixe hem and the folk of Troie town, / And of thassege he gan hir ek biseche / to telle hym what was hire opynyoun; / ffro that demaunde he so **descendeth *down*** To axen hire if that hire straunge thoughte / The Grekis gise and werkes that they wroughte (*a.*1425(*c.*1385) Chaucer *TC*)
'He started to talk first about the war between them and the people of Troy, and also entreated her to tell him her opinion about the siege, and from that question, he descends down to ask her if the ways and the deeds of the Greeks seemed strange to her.'

In line with the basic 'GOOD IS UP'-metaphor, there are also NON-LITERAL uses in which *descenden* refers to people descending into corruption, into poverty, or on the wheel of fortune, i.e. falling in status:

(311) a. ȝef i schal **descende *in to euerlastynge corrupcion*?** (*c.*1410 *St.Anselm Medit.*)
'If I shall descend into everlasting corruption?'

b. And who despendyth more than his Powere or his goodis strechyth, **descende** he moste ***in Powerte*** (*a*.1500(?*a*.1425) *Lambeth SSecr.*)
'And who spends more than his power or his goods reach, he must descend into poverty.'

c. þouȝ þou **descende** ***doun*** whanne resoun of my pleye axeþ it (?*a*.1425(*c*.1380) Chaucer *Bo.*)
'Though you descend down when the reason of my [i.e. Fortune's] game demands it.'

Another group among the OTHER NON-LITERAL attestations is the one of FICTIVE MOTION, with 33 attestations. In motion verb typology studies, the term 'fictive motion' is used to denote 'stationary spatial scenes that "contain" PATH-like configurations' (Matlock 2004: 223), such as 'the road runs along the coastline'. Some examples for FICTIVE MOTION uses of *descenden* are (312a–b).

(312) a. the site of that cyte, hauenge the mownte of Syon of the northe **descendenge** ***towarde the sowthe*** with a softe dependence (?*a*.1475(?*a*.1425) *Higd.(2)*)
'The site of that city, having Mount Zion from the North descending toward the South with slight inclination.'

b. LYBYE the lowe þat **descendeth** ***down lowe toward the grete see of Spayne*** (?*a*.1425(*c*.1400) *Mandev.(1)*)
'Lower Libya, which descends down low toward the great sea of Spain.'

Most of these come from descriptions of landscapes, which cluster in Higden's *Polychronicon* (312a) and in *Mandeville's Travels* (312b), or from anatomical descriptions in Lanfranc's *Chirurgie* (313a), Arderne's *Fistula*, and Metham's *Palmistry*. While these are all from translations, there are also two autonomous attestations in Chaucer, one of them in the *Astrolabe* (313b).

(313) a. Bi þe myddil seem which þat is in þe heed, þere **discendiþ** a ueyne comynge fro þe lyuere (*a*.1400 *Lanfranc*)
'By the middle seam that is in the head, there descends a vein, coming from the liver.'

b. a lyne, þat ***cometh*** **dessendinge** fro the ryng ***down*** to the nethereste bordure. (*a*.1450(1391) Chaucer *Astr.*)
'A line that comes descending from the ring, down to the lowest margin.'

It may be interesting to note that in this autonomous example of *descenden* used in a context of FICTIVE MOTION (313b), the verb is not the main verb in the phrase, but only modifies *cometh*. The simultaneous use of the satellite *down* even makes it redundant. In the other examples of fictive motion above, which all come from translated texts, *descenden* is the main verb, although some of them are also highly redundant because of their use of satellites (cf. *descendeth down lowe* in (312b)). Again, one could argue that a borrowed PATH verb is better suited for the description of FICTIVE MOTION than a MANNER or a NEUTRAL MOTION verb, since it is the FICTIVE PATH that is the salient element in the situation (cf. the frequent redundant PATH expressions) and since there is neither MANNER nor MOTION involved—after all, the situation is stationary here.

9.3.5 Complementation patterns

In the majority of its attestations, *descenden* is combined with a PATH satellite, as the examples in the preceding two sections have shown. This is its typical complementation pattern in the French donor language varieties (see 9.3.1), and also the typical complementation pattern of motion verbs in Middle English. In a fifth of the attestations, it is used as a bare verb. There are only a mere two in the almost 500 attestations of *descenden* in which it is used with a direct object GROUND:

(314) a. Thenne he wold **descende *the mountayne*** (*c*.1500 *Melusine*)
'Then he would descend the mountain.'

b. Thenne he departed & **descended *the mountayne*** with hys men of armes (*c*.1500 *Melusine*)
'Then he departed and descended the mountain with his men of arms.'

In 74 of the attestations (16 per cent), *descenden* combines with the additional PATH adverb *down* or *downward(s)*. This is interesting, because, different from prepositions, such as *(on)to*, *of*, or *from*, the adverb is not needed to introduce the GROUND NP; it might therefore just as well be left out. Furthermore, different from the prepositional phrases, it carries exactly the same information ($\text{PATH}^{\text{down}}$) that is carried by *descenden* itself. The adverb *down/downward(s)* might therefore serve as a kind of 'text-internal gloss' to the verb *descenden*. It is more frequent in autonomous texts, where it is used in 22 per cent of the attestations, than in translations, with only 13 per cent, as shown in Table 9.8.

In the contexts of LINEAGE and HERITAGE, by contrast, the adverb *down* is extremely rare, with only two examples out of 62 (3 per cent).[201] This is consistent with the fact that the LINEAGE and HERITAGE uses are the ones in which the spatial sense $\text{PATH}^{\text{down}}$ seems to be the least important. In the LITERAL uses, by contrast, in which *descenden* combines with *down* more often, the notion of $\text{PATH}^{\text{down}}$ is very prominent. The same is true for the NON-LITERAL FICTIVE and THEOLOGICAL uses of *descenden*, which also combine with *down*. Also in NON-LITERAL metaphorical uses such as 'decline, degenerate, descend in status/

[201] What is more, in one of them, the rhyme depends on *down*: *A goddes of ful gret* ***renoun*** */ And by lyne descended* ***doun*** */ Fro the goddys high kynrede [. . .]* (*c*.1450(?*c*.1408) Lydg.*RS*) 'A goddess of very great renown, and by lineage descended down from the gods' high kindred.'

TABLE 9.8. Complementation of ME *descenden* in CME translated and autonomous texts

	WITH *DOWN(WARD(S))*		WITHOUT *DOWN(WARD(S))*		TOTAL
Translated	13%	(48)	87%	(331)	379
Autonomous	22%	(26)	78%	(94)	120
	16%	(74)	84%	(425)	499

($\chi^2=5.8473$, df=1, $p<0.05$)

TABLE 9.9. Complementation of ME *descenden*, without LINEAGE/HERITAGE contexts

	WITH *DOWN(WARD(S))*		WITHOUT *DOWN(WARD(S))*		TOTAL
Translated	14%	(47)	86%	(296)	343
Autonomous	27%	(25)	73%	(69)	94
	16%	(72)	84%	(365)	437

($\chi^2=8.91239$, df=1, $p<0.01$).

fortune', the spatial metaphor of $\text{PATH}^{\text{down}}$ (BAD/LESS IS DOWN) is quite active, and also here, *descenden* is found in combination with *down*. With the concept of LINEAGE and HERITAGE, however, $\text{PATH}^{\text{down}}$, as well as the MOTION metaphor in general, is far less motivated—note that LINEAGE works just as well with PATH^{out}, as shown in section 9.2 on *ishen/issuen*. Correspondingly, the adverb *down* is less frequent in these attestations. If they are disregarded, the difference between translated and autonomous texts noted above (Table 9.8) becomes even more pronounced, with autonomous texts featuring *down(ward(s))* in more than a quarter of the attestations of *descenden* (Table 9.9):

The redundant *descend down(ward(s))* is thus more frequent in autonomous texts than in translated ones. Let us throw a quick glance on the development of *descend down* after the Middle English period. In Early Modern English, this combination apparently becomes extremely rare: In the Penn–Helsinki corpus for this period (*PPCEME*), it occurs only once in 92 attestations, and, what is more, in a rather early text from this period, Torkington's diary of his pilgrimage to Jerusalem from 1517. For Present-Day English, the very large *BNC* only has a mere three attestations of this combination. Thus it seems that once *descend* is better established, the redundant support of *down* falls away. Normative efforts, which often dislike redundancies, may have had an influence in this development as well.

9.3.6 Summary

This section has shown that ME *descenden* is used only rarely for LITERAL motion events in autonomous texts. In the translations, in which ME *descenden* may be triggered by a French *descendre* or a Latin *descendere* (as it definitely is the case in the Earlier Wycliffite Bible), by contrast, the verb is used more often in LITERAL contexts. The fact that the more idiomatic Later Version of the Wycliffite Bible discards each and every instance of *descenden*, is a good indicator that it was simply not idiomatic in Middle English to use this verb for LITERAL motion events.

However, we have seen that *descenden* is used for a range of specific contexts: first of all, for THEOLOGICAL contexts of divine figures and gifts descending from heaven, in which the verb is definitely influenced by the Latin *descendere* used in these same contexts. In these attestations, *descenden* can be regarded as a theological terminus technicus. Second, *descenden* is often used in contexts of LINEAGE and HERITAGE, in which it is equally a technical term and does not even necessarily involve an idea of MOTION. That the MOTION metaphor, which must lie at the origin of the LINEAGE and HERITAGE uses in the donor languages, is not really active in Middle English is supported by the fact that in these contexts, *descenden* almost never combines with the PATH adverb *down*. Other frequent uses of *descenden* are for all kinds of NON-LITERAL, metaphorical motion, such as for descending in status or fortunes, as well as for FICTIVE MOTION. As the motion here is not LITERAL, these 'motion events' do not involve (MANNER OF) MOTION in the first place, and the most important notion that must be conveyed when talking about them is $\text{PATH}^{\text{down}}$. Therefore these contexts represent a suitable semantic niche for *descenden*.

We can conclude that, just like *entren*, *descenden* does not seem to be well integrated as a normal motion verb in Middle English, i.e. one used for GENERAL LITERAL motion events (and, different from *entren*, does not acquire additional MANNER meaning), but is largely used for NON-LITERAL motion.

9.4 *Avalen*

9.4.1 Donor language portrait: OF/MF *avaler*

OF *avaler* is yet another verb meaning 'descend', yet different from OF *descendre*, it is not inherited from Latin, but is a new formation in Old French. It is formed from the particle *aval* 'down', which itself is a univerbation of the prepositional phrase *a val* 'to the valley' (on these types of new PATH verbs in the Romance languages, see 7.1.1 above).

A) Syntax

In Old and Middle French and in Anglo-Norman, *avaler* 'descend' is typically used with PATH satellites, either prepositional phrases (315a), adverbs (315b), or both (315c).

(315) a. **avalez** erent ***dedans un val*** (*Dermot2* 646, AND s.v.)
'They had descended into a valley.'

b. se commencerent les glasses à departir et **avaler** ***aval*** à grant impetuosité. (BAYE, I, 1400–10, 213, DMF s.v.)

'The frozen waters started to depart, and to descend down with great impetuousness.'

c. Et tuit li autres ensement ***De la tour aval*** **avalerent** (MACH., P. Alex., p.1369, 97, DMF s.v.)
'And all the others also descended down from the tower.'

In all these varieties of medieval French, *avaler* can also be used with MEDIUM GROUND encoded as direct object, such as the stairs and the mountain in (316).

(316) a. Par mal talent **avale *les degrez*** (*Nymes* 415, TL s.v.)
'Angrily, he descends the stairs.'

b. E Jesus ***le munt*** **avalat** (*Mirur* 74ra16, AND s.v.)
'And Jesus descended the mountain.'

Furthermore, in all the varieties *avaler* can be used with a reflexive object pronoun (317a), as well as in caused motion constructions (317b).

(317) a. et vindrent a la coste de la montaigne, et **s'avalerent** aval en une moult belle praierie. (ARRAS, *c.*1392–3, 198, DMF s.v.)
'And they came to the side of the mountain and descended REFL/themselves down into a very beautiful prairie.'

b. Dedans ma chartre ***l'*avalez** (*SMarg.* I 230, TL s.v.)
'Descend him into my prison.'

B) Semantics

As in the above examples, *avaler* is used for all kinds of downward motion, both self-motion and caused motion. Specific senses of *avaler* include the metaphorical use for 'diminishing, reducing', as in (318a–b), based on the common orientational metaphor LESS IS DOWN. This is also found for people being reduced in fortune and status (318c).

(318) **Diminish, reduce**

a. 'to diminish in quantity' (AND s.v., sense v.n.3)
La farine ne defaldrad ne l'olie ne **avalerad** (Latin: minuere) jesque Deu enveit pluie en terre (*Liv Reis* 156)
'Flour will not be lacking and oil will not diminish until God sends rain to the earth.'

b. *'diminuer'* ('diminish', DMF s.v., sense I.A.4.a)
selon ce que le pris du blé **avaleroit** ou monteroit en plain marchié (Mét. corp. Paris L., t.1, 1372, 201)
'According to whether the price of wheat fell or rose on the open market.'

c. *au figuratif: De la Fortune* ('fig.: of Fortune', DMF s.v., sense I.B.2.c)
Que Fortune par ses conduis Monte l'un, l'autre **avale** [. . .] (MACH., *R. Fort.*, *c.*1341, 36).
'That Fortune, in her ways, raises this one and brings down the other.'

Specific caused motion uses of *avaler* in the donor languages that will be found in Middle English are 'lower a drawbridge' and 'lower a vizor', 'take off one's headwear'. These do not appear to have been prominent uses in the donor languages: Only the detailed entry in *DMF* mentions the drawbridge use, together with 'to lower a machine' and other objects (319a). The *AND* does not specifically mention it, but has one attestation of AN *avaler* with a bridge (319b) as well:

(319) **Caused Motion: Lower a drawbridge**

a. *Baisser, abaisser qc: Un pont-levis, un engin . . .* ('lower something: a drawbridge, a machine, etc.', DMF s.v., sense I.B.2.b)
et ceulx de la ville **avalerent *le pont*** et ouvrirent portes et barrieres [. . .] (ARRAS, *c.*1392–3, 180)
'And those of the town lowered the bridge and opened ports and barriers [. . .].'

b. et crierent a les gardeyns del dit pount pur **avaler *le dit pount*** et les lesser entrer (*Anon Chr* 141.4, AND s.v., sense v.a.1)
'And they shouted to the guardians of the said bridge to lower the said bridge and let them enter.'

The *DMF* mentions the specific use for 'lower a vizor' (320) as one among many caused motion uses of *avaler*.

(320) **Caused Motion: Lower a vizor**
Baisser, abaisser qc: la visière: (DMF s.v., sense I.B.2.b)
de haucier et de **avaler *sa visiere*** (Cout. bourg. glosé P.M., *c.*1380–1400, 308)
'And to raise and to lower his vizor.'

(321) **Caused Motion: Take off headwear**

a. **Avaler *la coiffe*** du haubert (*Graal*, Vat. Chr. 1687, fol. 123a, Godefroy s.v.)
'Take off the hood of the hauberk.'

b. **Avaler *le chapperon*** sur les espaules (*Artus*, Vat. Chr. 738, fol. 1a, Godefroy s.v.)
'Take the hood down to the shoulders.'

c. Et la dame devant son vout Jusqu'au manton **sa guinple *avale*** (*Guil. d'A.* 2561, TL s.v.)
'And the lady takes the veil in front of her face down to her chin.'

TL and Godefroy do not explicitly mention any senses relating to vizors and headwear, but give the attestations in (321a–c). The *AND* does not have any attestation in this kind of context.[202] This use therefore seems to be just one among many caused motion uses of French *avaler*, but it will be prominent in the attestations of ME *avalen*, as we will see in the following sections.

9.4.2 ME *avalen*

Though *avaler* is widely used in medieval French, *avalen* is not very frequently attested as a loan word in Middle English. The search for its various spellings in the *CME* provided only 36 attestations. A problem was, however, that one of the attested spellings of *avalen* is *avail**, a search for which would have yielded an overwhelming majority of the frequent Middle English verb *availen* 'be helpful', i.e. too many false positives. Due to this common homograph, the precision of the results would have deteriorated disproportionately, so the slightly suboptimal recall was accepted. In order to compensate for a number of possibly missing occurrences with *avail**-spellings in the corpus, and for the generally small number of attestations, in a second step the quotations from the *Middle English Dictionary* entry were included as well. In this way, the total number of attestations increases to 75.[203] The kinds of uses in which *avalen* is attested in autonomous texts and translations are shown in Figure 9.7 and Table 9.10.

ME *avalen* is less often used for GENERAL LITERAL MOTION events in autonomous texts than it is in translated ones (see the black areas in Figure 9.7), and the proportion of NON-LITERAL contexts (see the white areas in Figure 9.7) is higher in the autonomous texts than in the translations. This observation is in line with what has been found for the other verbs above: borrowed PATH verbs seem to have a stronger tendency towards metaphorical or other NON-LITERAL use in genuinely Middle English texts and are only reluctantly used for GENERAL LITERAL MOTION. What makes *avalen* very different from the other borrowed PATH verbs, however, is that half of its uses are examples of CAUSED MOTION. In the case of all the other borrowed PATH verbs, CAUSED MOTION

[202] There is one attestation with helmets being *avalé* in the *AND*: *helmes et bacynetz availlez* (*Anon Chr* 27.17) 'helmets and basinets (were) brought down' (under sense v.a.2 'to bring down, lower'). The context, however, has nothing to do with the opening of helmets: It is the description of a battle, where 'lances are shattered, swords broken, hauberks ripped open, helmets and basinets brought down, and shields cut in four' (*En quele conflykte launces furrent frussez, espeis debrisez, haubergeouns desmaillez, helmes et bacynetz availlez et escues desquarterez* (*Anon Chr*, ed. Galbraith 1970: 27)).

[203] Just like the texts from *CME*, the texts from which the additional *MED* examples are taken were checked in the *Manual of the Writings in Middle English* (Severs et al. 1919–1951), to identify them as genuinely Middle English texts or as translations from a Latin or French original. Although on the one hand, the inclusion of the additional *MED* attestations adds to the representativeness of the study, on the other hand, it might be problematic that the attestations were taken from a dictionary entry, which by definition aims at giving attestations for each and every use, however infrequent. A comparison of the distribution of the respective uses of *avalen* in the *CME* alone and with those from the *MED* included, however, shows that the proportions stay very much the same.

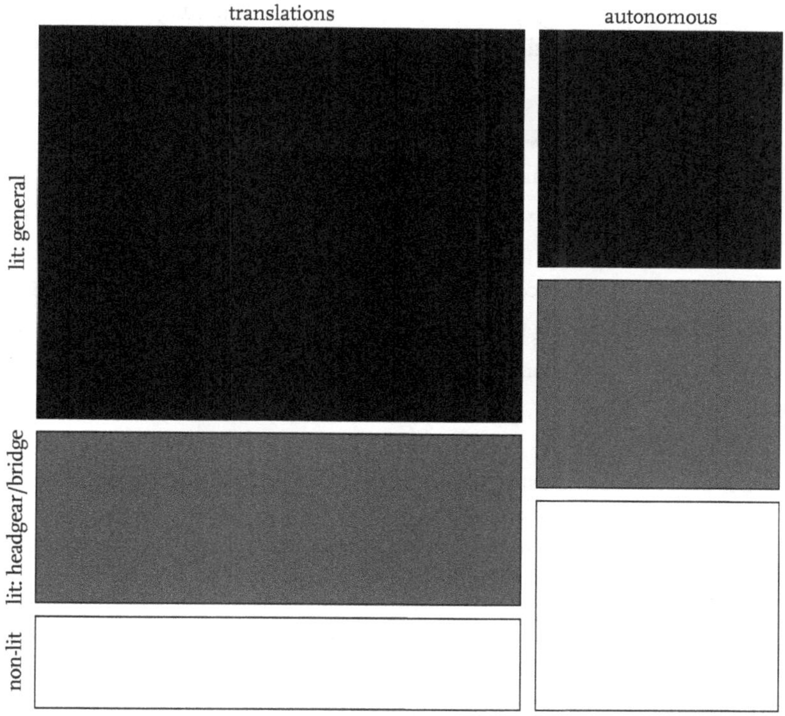

FIGURE 9.7. Usage contexts of ME *avalen* in CME translated and autonomous texts and MED quotations.

TABLE 9.10. Usage contexts of ME *avalen* in CME translated and autonomous texts and MED quotations.

	LITERAL: GENERAL	LITERAL: HEADGEAR	LITERAL: BRIDGE	NON-LITERAL: HUMBLE/ REDUCE	NON-LITERAL: OTHER	TOTAL
Translated	60% (30)	10% (5)	16% (8)	8% (4)	6% (3)	100% (50)
Autonomous	36% (9)	32% (8)	—	24% (6)	8% (2)	100% (25)
Total	52% (39)	17% (13)	11% (8)	13% (10)	7% (5)	100% (75)

attestations were only marginal. The kinds of LITERAL and NON-LITERAL uses are discussed in the following two sections.

9.4.3 LITERAL uses

The first thing one notices is that in hardly any of its LITERAL attestations does *avalen* refer to human FIGURES (only one attestation in the autonomous texts). This is different from the other PATH verbs. *Avalen* is used for rain (322a) and water (322b), for sails (322c) and other objects, such as food (322d). Note that in (322d), *avalen* is supported by the cooccurrence of the synonymous *goo dounne* 'go down.'

(322) a. That swych a ***reyn*** from heuen gan **auale** (*a*.1425(*c*.1385) Chaucer *TC*)
'That such a rain started to descend from heaven.'

b. The ***flod*** in such condicioun / **Avaleth**, that his drinke areche / He mai noght. ((*a*.1393) Gower *CA*)
'The flood descends in such a way that he may not reach his drink.'

c. Whan thei this strange vessel syhe / Come in, and hath his ***Seil*** **avaled** ((*a*.1393) Gower *CA*)
'When they see this strange vessel come in, which has its sail descended.'

d. And whan he supposed to haue etyn, þe ***mete*** myght not ***goo dounne*** ne **auale** in-to his stomake. (*c*.1482 *Brut*)
'And when he wanted to eat, the food could not go down nor descend into his stomach.'

In many of these attestations, particularly those with rain and flood like (322a–b), the similarity in sound to ME *afallen*, a prefixed form of ME *fallen* 'fall' may have supported the use of *avalen*.[204]

Among these LITERAL attestations with all sorts of inanimate FIGURES, two kinds of uses stick out as prominent (see Figure 9.7 and Table 9.10 above); both are instances of CAUSED MOTION. The first one, 'to bring down a draw bridge' (also present in French; see 9.4.1), is only found in translated texts, with most examples coming from Caxton's *Charles the Great*:

(323) For they lyfte vp the brydge, and entred, and after **aualed *the drawe brydge***. (1485 Caxton *Charles the Great*)
'For they lifted up the bridge, and entered, and then brought down the draw bridge.'

The second recurrent literal context is more interesting: Among the autonomous texts, the use of *avalen* to refer to 'removing one's headwear' is the most prominent one.[205] These attestations refer to taking off hats, hoods and other 'normal' headgear, such as in (324a–c), as well as military headgear, or to opening of helmets such as in (324d).

(324) a. He was asked yif he..dede..eny reuerence..he seid that..he, sitting stille, **avaled *his hode***. ((1439) *Cases before the King's Council* in *Seld.Soc.35*)

[204] According to the *MED*, one of the uses of ME *afallen* is specifically 'of a flood: subside; of a stream: flow downward' (s.v. *afallen*, sense 1.b).
[205] Cf. the *MED*'s sense 4 'to remove one's/somebody's headgear; to open the visor' (s.v. *avalen*).

'He was asked if he had done any reverence. He said that, sitting still, he had taken off his hood.'

b. He nolde **avalen** neither ***hood*** ne ***hat*** ((*c.*1390) Chaucer *CT.Mil.*)
'He did not want to take off hood or hat.'

c. Chyld whan þou gost yn eny strete, / Iff þou eny gode man or woman mete, / **Avale *thy hode*** to hym or to here, / And bydde, "god spede dame or sere!" (*c.*1475 Symon *Lesson*)
'Child, if you meet any respectable man or woman when you are walking in the street, take off your hood for him or her, and say, "Good luck, madam or sir!" '

d. thene the kyng **aualyd *his vyser*** (1485 Malory *Wks.*)
'Then the king opened his visor.'

As we have seen in 9.4.1, this use of *avaler* exists in the donor language, but is only one among many CAUSED MOTION uses of the verb: Among the different dictionaries of Old and Middle French, it is explicitly mentioned only by the *DMF*. The specialization of *avalen* for 'taking off headwear' thus seems to have taken place in English only, or rather, English speakers seem to have borrowed *avaler* in this meaning specifically, which is only a peripheral use in the donor languages. Note that all the attestations in (324a–d) are in a context of courtesy and politeness. This is a typical semantic area of borrowings from French in Middle English.

Just how prominent this sense of ME *avalen* is can be seen in (325): *Avaled* is used as an adjective meaning 'bareheaded' here. The lack of any other indication in the context (such as *hat, helmet*, etc.) that would point to this meaning indicates that 'taking off headwear' was a decontextualized meaning of *avalen* in Middle English.

(325) Priamus ran to halle **a-valed** / Ther these kynges to-gedur taled (*c.*1425(*c.*1400) *Ld.Troy*)
'Priamus ran bareheaded to the hall where these kings were talking together.'

It is probably through uses referring to opening of helmets, like (324d) above, that the meaning of PATH^{down} became backgrounded, and the—originally contextual—meaning of 'take off headwear, open helmet' became an independent, decontextualized meaning of *avalen* (cf. 2.2.3). This development, then, also allows uses like (326) with *up*.

(326) He **auaylet** [Thrn: lyfte] **vppe** his viserne fro his ventalle. (*c.*1475 *Awntyrs Arth.*)
'He lifted up his visor from his helmet's vent.'

Note that one of the manuscript alternatives for *avalen* in this attestation is *lyfte* 'lifted', i.e. a verb directly opposite to the original meaning of *avalen*. This, as well as the combination with *up*, suggests that the meaning 'take off headwear, open helmet' was strongly entrenched with *avalen*: It is highly unlikely that the writer had meant 'descend up the visor'. What may also have contributed to the backgrounding of PATH^{down} in *avalen* is that the 'headgear' attestations (just like the 'bridge' ones) mostly come without any directional adverb 'down', as seen in (324a–d), and the fact that there is no formally corresponding noun or Latin verb to 'stabilize' the PATH meaning.

The LITERAL uses of *avalen* thus suggest that it was not borrowed as a general PATH verb, but in very specific senses, such as for receding water, lowering sails, lowering bridges, and, particularly, taking off headwear.

9.4.4 NON-LITERAL uses

Most NON-LITERAL uses of *avalen* are in the meaning of 'to humble oneself or somebody, to reduce somebody (in rank or fortune)' (sense 5 in the *MED*), such as (327a–b), which make reference to Fortune bringing people down on her wheel. Note that also this use has a parallel in the similar sounding native *afallen* (see *MED* s.v. *afallen*, sense 2.b).

(327) a. For his fortune is to deceive / And forto change upon the ***whel*** / His wo with othre mennes wel: / Of that an other man **avaleth**, / His oghne astat thus up he haleth ((*a*.1393) Gower *CA*)
'For it is his destiny to deceive, and to change his bad luck with other men's good luck on the wheel [of Fortune]: By another man's descending he thus drags up his own state.'

b. This marchaunt whilom so worshipable..***Fortune*** did **avale**. (*c*.1460(*a*.1449) Lydg. *2 Merch.*)
'Fortune made descend this formerly so honorable merchant.'

In other attestations of *avalen*, it is not Fortune, but another person (328a), or God (328b) who humbles the FIGURE:

(328) a. There were made..Statutz ageinst Walshmen, to **avale** their pompe and pride ((1444) *RParl.*)
'Statutes against Welshmen were made to humble their pomp and pride.'

b. And myghty Tyrauntes fro her Riall see He hathe **avalede** and yputte dovne. (*c*.1450(?*a*.1422) * Lydgate, *LOL*)
'And he [God] has brought down and descended mighty tyrants from their throne.'

These specific uses in the context of the wheel of Fortune and of 'bringing someone down from their high horse' could also be characterized as belonging to courtly discourse domains. This is a register in which French loanwords are particularly common (cf. 6.1).

9.4.5 Summary

This section has shown that *avalen* is borrowed into English in a few very specific meanings, particularly 'taking off headwear', and in senses that relate to humiliation. In the 'headwear' meaning, it even seems to lose its original meaning PATH^{down}. These two specific uses, both characteristic of courtly contexts, account for more than half of the autonomous Middle English attestations of *avalen*.

GENERAL LITERAL uses of *avalen* are more widespread in translated texts, where they are twice as frequent as they are in autonomous Middle English texts (see Figure 9.7 and Table 9.10). The verb is very rarely used for animate FIGURES, but tends to refer to the descending of rain, water, sails, and other objects. This indicates that *avalen* was not a common verb for ordinary motion events in Middle English.

9.5 *Ascenden*

9.5.1 Donor language portrait: OF/MF *ascendre*, L *ascendere*

Like its antonym *descendre*, *ascendre* has developed from a Latin prefixed verb *ad-scandere* (> L *ascendere*) 'to to-climb'. The idea of 'climbing', however, has receded and given way to the pure meaning of PATH^{up}. *Ascendre* has apparently not had a continuous history from Latin to French, but has been reintroduced into French as a Latinism.[206] Tobler–Lommatzsch does not have an entry for it; the *TLF* dates it to 1270. Furthermore, the *AND* gives attestations from Anglo-Norman texts only from the second half of the thirteenth century.

A) Syntax

In Latin as well as in (Anglo-)French, *ascend(e)re* is used either bare or with prepositional PATH satellites, such as in (329a–b):

(329) a. fecit eum **ascendere** retro ***super equum*** (1211 *SelPl Crown* 58, DMLBS s.v., sense 1.a)
'He made him ascend behind onto the horse.'

b. [La soupe au vin] [. . .] prohibe les fumees caligineuses **adscendre *aulx cerveau*** (*Rég. santé corps* C., 1480, 103, DMF s.v.)
'[Wine soup] [. . .] prevents the steaming fumes from ascending to the brain.'

In Anglo-Norman and Latin, it is used also with GROUND as direct object, such as in (330a–b). The dictionaries do not list this structure for continental French.

(330) a. **ascendit *le mount*** (*Rich I* 18.9, AND s.v.)
'He ascended the mountain.'

[206] See also the relatinizing spelling with <d> in (329b).

b. **ascendit** ***equum*** et transivit pontem (1255 *SelPlForest* 115, DMLBS s.v., 3.c)
'He ascended the horse and traversed the bridge.'

B) Semantics

In Latin, the verb is used for all kinds of motion upwards, particularly on mountains and horses (as in (330b)). A specific use is the THEOLOGICAL one for ascending into heaven, as in (331a–b), supported by the name for the feast, Latin *ascensio*, medieval French *(la feste de) l'Ascension* 'Ascension Day'.

(331) **THEOLOGY**
a. 'to ascend into Heaven' (DMLBS s.v., sense 1.c)
ad caelestia regna [. . .] **ascendit** (BEDE HE II 20)
'He ascended [. . .] to the heavenly kingdoms.'

b. Jhesu Cryst **ascendist** ***a cels*** (*Bible* 59.1, AND s.v.)
'Jesus Christ ascended to heaven.'

Like the corresponding adjective and noun *ascendant*, the verb also is used as a technical term in ASTRONOMY in Latin and medieval French:

(332) **ASTRONOMY**
a. 'to ascend, be in the ascendant (astr.)' (DMLBS s.v., sense 4)
sol [. . .] venit in Cancro, quo non altius **ascendere** potest (*V. Osw.* 443)
'The sun [. . .] comes into Cancer, where it can ascend no higher.'

b. 'ASTR. *[D'un astre] S'élever, monter à l'horizon*' ('of stars: rise, ascend on the horizon', DMF s.v., sense II.B)
Une influence [. . .] Vient [. . .] par les extremitez Des infortunees planettes Ou des bonnes fines et nettes, **Ascendens** a l'eure que l'omme Naist de mere (CHR. PIZ., M.F., II, 1400–3, 84).
'An influence comes from the borders of the infortunate planets, or of the good, fine, and nice ones, ascending in the hour when a person is born from the mother.'

Thus, already in the varieties of medieval French, *ascendre*, being a Latinism, seems somewhat restricted to these specialized uses as a technical term in theology and astronomy. For general upward motion, *monter* seems to have been preferred, which is discussed in 9.6, together with its prefixed form *amonter*.

9.5.2 ME *ascenden*

Of all the borrowed PATH verbs investigated here, ME *ascenden* is the youngest: Its earliest attestations are from the Wycliffite Bible translations from the 1380s, while the other verbs are first attested around 50 years (ME *avalen*, *descenden, ishen, mounten*) or even a century before that (ME *entren, amounten*). I found 170 attestations of *ascenden* in the texts from the *CME*. Recurrent contexts in which the verb is used include astronomy and ascension to heaven, i.e. the ASTRONOMICAL and THEOLOGICAL uses also present in the donor languages (see 9.5.1). As shown in Figure 9.8 (the light grey areas) and Table 9.11, these account for as many as three quarters of the attestations of *ascenden* in autonomous Middle English texts.

This suggests that *ascenden* is very much a technical term in Middle English. The different types of use will be discussed in the following sections.

9.5.3 LITERAL uses

The first thing that is noticeable is how rare the GENERAL LITERAL uses of *ascenden* are in the autonomous Middle English texts: There are only seven of them, i.e., only a sixth of the attestations in the autonomous texts. In the translated texts, particularly those from Latin originals such as in the *Polychronicon*, the Earlier Wycliffite Bible or the *Prose Life of Alexander*, ME *ascenden* is far more frequently used to refer to general upward motion, such as in (333a–c). This

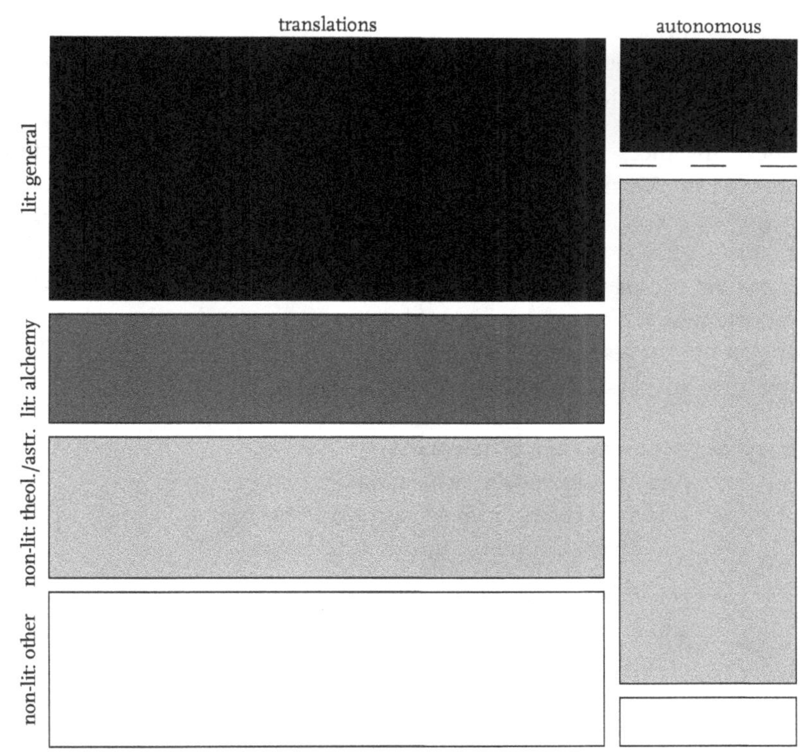

FIGURE 9.8. Usage contexts of ME *ascenden* in CME translated and autonomous texts.

TABLE 9.11. Usage contexts of ME *ascenden* in CME translated and autonomous texts.

	LITERAL: GENERAL		LITERAL: ALCHEMY		NON-LITERAL: THEOLOGY/ ASTRONOMY		NON-LITERAL: OTHER		TOTAL	
Translated	40%	(51)	16%	(21)	21%	(27)	23%	(30)	100%	(129)
Autonomous	17%	(7)	—		76%	(31)	7%	(3)	100%	(41)
Total	34%	(58)	12%	(21)	34%	(58)	19%	(33)	100%	(170)

is likely to reflect its use in Latin; the passage corresponding to (333b) in the Vulgate, for instance, also features L *ascendit.*

(333) a. Tullia [. . .] **ascendede *here chariette*** (?*a*.1475(?*a*.1425) *Higd.(2)*)
'Tullia [. . .] ascended her chariot.'

b. til he wente out in to the hil **in which** Moises **ascendide** (*a*.1425(*a*.1382) *WBible(1)*) (*usquequo exiit in montem in quo Moses ascendit* (2Macc 2:4))
'Until he went out onto the hill onto which Moses had ascended.'

c. Þis Mountayne hadd made of saphyres twa thowsande gree3 & a halfe, by þe whilke men **ascendid *to þe summit of þe Mountayne*** (*c*.1440 *PLAlex.*)
'This mountain has two and a half thousand steps made from sapphires, by which people ascend to the summit of the mountain.'

Of the 11 instances of ME *ascenden* in the Earlier Version of the Wycliffite Bible, the Later Version retains only one (333b), the others are all replaced with *stien (up)* or *gon up*, as in (334a–b), which give the Latin verse with the earlier and later Wycliffite translation. Similar to the replacement of ME *descenden* with *gon down/comen down* in the Later Version (see (299–300) in 9.3.3), the replacement of ME *ascenden* with combinations of neutral verbs and satellites (*stien/gon up*) suggests that ME *ascenden* was not an idiomatic word for literal 'going up to an elevated place'.

(334) a. *et Anna non* ***ascendit*** (1Sam 1:22)
And Anna **assendide** not (*a*.1425(*a*.1382) *WBible(1)*)
And Anna **stiede** not **up** (*a*.1425(*c*.1395) *WBible(2)*)
'And Hannah did not go up [i.e. to the temple].'

b. *post triduum* ***ascendit*** *Hierosolymam a Caesarea* (Acts 25:1)
aftir the thridde day he **assendide** to Jerusalem fro Cesarie. (*a*.1425(*a*.1382) *WBible(1)*)
aftir the thridde dai he **wente vp** to Jerusalem fro Cesarie (*a*.1425(*c*.1395) *WBible(2)*)
'After the third day, he went up to Jerusalem from Caesarea.'

Sixteen per cent of the attestations of *ascenden* in translated texts are in the context of ALCHEMY, all from *The Book of Quinte Essence*, another translation from Latin (cf. Figure 9.8 and Table 9.11):

(335) putte it þanne in a distillatorie. and first þe vynegre wole **ascende**; þanne after ȝe schal se merueilis (*a.*1475 *Bk.Quint.*)
'Then put it in a distilling apparatus, and first the vinegar will ascend; then afterwards, you will see marvels.'

In these, *ascenden* can be regarded as a technical term in the domain of distillery (cf. sense 6 in the *MED* '*Alch.* To rise as vapor, vaporize', s.v. *ascenden*). In the next section, we will see further uses of *ascenden* as a technical term.

9.5.4 NON-LITERAL uses

The most widespread senses of *ascenden* in autonomous texts, as shown in Figure 9.8 and Table 9.11 above, are ascension to heaven and ascension in the context of planets, i.e. the THEOLOGICAL and ASTRONOMICAL contexts in which the verb is also used in the donor languages (see 9.5.1). The THEOLOGICAL attestations of *ascenden* mirror the ones of its converse, *descenden* (see 9.3.4.1): While *descenden* is used for divine FIGURES coming down from the heavenly spheres, and for the Harrowing of Hell (*Descensus Christi ad Inferos*), *ascenden* is used for the reverse, mostly for Jesus ascending to heaven (cf. the corresponding noun and name of the feast, *Ascension*), as in (336a–b).

(336) a. Seþin Crist ***vp vnto heuyn*** **ascend** (*a.*1425(?*c.*1375) *NHom.(3) Leg.*)
'Since Christ ascended up into heaven.'

b. Jesus: ***To my God and my fadir dere***, To hym als-swithe I schall **assende** (*a.*1450 *Yk.Pl.*)
'Jesus: "To my dear God and father, to him I shall ascend soon."'

Ascenden replaces the older *stien up*, which was the verb used to refer to the Ascension in Old English (see Gneuss 1955; Käsmann 1961: 34–6, 130–1). According to Käsmann (1961: 130–1), the corresponding noun OE *upstige* 'Ascension' was lost in early Middle English. After a period of experimentation with different alternatives, such as *upastihung*, *astiung*, the terminological gap was finally filled with the French/Latin noun *ascension*. The verb *stien up*, by contrast, had stayed on after the loss of the noun *upstige* 'ascension'. Only after the noun *ascension* was established will the corresponding verb *ascenden* have gradually replaced the older *stien up*. This can be seen in the different translations of the *Polychronicon*: The earlier one by Trevisa from the late fourteenth century still has *steihe up* (337a), i.e. a native structure with MOTION verb and PATH satellite, while the fifteenth century anonymous translation in MS. Harl. 2261 has the PATH verb *ascendede* (337b).

(337) a. Out of þat mount Crist **steihe vp** into heuene ((*a*.1387) Trev.*Higd.*)
'From this mount Christ ascended into heaven.'

b. From whiche mownte Criste **ascendede** to heuyn (?*a*.1475(?*a*.1425) *Higd.(2)*)
'From which mount Christ ascended into heaven.'

To a minor extent, *ascenden* is also used for other divine FIGURES in THEOLOGICAL contexts, such as *goddes miht* in (338).

(338) Gabriel said, "leudy, nay, / Hou þat may be i sal þe say, / þe haly gast in þe sal lende, / And ***goddes miht in þe* ascende**" (*a*.1400 *Cursor*)
'Gabriel said, "No, lady, I will tell you how that [i.e. immaculate conception] may be: the Holy Ghost will land in you, and God's power will ascend in you." '

In these THEOLOGICAL uses, *ascenden* can be regarded as a technical term, just like in its use for 'rising above the horizon' in the domain of ASTRONOMY. The attestations from autonomous texts in contexts of ASTRONOMY are almost all from Chaucer's *Astrolabe*, a manual on how to use an astrolabe, such as (339).

(339) to knowe by nyht or by day the degree of any signe þat **assendith** on the est Orisonte, which þat is cleped communly the assendent (*a*.1450(1391) Chaucer *Astr.*)
'How to know by night or by day the degree of any sign that ascends on the east horizon, which is commonly called the ascendant.'

Three astronomical attestations, however, occur outside such specialist texts, in the *Canterbury Tales*, i.e. in narrative fiction, such as in (340).

(340) Phebus hath laft the angle meridional, And yet **ascendynge** was the beest roial, The gentil leon, Whan that this Tartre King, Cambyuskyan, roos fro his bord [. . .] ((*c*.1395) Chaucer *CT.Sq.*)
'Phebus had left the meridional angle, and the royal beast, the gentle lion, was ascending, when this Tartar king, Cambyuskyan, rose from his table [. . .].'

9.5.5 Summary

This section has shown that *ascenden* has only a very limited range of uses: It mostly occurs as a technical term in the domains of ASTRONOMY and ALCHEMY, as well as to refer to the Ascension in THEOLOGICAL contexts, where it starts to replace the older term *stien up* in the course of the Middle English period. In autonomous Middle English texts, it occurs only rarely in descriptions of GENERAL LITERAL MOTION. It is only in translations from Latin that *ascenden* is used more often for GENERAL LITERAL MOTION contexts (in 40 per cent of the attestations), which is most likely due to the influence of the Latin original.

9.6 *Mounten* and *amounten*

9.6.1 Donor language portrait: OF/MF *monter* and *amonter*

Both F *monter* and F *amonter* are derived from the basis *mont* 'mountain', and their basic meaning is 'to move up'.

A) Syntax

In Anglo-Norman as well as continental Old and Middle French, the unprefixed *monter* is found with prepositional (341a) and adverbial (*amunt* 'up' in (341b)) PATH satellites as well as with GROUND in the direct object (342a–b).

(341) a. Sire, vuillez vous approcher a la sale ou vuillez **monter** (MS mointer) ***a la chamber***? (*Man lang* ANTS 73.2, AND s.v.)
'Sir, do you want to approach yourself to the hall, or do you want to mount into the chamber?'

b. Hastivement a la nef vunt; ***Par l'eschiele*** **muntent** ***amunt*** (*MFce Lais G* 702, TL s.v.)
'Hastily they come to the ship; they mount up by the ladder.'

(342) a. cele damoisele **mounta** un jur **le mestre tour de cest chastel** (*Fouke* ANTS 46.15, AND s.v.)
'One day this damsel mounted the highest tower of this castle.'

b. car onques tel enfant n'avoit **monté ce lieu la**. (*Mir. st Guill., c.*1347, 5, DMF s.v.)
'Because never had such a child gone up to this place [i.e. the temple in Jerusalem].'

Also the prefixed *amonter*, which is often used figuratively, is found with prepositional PATH satellites (343) and direct object GROUND (344). *TL* and *DMF* also give numerous caused motion attestations with the FIGURE in the direct object in figurative senses like 'to raise, increase sth.', as in (345), which seem to be absent in Anglo-Norman according to the *AND*.

(343) ceo ke fet avons jeske a ce jur **Amunt'** ***a mut poy de profyt*** (*S Fran* ANTS 5935, AND s.v.)
'What we have done until this day amounts to very little profit.'

(344) A grant espleit ount **amonteiz** Trestoz **les monz** et avaleiz (*MSMich.* 557, TL s.v.)
'With great effort, they mounted and descended all the mountains.'

(345) tels cuide venger sa honte Qui l'acroist ançois et **amonte** (*Ferg.* 81, 29, TL s.v.)
'One wants to revenge his shame, who instead increases and augments it.'

B) Semantics

In medieval French, *monter* is used in a wide range of contexts—the entry in *TL*, for instance, consists of an entire 14 columns. The use of *monter* that is adopted most prominently in Middle English is the one referring to mounting horses and other riding animals, such as in (346a–c), in which it is the converse of *descendre* (see 9.3.1). The prefixed *amonter* is not found in this sense.

(346) *monter* HORSE

a. *'zu Pferde steigen, aufsitzen'* ('to mount a horse', TL s.v. *monter*)
Baruns, ne vus targiez, Eissez des nefs, **muntez**, si chevalchiez! (*Ch. Rol.* 2806)
'Barons, hurry up, exit from the ships, mount, and ride!'

b. (horse) 'to mount (a horse)' (AND s.v. *munter*, v.a. 1)
Lui quens **mounta un destrer abrivé** (*Boeve* 138)
'The count mounted a swift warhorse.'

c. *En partic. [À propos de monture]: monter (à cheval etc.)* ('with respect to riding animals: mount (a horse, etc.)', DMF s.v. *monter*, sense I.A.1.d)
le gentil chevalier fist tantost la dame et sa compaignie **monter** (LE BEL, *Chron.* V.D., t.1, 1352–6, 16)
'The noble knight immediately made the lady and her company mount [i.e. on horses].'

Two among the various figurative uses of *monter* are the senses 'amount to, add up to, be worth' and 'to mean, signify'. These are present in both insular and continental varieties of medieval French, see the respective senses from *TL*, *DMF*, and *AND* listed in (347a–c) and (348a–c). They occur with direct object and with prepositional phrase complementation.

(347) *monter* AMOUNT TO, ADD UP TO, BE WORTH

a. *'übertr. betragen, sich belaufen auf', 'wert sein'* ('fig. 'to amount to', 'be worth', TL s.v.)
quatre foiz nonante degrez, qui **montent trois cenz et soissante degrez** (*Brun.Lat.* 131)
'Four times ninety degrees, which amounts to 360 degrees.'

b. *'Atteindre une certaine valeur'* ('reach a certain value', DMF s.v., sense III.B.3)
il prindrent toute la robe qui troverent en icelle chambre, **montant *à la somme ou valeur de cinq cens frans*** (*Reg. crim. Chât.*, I, 1389–92, 249)

'They took all the clothes they found in this room, amounting to the sum or value of 500 Francs.'

c. 'to add up to, amount to' (AND s.v., v.a. 3)
le premer tiercz qi **mounte seize centz seissante sis livres** (*Treaty Rolls* ii 9.23)
'The first third [of a payment] that amounts to 1,666 pounds.'

(348) *monter* MEAN, SIGNIFY

a. *'zu bedeuten haben'* ('to signify', TL s.v.)
savez **que monte** Ceste chose que ci vëez? (*Erec* 5792)
'Do you know what this thing that you see here means?'

b. *'Signifier, vouloir dire'* ('signify, mean', DMF s.v., sense B.5)
Cheÿ tempeste, fourdre, espars En pluiseurs lieux, en pluiseurz pars, Et maintz homs s'en espoeunta ; Mais nul ne sçot **que** ce **monta** (*Renart contref.* R.L., t.1, 1328–42, 104).
'There fell a thunderstorm, flash, and lightning in various places, in various areas, and many people were frightened of it, but no one knew what it meant.'

c. 'to mean' (AND s.v., v.a. 3)
le dit Thomas demandra soun dit meistre **qe mounteroit** le parole (*King's Bench* vii 112)
'The said Thomas asked his said master what the word meant.'

The sense 'to amount to, be worth' is also present in the nouns *monte* and *montance* 'sum, amount, value' that are derived from the verb *monter* (*TL* and *DMF* s.vv., *AND* s.vv. *munte, muntance*).

In Anglo-Norman, the sense 'amount to, be worth' is available not only for *monter*, but for the prefixed *amonter* as well, and the corresponding nouns are *amunte* and *amuntance* 'sum, amount, value'. Like *monter, amunter* can take both a direct object and a prepositional phrase in this sense, as in (349a–b). Also in the sense of 'mean, signify', both complementation patterns are attested, see (350a–b).

(349) AN *amunter* AMOUNT TO, ADD UP TO, BE WORTH

a. 'to amount to' (AND s.v.)
diverses subsides, queles **amontont MLCCCC marcz** (*Windsor* 197)
'Various subsidies, which amount [i.e. amount to] MLCCCC marks.'

b. 'to amount to, be worth' (AND s.v., *amunter a*, sense 2)
les amendes [. . .] **amounteront *a treble & quatreble de lour coustages*** (*Rot Parl1* i 416)
'The compensations [. . .] will amount to three and four times their costs.'

(350) AN *amunter* **MEAN, SIGNIFY**

a. 'to mean, add up to' (AND s.v., v.a. sense 3)
lui roi P. demaunda **quoi** ceste chose devoit **amounter** (*Triv* 65.3)
'King P. asked what this thing should mean.'

b. 'to mean, signify' (AND s.v., *amunter a*, sense 3)
Car il meisme asez reconte ***A quei*** cele aventure **amonte** (*S Greg* 345)
'Because he himself reports adequately what this marvel means.'

While *monter* in the senses of 'to amount to, be worth' and 'to mean, signify' exists in all the varieties of medieval French, the prefixed *amonter* in these two senses appears to be typical for Anglo-Norman only: *TL* and *DMF* do not list any attestations of *amonter* in the sense of 'to mean, signify'. In the sense of 'to amount to, be worth', the only quotation in *DMF* is from an Anglo-Norman text (s.v. *amonter*, sense II.A). *TL* has a single attestation in this meaning, (351):

(351) Anceis fust el (*l'espee*) tote esgrunee Qu'el abatist de l'elme tant Ki **amontast un peis pesant** (*En.* 4434, TL s.v.)
'It [the spear] would rather be all broken in pieces than to cut off from the helmet as much as would amount the weight of a pea.'

Here, the image of a balance scale seems to be involved: The splinter of the helmet, put in one pan of the scale, would not even be heavy enough to 'amount' the pea in the other pan. This image might be at the origin of *(a)monter* + direct object in the sense of 'amount to, be worth', while *(a)monter à* in this sense may be based on the image of literally 'amounting to' a certain height on a measuring stick.

9.6.2 ME *mounten* and *amounten*

Together with ME *entren*, ME *amounten* is the verb that is attested earliest in Middle English among the borrowed PATH verbs analysed here: Both are first attested in the *Kentish Sermons*, a translation from French from *c.*1275. Different from the frequently used *entren*, however, *amounten* is considerably rarer, with only around a hundred attestations in the *CME*. The first attestation of the non-prefixed *mounten* is from *c.*1300 in the *South English Legendary*, and it was found 232 times in the *CME*.

The analysis of the attestations of the two verbs in the *CME* shows that there is a clear division of labour: *Mounten* is used almost exclusively for LITERAL motion, predominantly for 'mounting horses' (see Table 9.12 and the black and dark grey areas in Figure 9.9). The prefixed *amounten*, by contrast, is almost exclusively used NON-LITERALLY, very often in the sense of 'amounting to a sum (of money)' and 'signifying' (see Table 9.13 and the white and lighter grey areas in Figure 9.9). Thus, while in Anglo-Norman, the non-prefixed *munter* is found in all these uses ('mounting horses', 'amounting to, adding up to', 'signifying'), it specializes on the HORSE contexts in Middle English, and leaves the NON-LITERAL contexts to *amounten*.

The following two sections will discuss the LITERAL and NON-LITERAL uses of *mounten* and *amounten* in more detail.

TABLE 9.12. Usage contexts of ME *mounten* in CME translated and autonomous texts.

	LITERAL: GENERAL		LITERAL: HORSE		NON-LITERAL		TOTAL	
Translated	19%	(32)	75%	(126)	6%	(10)	100%	(168)
Autonomous	8%	(5)	82%	(53)	10%	(6)	100%	(64)
Total	16%	(37)	77%	(179)	7%	(16)	100%	(232)

TABLE 9.13. Usage contexts of ME *amounten* in CME translated and autonomous texts.

	LITERAL: HORSE		NON-LITERAL: SUM, VALUE		NON-LITERAL: SIGNIFICATION		NON-LITERAL: OTHER		TOTAL	
Translated	—		85%	(22)	—		15%	(4)	100%	(26)
Autonomous	1%	(1)	68%	(55)	27%	(22)	4%	(3)	100%	(81)
Total	1%	(1)	72%	(77)	21%	(22)	7%	(7)	100%	(107)

9.6.3 LITERAL uses

The vast majority of the attestations of *mounten* (77 per cent) refer to mounting a horse, a use already prominent in the donor languages (see 9.6.1), with the respective proportion in autonomous and translated texts being quite similar (82 and 75 per cent respectively, see Table 9.12 and the dark grey areas in Figure 9.9). Prototypical examples are (352a–b).

(352) a. she [. . .] **mounted** ***vpon a good cursere*** (?c.1450 *Knt.Tour-L.*)
'She [. . .] mounted on a good war horse.'

b. Soo thenne they dressid on their helmes / & putte on their sheldes / & **mounted** ***vpon horses*** (1485 Malory *Wks.*)
'So then they put on their helmets and their shields and mounted upon horses.'

However, all these HORSE attestations come only from a small number of texts from the end of the Middle English period (M4): Malory's *Morte Arthur*, *The Four Sons of Aymon*, *Melusine*, *Blanchardyn*, *Charles the Great*, and *The Book of the Knight de la Tour-Landry*. The lack of *mounten* in other texts is not necessarily due to 'mounting horses' not being written about: Before *mounten* became the usual verb for this sense, native verbs referring to mounting horses had been *worthen* (cf. *MED*, sense 7), *(bi)striden* (cf. *MED* svv. *striden*, sense 2; *bistriden*, sense a), *stien* (cf. *MED*, sense 2c), or *leven* (cf. *MED*, sense 1.b.b), and these are used also in earlier texts from the *CME*, such as in (353a–d).

(353) a. He **werth** ***opon is palefrai.*** (c.1300 *SLeg.Becket*, MED s.v. *werthen*)
'He mounted upon his palfrey.'

b. **Lyarde** he **bistrydeth**, And raped hym to-iherusalem-ward.
(c.1400(c.1378) *PPl.B*, MED s.v. *bistriden*)
'He mounts upon Lyard and hastens himself toward Jerusalem.'

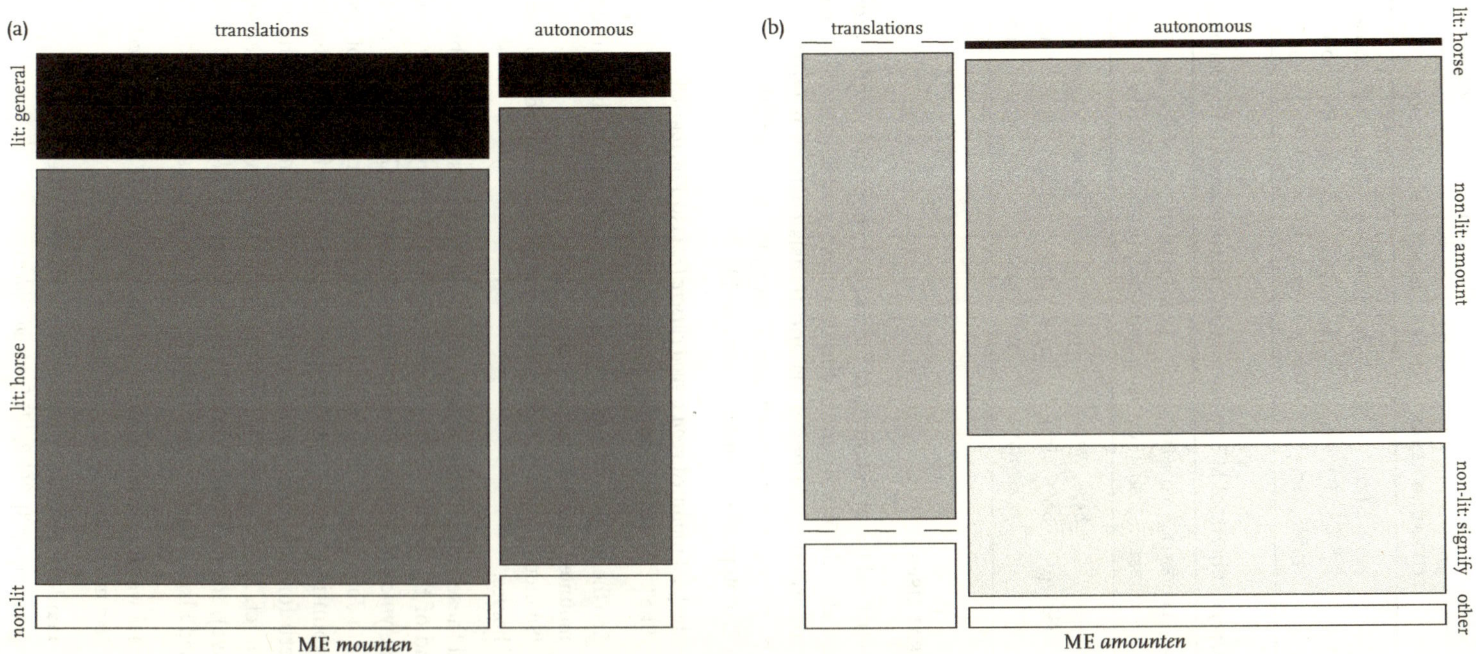

FIGURE 9.9. Usage contexts of ME *mounten* and *amounten* in CME translated and autonomous texts.

c. foure hundrid ȝonge men þat **steȝedyn** ***vp camailes*** & flowen. ((*a*.1382) *WBible(1)*, MED s.v. *stien*)
'Four hundred young men who mounted on camels and fled.'

d. Þay..***on hors*** **lepes**. (*c*.1400(?*c*.1380) *Cleanness*, MED s.v. *lepen*)
'They leap on horses.'

Later *mounten* became the usual term for this meaning. 'Mounting horses' is, of course, another activity related to courtly culture, which is one of the areas where the French influence on the lexicon is strongest.

The few other LITERAL attestations of *mounten* in autonomous texts mostly refer to flying FIGURES (Icarus, dragons) mounting in the air, such as in (354).

(354) This Icharus began to **monte** ((*a*.1393) Gower *CA*)
'This Icarus started to mount.'

Only one instance, (355), is peculiar in that *mounten* does not seem to have a prominent 'up'-meaning here—there is nothing in the context that would suggest that the narrator is located in a lower place before or that he moves upwards. The choice of word may of course be due to the limitations imposed by the alliterative verse.

(355) Þanne y **munte** me ***forþ*** þe mynstre to knowen, / And a-waytede a woon wonderlie well y-beld (*c*.1600(?*c*.1395) *PPl.Creed*)
'Then I went on (lit. 'mounted myself forth') to see the minster, and saw a building, wonderfully well constructed.'

In the translations, by contrast, the range of contexts in which *mounten* occurs is larger: The verb is used to refer to people mounting hills, mountains and towers, walls and ladders, into halls and basins. Most of these attestations (18 out of 32) are from translations by Caxton—*Blanchardyn and Eglantine*, *The Four Sonnes of Aymon*, and *Charles the Great*—and from the translation of Jean d'Arras' *Mélusine* (13 out of 32). This will often be due to the presence of *monter* in the original French, such as in (356), where the Middle English translation of *Mélusine* sticks to the same lexical material as its original.

(356) Les dames retournèrent et **montèrent** ***sur les eschafauds*** (*Mélusine* ed. Brunet (1972: 268))
the ladyes & damoyselles **mounted** ***vpon the scafoldes*** (*c*.1500 *Melusine*)
'The ladies and damsels mounted upon the scaffolds.'

The wider range of possible LITERAL GROUNDS that *mounten* is used for in the translated texts is likely to be due to the fact that it is used for all kinds of upward motion in French (see 9.6.1).

9.6.4 NON-LITERAL uses

Most of the occurrences of *mounten* thus are LITERAL. The rare NON-LITERAL attestations in autonomous texts are either in the meaning of 'increase' (i.e. an instance of the metaphor MORE IS UP) or of 'mounting to heaven' (357).

(357) He ȝyue vs grace so to acounte þat we may ***to heuen* mounte** (*a*.1400 *Cursor*)
'He gave us grace to confess thus, that we may mount to heaven.'

The word is used in the rhyme in (357), but note that, different from *ascenden*, in this case the FIGURE are people, not Christ or any other divine FIGURE rising to heaven. This is another indication that *ascenden* is a technical term for the *Ascension*, and not for just any FIGURE rising to heaven.

In three of its NON-LITERAL attestations, *mounten* is used in the meaning of 'add up to', a use which is, however, far more typical for the prefixed form *amounten*: Almost three quarters of the attestations of *amounten* are in the context of sums (mostly of money) and values and can be paraphrased as 'add up to' (cf. *MED* s.v. *amounten*, sense 3). (358a–b) are examples from administrative contexts:

(358) a. the money **amonthe *to the Summa x. li*** (1478 *Stonor*)
'The money amounts to the sum of ten pounds.'

b. [. . .] (of the said terme that is to say a yere and half therof to lerne gramer and the resydue of the said two yeres) which **amounteth *to half-a-yere*** to scole for to lerne to write (1450–54C1/19/492 Petition of Thomas Bodyn (1450 *Chanc.E*))
'[. . .] that is to say, of the abovementioned term a year and a half to learn grammar, and the rest of the abovementioned two years, which amounts to half a year, to school in order to learn to write.'

Although *amounten* in this sense usually appears in contexts of sums of money, as in (358a), it is used for other sums as well, such as for time in (358b). These uses of *amounten* are obviously most widespread in texts which deal with sales, rents, last wills, and financial transactions in general, i.e. the Paston and Stonor Letters, documents from the Chancery, Gilds, and Church Registers. However, it is also attested outside these administrative contexts, and occurs in religious texts, such as—rather early—in *Handlyng Synne* (359a), or in history writing like the *Brut Chronicle* (359b). ME *amounten* in this sense, therefore, is not restricted to administrative texts as a genre, but can be found in all sorts of texts.

(359) a. Whan þey hadde gadred and a-countede **ðrytty mark** hys gode **amounted** (*a*.1400(*c*.1303) Mannyng *HS*)
'When they had gathered and counted, his [i.e. the dead man's] goods amounted thirty marks.'

b. þe King [. . .] borowed a somme of gold þurghout þe Reame, of temporall peple, þat **amounted a c Ml marc of money** (?*c*.1450 *Brut-1447*)
'The King [. . .] borrowed a sum of gold from secular people throughout the realm, which amounted to 100,000 marks of money.'

The remaining NON-LITERAL uses of *amounten* are all related to the ones in contexts of sums and values, and like those, they are attested rather early (see, e.g., (360a)): There are a good twenty attestations in which *amounten* can be paraphrased as 'to signify', i.e. the sense is basically the same as in the SUMS and VALUES contexts, only that they do not refer to numbers, but to situations (360a–b), words in other languages (360c), or encrypted messages (360d).

(360) a. **Wat** may **amounti** þat he wyle / So by-come oure fode? (*c*.1350(*a*.1333) Shoreham *Poems*)
'What may this signify, that he [i.e. Christ] will thus become our food?'

b. For which this emelye astoned was, And seyde, **what amounteth** this, allas? ((*c*.1385) Chaucer *CT.Kn.*)
'Therefore Emily was astonished and said, "Alas, what does this mean?"'

c. *Ignis deuorabit tabernacula eorum qui libenter accipiunt munera.* [. . .] þis latin **Amounteþ, Þat Fuir schal falle and brenne atte laste Þe houses and þe homes** [. . .] (*c*.1390 *PPl.A(1)*)
'[. . .] this Latin [i.e. sentence] means that fire shall fall and finally burn the houses and the homes [. . .]'

d. The messager which forth it bar, **What** it **amonteth** is noght war ((*a*.1393) Gower *CA*)
'The messenger who carried it away is not aware of what it signifies.'

As we have seen in 9.6.1, in the donor language varieties these NON-LITERAL senses of 'add up to, be worth' and 'signify' are present for the prefixed *amunter* in Anglo-Norman only, and for the non-prefixed *monter* in both Anglo-Norman and continental French. In Middle English now, they hardly occur with the non-prefixed *mounten* but predominantly with the prefixed *amounten*. This will be discussed in the following section on complementation patterns.

9.6.5 Complementation patterns

In around two thirds of the SUM, VALUE, and SIGNIFICATION uses discussed above, *amounten* is used with a direct object (see examples (359) and (360) above). The rest has a prepositional phrase with *to*, such as in (358). The prepositional attestations, however, occur, for one thing, mostly in the more administrative texts (documents from the Gilds and Chancery, and *Paston* and *Stonor Letters*), and, for another, they are also attested slightly later than the ones with direct object. *Amounten* with direct object can be found in a wider range of texts and genres. How can this be interpreted? I think that an explanation for the prevalence of the direct object uses of *amounten* may lie in the argument structure of synonymous Middle English verbs: *Menen* 'to mean' and *toknen* 'to signify', for instance, are also used with direct objects, such as in (361), the same passage from *Piers Plowman* as (360c), only from the B-version:

(361) *Ignis devorabit tabernacula eorum qui libenter accipiunt munera* / [. . .] this Latyn is to **mene** / **That fir shall falle** [. . .] (*c.*1400(*c.*1378) *PPl.B* (LdMisc 581))
'[. . .] this Latin [sentence] is to mean that fire shall fall [. . .]'

ME *amounten* may simply be used in analogy to its synonyms *menen* and *toknen*. That the prepositional pattern is more frequent in administrative texts could be due to its being taken over from Anglo-Norman usage (*amonter à*) in this text type. By the beginning of the Early Modern English period, the prepositional pattern seems to have largely displaced the one with direct object (cf. *OED* s.v. *amount*, v.).

That the prefixed *amounten* is used for these NON-LITERAL meanings in England rather than the non-prefixed *mounten* (which is used in these meanings in continental French) could be related to the fact that native prefixed motion verbs often have metaphorical meanings (and take a direct object), as discussed in section 5.4. The prefixed *amounten* would fit this series of prefixed, transitive, metaphorical motion verbs better than the simple *mounten*.[207]

The non-prefixed *mounten*, which is predominantly used to refer to LITERAL motion, follows the typical complementation pattern of motion verbs: It is predominantly used with PATH satellites (cf. the attestations in 9.6.3). Actually, only in four per cent of its attestations does it combine with a direct object GROUND, and these ten attestations are all in texts translated from French.

9.6.6 Summary

This section has shown that *mounten* and *amounten* have almost no overlapping uses in Middle English: The non-prefixed form refers to LITERAL mounting horses in most attestations, the prefixed form is used in NON-LITERAL contexts of SUMS, VALUES, and SIGNIFICATION. The complementation patterns of ME *mounten* and *amounten*, at least in the beginning, reflect their meanings: *Mounten* is a MOTION verb and, consequently, combines with PATH satellites (usually *upon* N). *Amounten* is not used for LITERAL MOTION but in the senses of 'add up to' and 'signify' and combines with direct objects, like its native synonyms *menen* and *toknen*. The attestations in the *CME* suggest that the prepositional pattern of *amounten* (*amount to* N), which subsequently supersedes the direct object pattern, seems to have its origin in administrative texts.

[207] Note that the support of a corresponding prefixed noun is not likely to be involved: According to the *OED*, the noun *amount* is only attested from the eighteenth century onwards, and equivalent nouns in Middle English are far more frequent without the prefix (*mountaunce*) than with it (*amountance*, *amountement*).

9.7 *Conclusion*

This chapter has shown that each of the borrowed PATH verbs has a different story. To simply claim that the French and Latin pure PATH verbs are borrowed as an additional means of talking about MOTION, introducing a PATH conflating pattern into English, would be a gross reduction. On the contrary, the verbs are borrowed in specific senses, and in the beginning exactly *not* primarily as verbs to describe GENERAL LITERAL MOTION.

Most of the specific senses in which the verbs are borrowed pertain to discourse domains in which French and Latin influence on Middle English was particularly strong, such as administration, courtly life, warfare, law, business, and (especially for Latin) science and religion. These domains had predominantly been dealt with in French and Latin for an extended period of time after the Norman Conquest (cf., e.g., the contributions in Ingham 2010). When they started to be negotiated in English again, the established discourse traditions were in French or Latin, so parts of their vocabulary were taken over into English. As Schaefer puts it, this kind of borrowing is 'a matter of 'specific-purpose' (con)texts, and thus of discourse-traditional norm compliance' (2012: 529). Some examples we have seen in the preceding sections are the use of ME *descenden* (like F *descendre* and L *descendere*) in the sense of LINEAGE and HERITAGE, of ME *amounten* (like AN *amunter*) for SUMS and VALUES, the LEGAL and MILITARY uses of ME *entren* 'enter to take possession' (like AN *entrer*), the MILITARY sense of *ishen* 'make a sortie' (like AN *issir*), the THEOLOGICAL uses of *entren*, *descenden*, and *ascenden* (like their Latin equivalents), or the use of *ascenden* in ASTRONOMY. Also more generally, the comparatively high proportion of NON-LITERAL uses of the verbs (represented in the white and light grey areas in Figures 9.1–2 and 9.5–9) can be related to the extension of the functional range of Middle English to discourse domains hitherto dealt with in French or Latin, where these kinds of more abstract meanings, such as 'descend into details', 'enter into a treaty', or 'enter into membership' will have frequently occurred.

The analysis has shown that these NON-LITERAL and other specific uses of the verbs are always more frequent in autonomous, i.e. genuinely Middle English texts as compared to translations. This difference between autonomous texts and translations suggests that the senses that are borrowed predominantly are the 'peripheral' ones, often figurative, abstract, and technical, and exactly not the uses for GENERAL LITERAL MOTION events. In some of these senses, particularly for *amounten* 'add up to', 'signify', we might even doubt whether the verb can still with due right be called a motion verb.

In the translated texts, in turn, the proportion of attestations in which the verbs refer to GENERAL LITERAL MOTION is larger (see the black areas in Figures 9.1–2 and 9.5–9, which are generally more dominant in the left column representing the translated texts). As comparisons of translations to the original French or Latin text have shown, the reason for this is that Middle English translators often react to the etymological trigger of a French or Latin PATH verb in the original by using the same PATH verb in its Middle English form in the translation. Often, this triggering effect equally happens when the verb in the original is used to describe GENERAL LITERAL MOTION events, for which an autonomous Middle English text would rather use a NEUTRAL MOTION (such as *comen* 'come') or MANNER verb (such as *riden* 'ride') as shown in section 6.3. It seems logical that translators should react like this, when the verb has already been borrowed into the language anyway (although in different contexts), and when its meaning is transparent to the translators due to their knowledge of Latin and/or French. The wider range of application of the PATH verbs in translations may have opened up the way to a wider range of application in general.

The analysis of the usage contexts of the borrowed PATH verbs in this chapter furthermore suggests a difference between those PATH verbs that are formally transparent in Middle English (*entren, descenden, ascenden, mounten*) and those that are not (*avalen* and *ishen*). The PATH meaning in the non-transparent ones is more prone to being superseded by formerly contextualized meanings: *avalen* is so closely linked to 'removing one's headwear; opening the vizor' in Middle English that it can be used as an adjectival participle 'bareheaded', and even be combined with the PATH satellite *up* (probably because the vizor in this context happens to open upwards). *Ishen* is borrowed in the specific MILITARY meaning of 'making a sortie, sally forth', i.e. basically as a MOTION+PATH+MANNER verb. Also *ishen* has an attestation where somebody *ishes in* (see *MED* s.v. *ishen*, sense 1.d: *ishen in* 'to enter'; see Huber 2012: 342 for discussion). In this verb, therefore, the PATH meaning must have given way, by and large, to a MANNER meaning (see section 8.3), so that the verb had been adapted to the Middle English system before it eventually died out.

The PATH meaning of the transparent verbs, by contrast, is more stable: Some of them are transparent (for most literate speakers) by virtue of their very close Latin equivalent (*ascenden, descenden*, and, to some extent, *entren*), so that the continuing influence of Latin will have supported their PATH meaning. Moreover, the existence of borrowed nouns formed from the same base as the verbs and having a salient PATH component in their meaning (such as *ascenden—ascension, entren—entry*, or *mounten—mount, mountain*) will have added to their transparency. This word family support will have stabilized the PATH meaning of the verbs, irrespective of whether the speaker had any knowledge of French or Latin.

CHAPTER 10 | General conclusion

THE PRESENT STUDY SET out with the goal of investigating the expression of intransitive motion in Old and Middle English, with a focus on the contact between PATH-conflating medieval French and MANNER-conflating Middle English. This chapter will provide a general conclusion to the results.

10.1 *Motion encoding in Old and Middle English*

After a presentation of the theoretical framework—a constructionist approach to the typology of motion encoding—and relevant previous work (chapters 2 and 3), I show that medieval English is indeed a MANNER-salient, satellite-framing language, much like Present-Day English (chapters 4–6). This issue is investigated essentially from two complementary perspectives: inventory and usage.

Since MANNER-salient languages are expected to have a large inventory of MANNER OF MOTION verbs (see the studies by Slobin and others reported in section 2.1.4.3), the first means to investigate the typological status of Old and Middle English is to collect an inventory of motion verbs (sections 5.1 and 6.2). For Old English, this is done on the basis of the *TOE*, supplemented by the information in the dictionaries *DOE*, *BT*, and *CH*, and the lists of motion verbs in Ogura (2002: 161–4) and Fanego (2012: 41–2). For Middle English, the basis of the inventory is a definition search in the *Middle English Dictionary* and the list of verbs in Fanego (2012: 58–63). The motion verb inventories are presented in Tables 5.1 and 6.1; Appendices A and B give more detailed information.

The result of the analysis of the inventories confirms earlier research by Fanego (2012) in showing that both Old and Middle English have a large inventory of MANNER OF MOTION verbs at their disposal and that, throughout the medieval period, the language keeps acquiring new MANNER verbs. During the Middle English period there are several such verbs from French among them, some of which are not even used to express INTRANSITIVE MOTION in the donor languages (e.g. ME *butten*, *frounten*, *halen*, see section 6.2.4.4), but only start to be used for INTRANSITIVE MOTION in Middle English. More importantly, most of the MANNER verbs are attested with PATH satellites (e.g. *hie swumman ofer to ðæm eglande* 'they swam over to the island'; see sections 5.1.5 and 6.2.6), as expected in a satellite-framing, MANNER-conflating language.

The characterization of medieval English as strongly satellite-framing is further buttressed by the fact that the inventory identifies several non-MOTION verbs occurring in contextual motion meanings brought about by their use in an intransitive motion construction (sections 5.1.4 and 6.2.5). These include verbs meaning 'to labour, struggle, fight, toil' (OE *feohtan, hīgian, winnan*; ME *labouren, swinken, travailen, winnen*) which can be construed as metonymically related to MOTION as MEANS OF MOTION (similar to PDE *they toiled up the hill*, a use attested from the sixteenth century onwards, cf. *OED* s.v. *toil*). For OE *hīgian* 'to labour, strive, endeavour' (*TOE* s.v.), this contextual motion meaning must have been frequent enough to become entrenched as a decontextualized meaning of the verb in ME *hien* 'to go quickly' (*MED* s.v.); the same must have happened later to ME *travailen* (> PDE *travel*).

That the presence of a satellite is enough to convey the idea of motion is illustrated in an extreme way by the verbless Old English *Hingwar mid eallre his fyrde fuse to Eadmunde* 'Hinguar with all his army quickly to Edmund' (see 5.2.1). This diagnosis is also supported by the fact that the modal verbs, some of which can still function as main verbs in medieval English, as well as other verbs denoting intention, desire, and so forth can occur as main verbs in intransitive motion constructions, yielding a meaning of 'modally broken motion' (section 6.2.5.4, e.g., *Criseyde moste out of the town* 'Criseyde must [i.e. must go] out of the town'; see also section 6.3.1 on ME *longen* 'to desire'). Their use in the intransitive motion construction can be characterized with Berthele (2007: 247 on Swiss German dialects) as sanctioned by an r-relation of MODAL PRECONDITION ('modale Rahmenbedingungen'), for instance with 'being obliged to go somewhere' as a precondition to going somewhere.

The satellite-framing pattern is confirmed as being the most frequent structure by the two usage studies in sections 5.2 and 6.3, which aim at examining which verbs and structures are typical in medieval English motion encoding: 84 per cent of the motion descriptions in the Old English texts and 76 per cent in the Middle English ones are instances of the intransitive motion construction (sections 5.2.4 and 6.3.4). Section 5.4 addresses its changing form, with the realization of PATH satellites as prefixes (e.g. OE *oð-* 'away' in *oð-windan* 'get away') disappearing. Already in the Old English motion descriptions analysed in 5.2, PATH is hardly ever (in merely 11 out of more than 500 instances) conveyed by prefixes. By contrast, the satellite contains at least one prepositional phrase in three quarters of the clauses (e.g. *unto Holande* in ME *saylyde unto Holande* 'sailed to Holland'), and at least one adverb in around 40 per cent (e.g. *out* in ME *out renneth blood* 'the blood runs out'); the latter is slightly more frequent if the verb is a MANNER verb. These results meet the expectations for a satellite-framing, MANNER-salient language.

The most frequent verbs in the texts in the usage studies are the generic or deictic NEUTRAL MOTION verbs OE *cuman* and *faran/fēran* and ME *comen* and *gon*. These account for more than half of the motion descriptions, ranging between 33 per cent (in the Middle English *Knight's Tale*) and 80 per cent (in the Old English *Chronicle E*). MANNER verbs are the second most frequent group. Ranging between 5 per cent (*Chronicle E*) and 54 per cent (in the *Knight's Tale*), their proportion, like that of the NEUTRAL MOTION verbs, strongly depends on text type, subject matter, and style (see the discussion in 5.2.2 and 6.3.2). Similar differences in the proportion of MANNER verbs depending on text type have been found by Cappelle (2012) for Present-Day English fictional and biographical texts. In view of the limitations entailed by the nature of the Old English texts (see section 3.1.2.1 in general and 4.2.2 on the individual texts) and the fact that the three texts from the *Canterbury Tales* analysed for Middle English (with an average of 43 per cent MANNER verbs) are probably most comparable to Present-Day English novels, in which Slobin (2004a: 231) finds 41 per cent MANNER verbs, we can conclude that medieval English exhibits a similar degree of MANNER-conflation as Present-Day English.

10.2 *Motion verbs in intertypological language contact*

Chapter 7 shows that, like medieval English, medieval French typically uses satellite-framing structures. With respect to the typical verb semantics, however, medieval French must be characterized as predominantly PATH-conflating, since PATH verbs occur far more frequently in motion descriptions than MANNER verbs (see Kopecka 2013 and the case study in 7.2, which finds PATH verbs in 33 per cent and MANNER verbs in 17 per cent of the Old French motion descriptions). Furthermore, MANNER verbs are combined with PATH satellites less frequently than other motion verbs in medieval French. This emphasizes yet again that a distinction between MANNER- and PATH-conflating languages is not to be understood as absolute but as a continuum (see section 2.1.2 and the contributions in Goschler & Stefanowitsch 2013)—medieval French is more PATH-conflating and less MANNER-conflating than medieval English.

In the Middle English period, therefore, we have a situation of intense language contact (see 6.1) between a predominantly MANNER-conflating and a predominantly PATH-conflating language. Both of them, however, typically use satellite-framing structures, although the realization of satellites as adverbs (e.g. ME *out*, OF *fors*) is far more frequent in the Middle English texts (about 40 per cent) than in the Old French ones (about 11 per cent). The borrowing of French pure PATH verbs into Middle English as a consequence of this intertypological contact is investigated in chapters 8 and 9.

For medieval English, the usage studies show that PATH verbs are rarely used (in four and six per cent of the motion descriptions in Old and Middle English respectively, see sections 5.2.3 and 6.3.3). Moreover, the PATH verbs that are used most often are verbs like OE *forlǣtan* 'to leave, abandon' and ME *londen* 'to land', which, as pointed out in section 5.1.3, cannot be considered pure PATH verbs. In fact, the inventory study shows that there are hardly any native pure PATH verbs apart from OE *genēalǣcan*/ME *neighlechen* 'to approach' (sections 5.1.3 and 6.2.2). This diagnosis is supported by an analysis of the translational equivalents of the Latin PATH verbs *intrare/inire/ingredi, exire, descendere, ascendere,* and *adpropinquare* in the Old English versions of the Gospels (section 5.3): only L *adpropinquare* 'to approach' is consistently rendered with the Old English pure PATH verb *genēalǣcan,* while the other Latin PATH verbs are mostly translated by a combination of a NEUTRAL MOTION verb and a PATH satellite (such as *gān in* 'go in').

The pure PATH verbs *entrer, issir, avaler/descendre, ascendre/monter/amonter* coming from French, whose meanings correspond to the Latin ones cited above, are thus indeed semantically novel in Middle English: they do not have any native equivalents. The typical syntactic environment in which they are encountered in French, though—the combination with a directional prepositional phrase such as in OF *ist del castel* 'exited from the castle', *entra en le loge* 'entered into the hut'—does not mark them as any different from English motion verbs.

The investigation of these French borrowed PATH verbs in Middle English on the basis of the *Corpus of Middle English Prose and Verse* in chapter 9 suggests that they are borrowed not primarily as verbs to express GENERAL LITERAL MOTION, i.e. not in their core meaning in the donor languages. Rather, they are borrowed in specific, often technical and figurative senses, many of which are related to discourse domains that had been dealt with in French and Latin for a long time in the Middle English period (e.g. administration, business, religion, warfare, etc.). With the functional extension of Middle English to these domains, the French and Latin vocabulary established in the respective discourse tradition must have been transferred to English, such as ME *entren* in the legal sense of 'enter into possession', *ishen* in the sense of 'make a sortie', or *descenden* in the sense of 'lineage'.

In many of these specific senses, the original MOTION meaning is no longer very prominent. For some verbs, this lack of a prominent MOTION meaning is reflected in their complementation patterns: ME *amounten* in the sense of 'to mean', for instance, combines with direct objects in the beginning (e.g. *what amounteth this* 'what does this mean'); ME *descenden* in the sense of LINEAGE and HERITAGE usually does not take an additional PATH adverb such as *down* (e.g. *Charles, discended off Carolus Magnus* 'Charles, descended from Charlemagne'); ME *entren* in the HOSTILE and POSSESSION contexts develops a direct object pattern next to the prepositional one (e.g. *brak & entrid þe hows* 'broke and entered the house').

However, the analyses in chapter 9 show that translators often react to the trigger of a French or Latin PATH verb in the original by rendering it with the respective borrowed PATH verb in their Middle English translation. This is also the case in contexts of GENERAL LITERAL MOTION EVENTS: The proportion of GENERAL LITERAL MOTION contexts is larger in the attestations from translated texts than in those from autonomous ones in Middle English. In the course of time, this may have encouraged a wider use of borrowed PATH verbs in GENERAL LITERAL MOTION contexts including in English texts not translated from French or Latin.

Finally, it seems that the PATH reading of the borrowed verbs is supported by (a) the existence of a borrowed word family that shares a prominent notion of PATH (e.g. ME *entre*, n.—*entren*, v.; *ascension*, n.—*ascenden*, v.) and (b) a formal similarity to the corresponding Latin verb (L *intrare*, *descendere*, *ascendere*). This support is less strong for *ishen/issuen* and *avalen*: The link of *ishen* to its Latin etymon *exire* is no longer transparent, the only 'family member' it has is the noun *issue*; *avalen* has neither a Latin etymon nor a borrowed word family. Both verbs undergo a fading of their original pure PATH meaning once they are borrowed into Middle English, or perhaps, in line with native Middle English means of expressing MOTION, they were not interpreted very strongly as PATH verbs by Middle English speakers in the first place: ME *ishen* when used LITERALLY—which is its only use in autonomous texts—is basically a MOTION+PATH+MANNER verb ('to make a sortie') and is even found in combination with the PATH satellite *in*, whose meaning is opposite to the PATH meaning 'out' that the verb originally has. ME *avalen* (originally 'to descend') is very strongly tied to the meaning of 'taking off one's headwear, opening one's helmet' and thus also found in combination with the PATH satellite *up*. The hypothesis gleaned from the cognitive linguistic studies reported in section 8.3, namely that speakers of a MANNER-conflating language are likely to interpret unknown or borrowed PATH verbs as carrying additional MANNER senses, is thus only borne out for those verbs whose PATH meaning is not buttressed by related loans in Middle English or by transparent Latin etyma.

In the question of intertypological contact between medieval French and English, the present study primarily focuses on the English side of the contact situation. However, we also see that Anglo-French appears to have been influenced by Middle English habits of talking about MOTION, at least with respect to the contextual motion use of non-MOTION verbs: Several non-MOTION verbs are attested in intransitive motion uses in Anglo-French (AN *travailler*, *labourer*, *frusher*, *purposer*), but not in continental varieties of medieval French. These are likely to represent cases of contact semantics due to Middle English influence, as argued in section 6.2.3.2—bilingual speakers may either have transferred their expressional habits from Middle English and started to use these verbs in contextual motion meanings in Anglo-French, or the verbs may have developed motion meanings in Middle English, which then came into Anglo-French as semantic loans.

Further influences of typical Middle English ways of motion expression on Anglo-French are yet to be investigated: The incidence of PATH satellites as adverbs (e.g. *fors* 'out', *enz* 'in') could be expected to be higher in Anglo-French than in weak-contact continental French varieties (cf. Schøsler 2008: 207), as could the use of MANNER verbs, especially in combination with (boundary-crossing) satellites. Since all of these are attested but—compared to Middle

English—not extremely prominent traits in medieval French, they might have been intensified in Anglo-French no matter whether one assumes an acquisition of Anglo-French in a naturalistic setting (in the scenario of second language maintenance suggested by Ingham 2012) or as an instructed L2.

To conclude, with the detailed examination of the early use of the borrowed PATH verbs in Middle English, the present study sheds light on the cognitive and contact-linguistic aspects involved in the process of their borrowing and subsequent development. This makes it a contribution to a neglected field in the research on loanwords in English, which has hitherto focused primarily on the impact of historical borrowing on the vocabulary of Present-Day English (e.g. Grant 2009). The study finds that initially the verbs are not borrowed as pure PATH verbs for GENERAL LITERAL MOTION events (i.e. not in their core meanings from the donor languages), but in specific contexts mostly compatible with the typical ways of expressing motion in medieval English. The extensive analysis of motion encoding in medieval English shows that, like other Germanic languages as well, it typically encodes PATH in satellites (adverbs and prepositional phrases), while the verb in the intransitive motion construction, depending on text type, mostly carries a meaning of NEUTRAL (or deictic) MOTION or of MANNER OF MOTION, but may also be a non-MOTION verb or even absent, which testifies to the semantic content of the medieval English intransitive motion construction.

APPENDIX C | Texts from the *CME* used in the analysis of the borrowed PATH verbs

1. Autonomous Middle English texts

TEXT	DATE/MED STENCIL	EDITION	SUBJECT MATTER
"Nou ihc for þi..." (incipit)	?*a*.1325 *Nou ihc for þi*	*Die Kildare-gedichte*, ed. W. Heuser, *EBA* 14 (1904). 114–16.	religion
Against Boasting and Pride	*c*.1475(*a*.1400) *Boasting & P.*	*Yorkshire Writers*, ed. C. Horstmann, vol. 1 (1895). 122–4.	religion
An ABC Poem on the Passion of Christ	*a*.1475 *In place*	*Political, Religious, and Love Poems*, ed. F. J. Furnivall, *EETS* 15 (1866; re-ed. 1903). 271–8.	religion
An Apology for Lollard Doctrines (Wycliffite tract)	*c*.1475(?*c*.1400) *Wycl.Apol.*	*An Apology for Lollard Doctrines, Attributed to Wicliffe*, ed. J. H. Todd, *Camd.* 20 (1842). 1–113.	religion
Anthology of Chancery English	1417–1462 *Chanc.E*	*An Anthology of Chancery English*, eds. J. H. Fisher, M. Richardson, & J. L. Fisher (1984).	administration
Assembly of Gods	*c*.1500(?*a*.1475) *Ass.Gods*	*The Assembly of Gods*, ed. O. L. Triggs, *EETSES* 69 (1896).	fiction:allegory

(*continued*)

TEXT	DATE/MED STENCIL	EDITION	SUBJECT MATTER
Brome *Abraham and Isaac* (non-cycle mystery play)	*c.*1475 *Brm. Abraham*	*A Common-Place Book of the Fifteenth Century*, ed. L. T. Smith (1886). 49–69.	religion:drama
Buke of the Howlat	(*c.*1450–2) *Howlat*	*Scottish Alliterative Poems in Riming Stanzas*, ed. F. J. Amours, *STS* 27, 38 (1892, 1897).	fiction
Charter of the Abbey of the Holy Ghost	(*c.*1390) *Chart. Abbey HG*	*Yorkshire Writers*, ed. C. Horstmann, vol. 1 (1895). 337–56 (line 14).	religion
Chaucer, *Astrolabe*	*c.*1400 Chaucer *Astr.*	*A Treatise on the Astrolabe*, ed. W. W. Skeat, *EETSES* 16 (1872).	astronomy
Chaucer, *Canterbury Tales*	*c.*1385–1395 Chaucer *CT*	*The Riverside Chaucer*, 3rd ed., ed. L. D. Benson (1987).	fiction
Chaucer, *Troilus and Criseyde*	*a.*1425(*c.*1385) Chaucer *TC*	*Troilus & Criseyde: A new edition of 'The book of Troilus'*, ed. B. A. Windeatt (London 1984).	fiction
Cleanness	*c.*1400(?*c.*1380) *Cleanness*	*Purity*, ed. R. J. Menner, *YSE* 61 (1920).	religion
Contemplations of the Dread and Love of God	*c.*1506	*Yorkshire Writers*, ed. C. Horstmann, vol. 2 (1895). 72–105.	religion
Cursor Mundi	*a.*1400 *Cursor*	*Cursor Mundi*, ed. R. Morris, *EETS* 57 (1874), 59 (1875), 62 (1876), 66 (1877), 68 (1878).	history
Dunbar, *The tretis of the twa mariit women and the wedo*	*a.*1508 Dunb. *TuaMar.W.*	*The works of William Dunbar*, ed. J. Kinsley (Oxford, 1979).	fiction
Earliest English Wills in the Court of Probate	(1387–1454) *EEWills*	*The Fifty Earliest English Wills in the Court of Probate, London*, ed. F. J. Furnivall, *EETS* 78 (1882).	administration

TEXT	DATE/MED STENCIL	EDITION	SUBJECT MATTER
Form of Confession (1)	c.1390 *Form Conf.(1)*	*Yorkshire Writers*, ed. C. Horstmann, vol. 2 (1896). 340–5.	religion
Fortescue, *The Governance of England*	c.1500(a.1475) Fortescue *Gov.E.*	*The Governance of England*, ed. C. Plummer (1885).	politics
Golagros and Gawane	a.1500 *Gol.&Gaw.*	*Scottish alliterative poems in riming stanzas*, ed. F. J. Amours, Scottish Text Society, 1891–2 and 1896–7, 1–46.	fiction: romance
Gower, *Confessio Amantis*	(a.1393) Gower *CA*	*The English Works of John Gower*, ed. G. C. Macaulay, 2 vols., *EETSES* 81 (1900), 82 (1901).	fiction
Guild documents	(1389–1503) *Doc.in EGilds*	*English Gilds*, eds. L. T. Smith & L. Brentano, *EETS* 40 (1870, rev. 1892).	administration
Henryson, *Fables*	a.1500 Henr. *Fab.*	*The Poems of Robert Henryson*, ed. G. G. Smith, 3 vols. Scottish Text Society 64, 55, 58 (1906–14).	fiction: fables
Henryson, *Orpheus and Eurydice*	a.1500 Henr. *Orph.*	*The Poems of Robert Henryson*, ed. G. G. Smith, 3 vols. Scottish Text Society 64, 55, 58 (1906–14).	fiction
Henryson, *The Testament of Cresseid*	a.1500 Henr. *Test.Cress.*	Oxford Text Archive, U-243-A.	fiction
Hilton, *Of Angels' Song*	a.1450 ?Hilton *Angels' Song*	*Yorkshire Writers*, ed. C. Horstmann, vol. 1 (1895). Middle section p. 175; upper section pp. 176–82.	religion
Hilton, *On the Mixed Life*	c.1390 Hilton *ML*	*Yorkshire Writers*, ed. C. Horstmann, vol. 1 (1895). 264–9; lower half of pp. 270–92.	religion

(*continued*)

TEXT	DATE/MED STENCIL	EDITION	SUBJECT MATTER
?Hoccleve, *Songs in a ME translation of a prose version of De Guileville's Pilgrimage of the Soul*	*a*.1500(1413) ?Hoccl.*Poems PS*	*Hoccleve's Works: The Regement of Princes*, ed. F. J. Furnivall, *EETSES* 72 (1897). Pp. xxiii–lxii.	religion
Hoccleve, *Regement of Princes*	*a*.1450(1412) Hoccl.*RP*	*Hoccleve's Works: The Regement of Princes*, ed. F. J. Furnivall, *EETSES* 72 (1897). 1–197.	politics: mirror for princes
How Antichrist and his Clerks Travail to Destroy Holy Writ (Wycliffite tract)	?*c*.1430(*c*.1400) *Wycl.Antichr. (1)*	*The English Works of Wyclif*, ed. F. D. Matthew, *EETS* 74 (1880). 255–62.	religion: treatise
How Satan and His Children Turn Works of Mercy Upside-Down (Wycliffite tract)	?*c*.1430(*c*.1400) *Wycl.Satan & C.*	*The English Works of Wyclif*, ed. F. D. Matthew, *EETS* 74 (1880). 210–18.	religion: treatise
How the Office of Curates is Ordained by God (Wycliffite tract)	?*c*.1430(*c*.1400) *Wycl.Curates*	*The English Works of Wyclif*, ed. F. D. Matthew, *EETS* 74 (1880). 143–63.	religion: treatise
Joseph of Aramathie	*c*.1390(?*c*.1350) *Jos.Arim.*	*Joseph of Aramathie*, ed. W. W. Skeat, *EETS* 44 (1871).	fiction: romance
Langland, *Piers Plowman*, A, B, C	*c*.1400(?*a*.1387) *PPl.C*	*The Vision of William Concerning Piers the Plowman*, part 3, ed. W. W. Skeat, *EETS* 54 (1873).	fiction
Lincoln Diocese Documents	1450–1544 *Lin. DDoc.*	*Lincoln Diocese Documents*, ed. A. Clark, *EETS* 149 (1914).	administration

TEXT	DATE/MED STENCIL	EDITION	SUBJECT MATTER
Lydgate, *Debate of the Horse, Goose and Sheep*	*a*.1500 Lydg.*HGS*	*Political, Religious, and Love Poems*, ed. F. J. Furnivall, *EETS* 15 (1866; re-ed. 1903). 15–22.	fiction
Lydgate, *Banner of Saint Edmund*	*a*.1450(*c*.1433) Lydg.*St.Edm.*	*Altenglische Legenden n. F.*, ed. C. Horstmann (1881). 376–440.	religion: saints' lives
Malory, *Le Morte Darthur*	1485 Malory *Wks.*	*Le Morte Darthur, by Syr Thomas Malory*, ed. H. O. Sommer (1889).	fiction: romance
Material from the Book of Brome in the hand of Robert Melton	?*a*.1509 *Doc. Melton* in *Bk. Brome*	*A Common-Place Book of the Fifteenth Century*, ed. L. T. Smith (1886). 124–28, 154–6, 176.	administration
Metham, *Amoryus and Cleopes*	(1449) Metham *AC*	*The Works of John Metham*, ed. H. Craig, *EETS* 132 (1916). 1–81.	fiction
Morte Arthure	*c*.1440(?*a*.1400) *Morte Arth.(1)*	*The Alliterative Morte Arthure: A Critical Edition*, ed. V. Krishna (1976).	fiction: romance
Nine Points	*c*.1475 *9 PPGod(2)*	*Yorkshire Writers*, ed. C. Horstmann, vol. 1 (1895). Upper pp. 110–112.	religion
Northern Homily Cycle: Legendary, part 1	*a*.1425(?*c*.1375) *NHom.(3) Leg.*	*Altenglische Legenden, n.f.*, ed. C. Horstmann (1881). 3–16, 19–24, 52–76, 96–142, 152–159, 165–173.	religion: saints' lives
Of Clerks Possessioners (Wycliffite tract)	?*c*.1430(*c*.1400) *Wycl. Possessioners*	*The English Works of Wyclif*, ed. F. D. Matthew, *EETS* 74 (1880). 116–140.	religion: treatise
Of Prelates (Wycliffite tract)	?*c*.1430(*c*.1400) *Wycl.Prelates*	*The English Works of Wyclif*, ed. F. D. Matthew, *EETS* 74 (1880). 55–107.	religion: treatise

(continued)

TEXT	DATE/MED STENCIL	EDITION	SUBJECT MATTER
Paston Letters	1425–1495 Paston	*Paston Letters and Papers of the Fifteenth Century*, ed. N. Davis (1971).	letters
Pearl	*c.*1400(?*c.*1380) *Pearl*	*Pearl*, ed. E. V. Gordon (Oxford, 1953).	fiction: allegory
Pecock, *Book of Faith*	(*c.*1456) Pecock *Faith*	*Reginald Pecock's Book of Faith*, ed. J. L. Morison (1909). 109–305.	religion
Pecock, *The Repressor of Over Much Blaming of the Clergy*	(*c.*1449) Pecock *Repr.*	*The Repressor of Over Much Blaming of the Clergy, by Reginald Pecock*, ed. C. Babington, *RS* 19 (1860–1). 1–565.	religion: treatise
Piers the Plowman's Creed	*c.*1600(?*c.*1395) *PPl.Creed*	*Pierce the Ploughmans Crede*, ed. W. W. Skeat, *EETS* 30 (1867).	fiction
Prick of Conscience	*a.*1425(*a.*1400) *PConsc.*	*The Pricke of Conscience*, ed. R. Morris, *PST* (1863).	religion
Prose Brut (after 1333)	?*a.*1425(*a.*1400) *Brut-1377*, *c.*1450(*c.*1425) *Brut-1419*, etc.	*The Brut or the Chronicles of England*, ed. F. W. D. Brie, vol. 2, *EETS* 136 (1908).	history
Prose Brut (second ME version concluding in 1461)	?*a.*1500(*a.*1471) *Brut-1461(2)*	*An English Chronicle of the Reigns of Richard II, Henry IV [. . .]*, ed. J. S. Davies, *Camd.* 64 (1856).	history
Records of the Parish Church of St. Mary-at-Hill, London	v.d.(v.d.) *Rec. St.Mary at Hill*	*The Medieval Records of a London City Church [. . .]*, ed. H. Littlehales, *EETS* 128 (1905).	administration
Robert of Gloucester, *Chronicle*, Version A	*c.*1325(*c.*1300) *Glo.Chron.A*	*The metrical chronicle of Robert of Gloucester*, ed. W. Aldis Wright, Rolls Series 86 (1887).	history
Rolle, *Desire and Delight*	*c.*1440(*a.*1349) Rolle *Desire & D.*	*Yorkshire Writers*, ed. C. Horstmann, vol. 1 (1895). 194.	religion

TEXT	DATE/MED STENCIL	EDITION	SUBJECT MATTER
Rolle, *Form of Living*	*a*.1425 Rolle *FLiving*	*Yorkshire Writers*, ed. C. Horstmann, vol. 1 (1895). 3–49.	religion
Rolle, *Meditations on the Passion (2)*	*a*.1500 Rolle *MPass.(2)*	*Yorkshire Writers*, ed. C. Horstmann, vol. 1 (1895). 92–103.	religion
Sayings of the Fathers	*c*.1475(*a*.1400) *Sayings Fathers*	*Yorkshire Writers*, ed. C. Horstmann, vol. 1 (1895). 125–28.	religion
Shoreham, *Poems*	*c*.1350(*a*.1333) Shoreham *Poems*	*The Poems of William of Shoreham*, ed. M. Konrath, *EETSES* 86 (1902).	religion
Sir Gawain and the Green Knight	*c*.1400(?*c*.1390) *Gawain*	*Sir Gawain and the Green Knight*, ed. J. R. R. Tolkien & E. V. Gordon; 2nd ed., rev. N. Davis (1967).	fiction: romance
South English Legendary: Thomas Becket, Archbishop of Canterbury	*c*.1300 *SLeg. Becket*	*The Early South-English Legendary from Bodleian MS. Laud Misc. 108*, ed. C. Horstmann, *EETS* 87 (1887).	religion: saints' lives
Stonor Letters	1290–1483 *Stonor*	*The Stonor Letters and Papers, 1290–1483*, ed. C. L. Kingsford, *Camd.*, ser. 3, vols. 29 and 30 (1919).	letters
The Holy Book Gratia Dei	*c*.1440 *HBk. GDei*	*Yorkshire Writers*, ed. C. Horstmann, vol. 1 (1895). 300–305(item 7), 305–310 (item 8), 310–321(item 9).	religion
The Lantern of Light (Wycliffite tract)	?*a*.1425(*a*.1415) *Wycl.Lantern*	*The Lanterne of Liȝt*, ed. L. M. Swinburn, *EETS* 151 (1917).	religion
Northern Homily Cycle: Sermon for the Feast of Corpus Christi	*c*.1390(*c*.1350) *NHom.(2) Corp.Chr.*	*The minor poems of the Vernon ms. [. . .]*, ed. C. Horstmann (part 1) & F. J. Furnivall (part 2), *EETSOS* 98, 117 (1892, 1901).	religion: sermon

(continued)

TEXT	DATE/MED STENCIL	EDITION	SUBJECT MATTER
The Stations of Rome (1)	*a*.1500 *Stations Rome(1)*	*Political, Religious, and Love Poems*, ed. F. J. Furnivall, *EETS* 15 (1866; re-ed. 1903). 143–73.	travel/religion
Towneley Plays	*a*.1500(*a*.1460) *Towneley Pl.*	*The Towneley Plays*, eds. G. England & A. Pollard, *EETSES* 71 (1897).	religion: drama
Tractatus de Pseudo-Freris (Wycliffite tract)	c.1475(*a*.1400) *Wycl.Pseudo-F.*	*The English Works of Wyclif*, ed. F. D. Matthew, *EETS* 74 (1880). 296–324.	religion: treatise
Treatise of Ghostly Battle	*a*.1500 *Treat. GBattle*	*Yorkshire Writers*, ed. C. Horstmann, vol. 2 (1896). 420–36.	religion: treatise
Wycliffite Sermons	*a*.1425 *Wycl. Serm.*	*Select English Works of John Wyclif*, ed. T. Arnold, vol. 1 (1869).	religion: sermon
York Plays	*a*.1450 *Yk.Pl.*	*York Plays: The Plays Performed by the Crafts or Mysteries of York*, ed. L. T. Smith (1885).	religion: drama

2. Translated texts

TEXT	DATE/MED STENCIL	EDITION	SUBJECT MATTER
Alphabet of Tales	c.1450 *Alph. Tales*	*An Alphabet of Tales*, ed. M. M. Banks, *EETS* 126, 127 (1904, 1905).	fiction
Arderne, *Practica de Fistula in Ano*	c.1425 *Arderne Fistula*	*Treatises of Fistula [. . .] by John Arderne*, ed. D. Power, *EETS* 139 (1910).	science: medicine
Ayenbite of Inwyt	(1340) *Ayenb.*	*Don Michel's Ayenbite of Inwyt*, ed. R. Morris, rev. P. Gradon, *EETS* 23 (1866).	religion
Benedictine Rule (2)	a.1450 *Ben. Rule(2)*	*Three Middle-English Versions of the Rule of St. Benet*, ed. E. A. Kock, *EETS* 120 (1902). 48–118.	religion: rule
Bible	c.1425 *Bible* SNT(1)	*A Fourteenth Century English Biblical Version*, ed. A. C. Paues (1904).	religion: Bible
Book of Quintessence	a.1475 *Bk. Quint.*	*The Book of Quinte Essence*, ed. F. J. Furnivall, *EETS* 16 (1866; rev. 1889).	science: alchemy
Caxton Abstract of the Rule of St. Benet	c.1490	*Three Middle-English Versions of the Rule of St. Benet*, ed. E. A. Kock, *EETS* 120 (1902). 119–40.	religion: rule
Caxton, *Blanchardyn and Eglantine*	c.1489	*Caxton's Blanchardyn and Eglantine [. . .]*, ed. L. Kellner, *EETSOS* 58 (1890).	fiction: romance
Caxton, *Charles the Great*	c.1485	*Lyf of the noble and Crysten prynce, Charles the Grete [. . .]*, ed. S. J. H. Herrtage, *EETSOS* 36–37 (1880–81).	fiction: romance

(continued)

TEXT	DATE/MED STENCIL	EDITION	SUBJECT MATTER
Caxton, *Gilte Legend: The invention of the holy cross; the exaltation of the cross*	1493 Caxton *Gold.Leg.*	*Legends of the holy rood; symbols of the passion and cross poems*, ed. R. Morris, *EETSOS* 46 (1871).	fiction
Caxton, *The Eighth Book of the Polychronicon*	*c.*1482	*Polychronicon Ranulphi Higden [. . .]*, eds. C. Babington & J. R. Lumby, *RS* 41 (1865–86), vol. 8, 522–87.	history
Caxton, *The right plesaunt and goodly historie of the foure sonnes of Aymon*	*c.*1489	*The right plesaunt and goodly historie of the foure sonnes of Aymon. Englisht from the French by William Caxton, and printed by him about 1489*, ed. O. Richardson, *EETSES* 45 (1884).	fiction
Chaucer, *Boece*	?*c.*1400 Chaucer *Bo.*	*Chaucer's Translation of Boethius's De Consolatione Philosophiae*, ed. R. Morris, *EETSES* 5 (1868).	philosophy
Destruction of Troy	*c.*1540(?*a.*1400) *Destr.Troy*	*The Gest Hystoriale of the Destruction of Troy*, eds. G. A. Panton & D. Donaldson, *EETS* 39, 56 (1869, 1874).	fiction
English Conquest of Ireland	*a.*1500 *Conq. Irel.*	*The English Conquest of Ireland*, ed. F. J. Furnivall, *EETS* 107 (1896).	history
English Register of Godstow Nunnery, Oxfordshire	*a.*1475 *Godstow Reg.*	*The English Register of Godstow Nunnery, Near Oxford*, ed. A. Clark, vols. 2–3, *EETS* 130, 142 (1906, 1911).	administration
English Register of Oseney Abbey	*c.*1460 *Oseney Reg.*	*The English Register of Oseney Abbey, by Oxford*, ed. A. Clark, *EETS* 133, 144 (1907, 1913).	administration

TEXT	DATE/MED STENCIL	EDITION	SUBJECT MATTER
Flete, *Remedies Against Temptations*	1508(*c.*1400) *Remedy TTempt.*	*Yorkshire Writers*, ed. C. Horstmann, vol. 2 (1896). 106–123(item 2).	religion
Generides (2)	*a.*1500(*a.*1450) *Gener.(2)*	*Generydes*, ed. W. A. Wright, *EETS* 55, 70 (1873, 1878).	fiction: romance
Gesta Romanorum	*a.*1500(?*a.*1450) *GRom.*	*The Early English Version of the Gesta Romanorum*, ed. S. J. H. Herrtage, *EETSES* 33 (1878).	fiction
Guy of Warwick (1)	*c.*1330(?*c.*1300) *Guy(1)*	*The Romance of Guy of Warwick*, ed. J. Zupitza, *EETSES* 42, 49, 59 (1883, 1887, 1891).	fiction: romance
Higden's Polychronicon (2) (anonymous fifteenth-century translation)	?*a.*1475(?*a.*1425) *Higd.(2)*	*Polychronicon Ranulphi Higden [. . .]*, eds. C. Babington & J. R. Lumby, 8 vols., *RS* 41 (1865–86). Vols. 1–7, lower half right hand pp.	history
John Trevisa, *Higden's Polychronicon* (ME translation)	(*a.*1387) Trev.*Higd.*	*Polychronicon Ranulphi Higden [. . .]*, eds. C. Babington & J. R. Lumby, 8 vols., *RS* 41 (1865–86). Upper right pp.	history
Katherine of Rome	*c.*1508	*Companion to Rolle* [Works not by Rolle extracted from: *Yorkshire writers: Richard Rolle of Hampole . . . and his followers*], ed. C. Horstmann (London, 1895–96). Vol. 2, 123–128.	religion: saints' lives
Kentish Sermons	*c.*1275 *Ken. Serm.*	*An Old English miscellany containing a bestiary, Kentish sermons, Proverbs of Alfred, religious poems of the thirteenth century*, ed. R. Morris, *EETSOS* 49 (1872), 26–36.	religion: sermons

(continued)

TEXT	DATE/MED STENCIL	EDITION	SUBJECT MATTER
Lanfranc's Complete Art of Surgery (ME version) [*Lanfranc*]	*a*.1400 *Lanfranc*, ?*a*.1450 *Lanfranc*	*Lanfrank's 'Science of Cirurgie'*, ed. R. V. Fleischhacker, *EETS* 102 (1894).	science: surgery
Laud Troy Book	*c*.1425(*c*.1400) *Ld.Troy*	*The Laud Troy Book*, ed. J. E. Wülfing, *EETS* 121, 122 (1902, 1903).	fiction
Life of Soul	*c*.1400 *Life Soul*	*Life of soul: An edition from MS Laud Misc. 210*, ed. P. Schaffner (Ithaca, 1999).	religion
Love, *Mirror of the Blessed Life of Jesus Christ*	*c*.1430(*a*.1410) Love *Mirror*	*The Mirrour of the Blessed Lyf of Jesu Christ [. . .] Made Before the Year 1410 by Nicholas Love*, ed. L. F. Powell (1908).	religion
Lydgate, *Pilgrimage of the Life of Man*	*a*.1475(?*a*.1430) Lydg.*Pilgr.*	*The Pilgrimage of the Life of Man*, eds. F. J. Furnivall & K. B. Lockcock, *EETSES* 77, 83, 92 (1899, 1901, 1904).	religion
Lydgate, *Reason and Sensuality*	*c*.1450(?*c*.1408) Lydg.*RS*	*Lydgate's Reson and Sensuallyte*, ed. E. Sieper, *EETSES* 84, 89 (1901, 1903).	fiction
Mandeville's Travels (1)	?*a*.1425(*c*.1400) *Mandev.(1)*	*Mandeville's Travels*, ed. P. Hamelius, *EETS* 153 (1919).	travel
Mandeville's Travels (2)	?*a*.1425 *Mandev.(2)*	*Mandeville's travels: The Egerton version*, ed. G. F. Warner (1889).	travel
Melusine	*c*.1500 *Melusine*	*Melusine*, ed. A. K. Donald, *EETSES* 68 (1895).	fiction: romance
Metham, *Palmistry*	*c*.1450 Metham *Palm.*, *a*.1500 Metham *Palm.*	*The Works of John Metham*, ed. H. Craig, *EETS* 132 (1916). 84–117.	science: palmistry

TEXT	DATE/MED STENCIL	EDITION	SUBJECT MATTER
Metham, *Prognostications Based on Christmas Day*	*c*.1450 Metham *Progn.CD*	*The Works of John Metham*, ed. H. Craig, *EETS* 132 (1916). 146–7, 157–58.	prognostications
Midland Prose Psalter	*c*.1350 *MPPsalter*, *c*.1400 *MPPsalter*	*The Earliest Complete English Prose Psalter*, ed. K. D. Bülbring, *EETS* 97 (1891).	religion: Bible
Mirror of St. Edmund (3)	*c*.1390 *Mirror St.Edm.(3)*	*Yorkshire Writers*, ed. C. Horstmann, vol. 1 (1895). 240–61.	religion
Privity of the Passion (ME version of chapters 74–92 of *Meditationes Vitae Christi*)	*c*.1440 *Bonav. Medit.(3)*	*Yorkshire Writers*, ed. C. Horstmann, vol. 1 (1895). 198–218.	religion
Prose Alexander	*c*.1440 *PLAlex.*	*The Prose Life of Alexander*, ed. J. S. Westlake, *EETS* 143 (1913).	fiction
Prose Brut (ME version concluding in 1333)	*c*.1400 *Brut-1333*	*The Brut or the Chronicles of England*, ed. F. W. D. Brie, vol. 1, *EETS* 131 (1906). 1–286.	history
Prose Merlin	*a*.1500(?*c*.1450) *Merlin*	*Merlin*, ed. H. B. Wheatley, 4 vols., *EETS* 10, 21, 36, 112 (1865, 1866, 1869, 1899).	fiction
Robert Mannyng of Brunne, *Handling Sin*	*a*.1400(*c*.1303) Mannyng *HS*	*Robert of Brunne's Handlyng Synne [. . .]*, ed. F. J. Furnivall, 2 vols., *EETS* 119, 123 (1901, 1903).	religion
Robert Mannyng of Brunne, *The Chronicle of England, Part 1*	*a*.1450(*a*.1338) Mannyng *Chron.Pt.1*	*The Story of England by Robert Manning of Brunne*, ed. F. J. Furnivall, *RS* 87 (1887).	history

(*continued*)

TEXT	DATE/MED STENCIL	EDITION	SUBJECT MATTER
Robert Mannyng of Brunne, *The Chronicle of England, Part 2*	?*a.*1400(*a.*1338) Mannyng *Chron.Pt.2*	*Peter Langtoft's Chronicle*, ed. T. Hearne, 2 vols. (1725; reprint 1810). 1–341.	history
Rolle, *English Psalter and Commentary*	*a.*1500(*c.*1340) Rolle *Psalter*	*The Psalter [. . .] and Certain Canticles [. . .] in English by Richard Rolle of Hampole*, ed. H. R. Bramley (1884).	religion: Bible
Rule of St. Clare (Second Order of St. Francis)	*a.*1500 *Rule Minoresses*	*Two Fifteenth-Century Franciscan Rules*, ed. W. W. Seton (bound with *A Fifteenth-Century Courtesy Book*, ed. R. W. Chambers) *EETS* 148 (1914). 81–116.	religion: rule
Secreta Secretorum (anonymous ME version in Lambeth MS 501)	*a.*1500(?*a.*1425) *Lambeth SSecr.*	*Three Prose Versions of the Secreta Secretorum*, eds. R. Steele & T. Henderson, *EETSES* 74 (1898). 41–118.	politics: mirror of princes, sciences
Secreta Secretorum (ME version in Royal 18.A.7)	*c.*1450 *Royal SSecr.*	*Three Prose Versions of the Secreta Secretorum*, eds. R. Steele & T. Henderson, *EETSES* 74 (1898). 1–39.	politics: mirror of princes, sciences
Shirley, *The Death of James Stewart, King of Scots*	*a.*1475(*a.*1456) Shirley *Death Jas.*	John Shirley, *Here Folowing Begynnythe a Full Lamentable Cronycle of the Dethe and False Murdure of James Stewarde, Kyng of Scotys, Miscellanea Scotica* 2 (1818). 5–29.	history
Siege of Jerusalem (1)	?*a.*1450(*a.*1400) *Siege Jerus.(1)*	*The Siege of Jerusalem*, eds. E. Kölbing & M. Day, *EETS* 188 (1932).	fiction
Smaller Vernon Legendary	*c.*1390(?*c.*1350) *SVrn.Leg.*	*Sammlung altenglischer Legenden*, ed. C. Horstmann (1878). 3–97.	religion: saints' lives

TEXT	DATE/MED STENCIL	EDITION	SUBJECT MATTER
The Babees Book	c.1475 *Babies' Bk.*	*The babees book [. . .]*, ed. F. J. Furnivall, *EETSOS* 32 (1868). 1–9.	education
The Book of the Knight of La Tour-Landry	?c.1450 *Knt. Tour-L.*	*The Book of the Knight of La Tour-Landry*, ed. T. Wright, *EETS* 33 (1868; rev. 1906).	fiction, education
Three Kings of Cologne	c.1450(?c.1400) *3 KCol.(1)*, a.1450 *3 KCol.(1)*	*The Three Kings of Cologne*, ed. C. Horstmann, *EETS* 85 (1886).	fiction
Three Kings' Sons	c.1500	*The Three Kings's sons [. . .]*, ed. F. J. Furnivall, *EETSES* 67 (1895).	fiction
Translation of Saint Anselm of Canterbury's *Meditatio I*	c.1410 *St. Anselm Medit.*	*Yorkshire Writers*, ed. C. Horstmann, vol. 2 (1896). 443–5.	religion
Vision of Tundale	a.1500 *Tundale*	*Tundale: das mittelenglische Gedicht über die Vision des Tundalus*, ed. A. Wagner (1893).	religion
Wars of Alexander	c.1450(?a.1400) *Wars Alex.*	*The Wars of Alexander, an Alliterative Romance*, ed. W. W. Skeat, *EETSES* 47 (1886).	fiction: romance
Wycliffite Bible (early version) [*WBible(1)*]	a.1425(a.1382) *WBible(1)*, a.1425(c.1384) *WBible(1)*, etc.	*The Holy Bible [. . .] by John Wycliffe and His Followers*, eds. J. Forshall & F. Madden, 4 vols. (1850).	religion: Bible
Wycliffite Bible (later version) [*WBible(2)*]	a.1425(c.1395) *WBible(2)*, a.1450 *WBible(2)*, etc.	*The Holy Bible [. . .] by John Wycliffe and His Followers*, eds. J. Forshall & F. Madden, 4 vols. (1850).	religion: Bible
Yonge, *Secreta Secretorum* (ME version)	a.1500(1422) Yonge *SSecr.*	*Three Prose Versions of the Secreta Secretorum*, eds. R. Steele & T. Henderson, *EETSES* 74 (1898). 121–91(line 2), 196 (line 5)–248.	politics: mirror of princes, sciences

REFERENCES

7FMR = *Corpus de Textes de Français Médiéval et de la Renaissance*. http://atilf.fr/dmf/

AND = *Anglo-Norman Dictionary*. Ed. William Rothwell, Louise W. Stone, & T. B. W. Reid. 1977–92. London: Modern Humanities Research Association. http://www.anglo-norman.net/gate/

Ameka, Felix K. 1996. Semantics. In Hans Goebl, Peter H. Nelde, Zdenek Stáry, & Wolfgang Wölck (eds.), *Contact linguistics: An international handbook of contemporary research. Vol. I*, 130–8. Berlin: De Gruyter.

Archibald, Elizabeth & A. S. G. Edwards. 1996. Introduction. In Elizabeth Archibald & A. S. G. Edwards (eds.), *A companion to Malory*, xiii–xv. Cambridge, U.K.: Brewer.

Aske, Jon. 1989. Path predicates in English and Spanish: A closer look. *Proceedings of the Annual Meeting of the Berkeley Linguistics Society* 15, 1–14.

Attenborough, Frederick L. 1922. *The laws of the earliest English kings*. Cambridge, U.K.: Cambridge University Press.

Banks, Mary Macleod. 1904. *An alphabet of tales: An English 15th century translation of the* Alphabetum Narrationum *of Etienne de Besançon; from Additional MS 25,719 of the British Museum* (Early English Text Society: Original Series 126/127). London: Kegan Paul, Trench, Trübner.

Bally, Charles. [1925] 1952. *Le langage et la vie*. Geneva: Droz.

Bally, Charles. [1932] 1965. *Linguistique générale et linguistique française*. Bern: Francke.

Bammesberger, Alfred. 2003. Zu *fusæ* in der Runeninschrift auf dem Ruthwell-Kreuz. In Wilhelm Heizmann & Astrid van Nahl (eds.), *Runica—Germanica—Mediaevalia* (Ergänzungsbände zum Reallexikon der Germanischen Altertumskunde 37), 28–34. Berlin: De Gruyter.

Barlow, Michael. 2004. *MonoConc Pro 2.2*. Houston, TX: Athelstan Publications.

Bately, Janet. 1980. *The Old English Orosius* (Early English Text Society: Supplementary Series 6). Oxford: Oxford University Press.

Bately, Janet. 1990. The compilation of the Anglo-Saxon Chronicle, 60 BC to AD 890: Vocabulary as evidence. In E. G. Stanley (ed.), *British Academy papers on Anglo-Saxon England*, 261–97. Oxford: Oxford University Press.

Baugh, Albert C. & Thomas Cable. 2013. *A history of the English language*, 6th edn. London: Routledge.

Beavers, John, Beth Levin, & Shiao W. Tham. 2010. The typology of motion expressions revisited. *Journal of Linguistics* 46(2), 331–77.

Benson, C. D. 1986. *Chaucer's drama of style: Poetic variety and contrast in the* Canterbury Tales. Chapel Hill, NC: University of North Carolina Press.

Benson, Larry D. & F. N. Robinson. 1988. *The Riverside Chaucer*, 3rd edn. Oxford: Oxford University Press.

Bergs, Alexander & Janne Skaffari. 2007. Invitation to the *Peterborough Chronicle* and its language. In Alexander Bergs & Janne Skaffari (eds.), *The language of the Peterborough Chronicle* (Studies in English Medieval Language and Literature 20), 5–12. Frankfurt: Peter Lang.

Berman, Ruth & Dan I. Slobin (eds.). 1994. *Relating events in narrative: A cross-linguistic developmental study.* Hillsdale, NJ: Lawrence Erlbaum Associates.

Bernini, Giuliano, Lorenzo Spreafico, & Ada Valentini. 2006. Acquiring motion verbs in a second language: The case of Italian L2. *Linguistica e Filologia* 23, 7–26.

Berthele, Raphael. 2006. *Ort und Weg: Die sprachliche Raumreferenz in Varietäten des Deutschen, Rätoromanischen und Französischen* (Linguistik—Impulse und Tendenzen 16). Berlin: De Gruyter.

Berthele, Raphael. 2007. *Sein*+Direktionalergänzung: Bewegung ohne Bewegungsverb. In Ljudmila Geist & Björn Rothstein (eds.), *Kopulaverben und Kopulasätze: Intersprachliche und intrasprachliche Aspekte*, 229–52. Tübingen: Niemeyer.

Blake, N.F. 2002. The early history of, and its impact upon, the Middle English Dictionary. *Dictionaries: Journal of The Dictionary Society of North America* 23, 48–75.

Blake, Norman. 1992. The literary language. In Norman Blake (ed.), *The Cambridge history of the English language, Vol. 2: 1066–1476*, 500–41. Cambridge, U.K.: Cambridge University Press.

Blake, Norman F. 1969. *Caxton and his world.* London: Deutsch.

Bloom, Paul, Mary A. Peterson, Lynn Nadel, & Merril F. Garrett (eds.). 1996. *Language and space.* Cambridge, MA: MIT Press.

Blumenthal, Peter. 1987. *Sprachvergleich Deutsch–Französisch* [Romanistische Arbeitshefte 29]. Tübingen: Niemeyer.

BNCweb = *British National Corpus CQP-edition.* Sebastian Hoffmann & Stefan Evert. Lancaster University. http://bncweb.lancs.ac.uk

Boffey, Julia. 2002. Middle English lives. In David Wallace (ed.), *The Cambridge history of medieval English literature*, 610–39. Cambridge, U.K.: Cambridge University Press.

Booij, Geert & Ans van Kemenade. 2003. Preverbs: An introduction. In Geert Booij & Jaap van Marle (eds.), *Yearbook of morphology 2003*, 1–11. Dordrecht, Netherlands: Kluwer.

Bosworth–Toller = Joseph Bosworth & T. Northcote Toller. 1882–98. *An Anglo-Saxon dictionary*. Oxford: Oxford University Press. With supplements by T. Northcote Toller 1908–21 and A. Campbell 1972. http://bosworth.ff.cuni.cz/

Braet, Herman & Guy Raynaud de Lage. 1989. *Béroul: Tristran et Iseut.* Paris: Peeters.

Brinton, Laurel & Elizabeth Closs Traugott. 2005. *Lexicalization and language change* (Research Surveys in Linguistics). Cambridge, U.K.: Cambridge University Press.

Broz, Vlatko. 2013. Aspectual functions of the verbal prefix *a-* in Old English with reference to Gothic. In Gabriele Diewald, Leena Kahlas-Tarkka, & Ilse Wischer (eds.), *Comparative studies in early Germanic languages: With a focus on verbal categories* (Studies in Language Companion Series 138), 235–62. Amsterdam: John Benjamins.

Brucale, Luisa, Claudio Iacobini, & Egle Mocciaro. 2011. Typological change in the expression of motions events from Latin to Romance languages. Paper read at the 44th *Annual Meeting of the Societas Linguistica Europaea*, Logroño, Spain, 8–11 September 2011.

Brunner, Karl. 1962. *Die englische Sprache. II.* Tübingen: Niemeyer.

BT: See Bosworth–Toller.

Bülbring, Karl D. 1902. *Altenglisches Elementarbuch. 1. Teil: Lautlehre.* Heidelberg: Winter.

Buridant, Claude. 2000. *Grammaire nouvelle de l'ancien français.* Paris: SEDES.

Burnett, Heather & Mireille Tremblay. 2009. Variable-behavior Ps and the location of PATH in Old French. In Enoch O. Aboh, Elisabeth van der Linden, Josep Quer, & Petra Sleeman (eds.), *Romance languages and linguistic theory: Selected papers from 'Going Romance' Amsterdam 2007*, 25–50. Amsterdam: John Benjamins.

Burnley, David. 1983. *A guide to Chaucer's language.* London: Macmillan.

Burnley, David. 1992. Lexis and semantics. In Norman Blake (ed.), *The Cambridge history of the English language, Vol. 2: 1066–1476*, 409–99. Cambridge, U.K.: Cambridge University Press.

Bybee, Joan L. 2013. Usage-based theory and exemplar representations of constructions. In Thomas Hoffmann & Graeme Trousdale (eds.), *The Oxford handbook of construction grammar*, 49–69. Oxford: Oxford University Press.

Cadierno, Teresa. 2004. Expressing motion events in a second language: A cognitive typological perspective. In Michel Achard & Susanne Niemeier (eds.), *Cognitive linguistics, second language acquisition, and foreign language teaching* (Studies on Language Acquisition 18), 13–49. Berlin: Mouton de Gruyter.

Cadierno, Teresa & Lucas Ruiz. 2006. Motion events in Spanish L2 acquisition. *Annual Review of Cognitive Linguistics* 4, 183–216.

Cappelle, Bert. 2001. Is *out of* always a preposition? *Journal of English Linguistics* 29(4), 315–28.

Cappelle, Bert. 2012. English is less rich in manner-of-motion verbs when translated from French. *Across Languages and Cultures* 13(2), 173–95.

Cardini, Filippo-Enrico. 2008. Manner of motion saliency: An inquiry into Italian. *Cognitive Linguistics* 19(4), 533–69.

Cardini, Filippo-Enrico. 2010. Evidence against Whorfian effects in motion conceptualisation. *Journal of Pragmatics* 42(5), 1442–59.

Choi, Soonja & Melissa Bowerman. 1991. Learning to express motion events in English and Korean: The influence of language-specific lexicalization patterns. *Cognition* 41, 83–121.

Cifuentes-Férez, Paula. 2007. Human locomotion verbs in English and Spanish. *International Journal of English Studies* 7(1), 117–36.

Cifuentes-Férez, Paula. 2008. *Motion in English and Spanish: A perspective from cognitive linguistics, typology, and psycholinguistics.* PhD dissertation, Universidad de Murcia, Spain.

Cifuentes-Férez, Paula & Dedre Gentner. 2006. Naming motion events in Spanish and English. *Cognitive Linguistics* 17(4), 443–62.
Cini, Monica (ed.). 2008. *I verbi sintagmatici in italiano e nelle varietà dialettali: Stato dell'arte e prospettive di ricerca; atti delle giornate di studio Torino, 19–20 febbraio 2007*. Frankfurt: Peter Lang.
Claridge, Claudia. 2000. *Multi-word verbs in Early Modern English: A corpus-based study*. Amsterdam: Rodopi.
Clark, Cecily. 1986. Ælfric and Abbo. *English Studies* 49, 30–6.
Clark, Cecily. 1995. The narrative mode of *The Anglo-Saxon Chronicle* before the conquest. In Peter Jackson (ed.), *Words, names and history: Selected papers by Cecily Clark*, 3–19. Cambridge, U.K.: Brewer.
Clark Hall, John R. 1916. *A concise Anglo-Saxon dictionary*, 2nd edn. Cambridge, U.K.: Cambridge University Press. http://www.ling.upenn.edu/~kurisuto/germanic/oe_clarkhall_about.html
Cleasby, Richard & Gudbrand Vigfusson. 1874. *An Icelandic-English dictionary*. Oxford: Clarendon.
CME = *Corpus of Middle English prose and verse*. University of Michigan. http://quod.lib.umich.edu/c/cme/
Cooper, Helen. 1996. *The Canterbury Tales*, 2nd edn. (Oxford Guides to Chaucer). Oxford: Oxford University Press.
Corona, Gabriella. 2009. Ælfric's schemes and tropes: *Amplificatio* and the portrayal of the persecutors. In Hugh Magennis & Mary Swan (eds.), *A companion to Ælfric* (Brill's Companions to the Christian Tradition 18), 297–320. Leiden, Netherlands: Brill.
Croft, William & D. Alan Cruse. 2004. *Cognitive linguistics*. Cambridge, U.K.: Cambridge University Press.
Croft, William, Jóhanna Barðdal, Willem Hollmann, Violeta Sotirova, & Chiaki Taoka. 2010. Revising Talmy's typological classification of complex event constructions. In Hans C. Boas (ed.), *Contrastive construction grammar* (Constructional Approaches to Language 10), 201–35. Amsterdam: John Benjamins.
Cruse, D. Alan. 2004. *Meaning in language*. Oxford: Oxford University Press.
Crystal, David. 2005. *The stories of English*. London: Penguin.
Curzan, Anne. 2012. Interdisciplinarity and historiography: Periodization in the history of the English language. In Alexander Bergs & Laurel J. Brinton (eds.), *English historical linguistics. An international handbook*, 1233–56. Berlin: De Gruyter.
Davis, Norman. 1971. *Paston letters and papers of the fifteenth century*. Oxford: Clarendon Press.
de Knop, Sabine & René Dirven. 2008. Motion and location events in German, French and English: A typological, contrastive and pedagogical approach. In Sabine de Knop & Teun de Rycker (eds.), *Cognitive approaches to pedagogical grammar: A volume in honour of René Dirven* (Applications of Cognitive Linguistics 9), 295–324. Berlin: Mouton de Gruyter.
de la Cruz, Juan M. 1975. Old English pure prefixes: Structure and function. *Linguistics* 145, 47–81.
De Smet, Hendrik. 2010. Grammatical interference: Subject marker *for* and the phrasal verb particles *out* and *forth*. In Elizabeth Closs Traugott & Graeme

Trousdale (eds.), *Gradience, gradualness and grammaticalization* (Typological Studies in Language 90), 75–104. Amsterdam: John Benjamins.

Denison, David. 1981. *Aspects of the history of English group-verbs: With particular attention to the syntax of the Ormulum.* Oxford: Oxford University unpublished PhD thesis.

Denison, David. 1993. *English historical syntax: Verbal constructions.* London: Longman.

Denison, David & Richard M. Hogg. 2006. Overview. In Richard M. Hogg & David Denison (eds.), *A history of the English language*, 1–42. Cambridge, U.K.: Cambridge University Press.

Dietz, Klaus. 2004. Die altenglischen Präfixbildungen und ihre Charakteristik. *Anglia* 122(4), 561–613.

DMF = *Dictionnaire du Moyen Français (1330–1500).* Ed. by Robert Martin et al. Université de Nancy 2. Version 5 (2012). http://www.atilf.fr/dmf

DMLBS = *Dictionary of Medieval Latin from British sources.* Ed. by R.E. Latham et al., 1997–2013. Oxford: Oxford University Press.

DOE = *Dictionary of Old English.* 1986–. Ed. by Angus Cameron, Ashley Crandell Amos, Antonette diPaolo Healey et al. Toronto: Dictionary of Old English Project. http://www.doe.utoronto.ca *Dictionary of Old English: A–G* online, 2007.

DOEC = *Dictionary of Old English web corpus.* 2009. Compiled by Antonette diPaolo Healey with John Price Wilkin and Xin Xiang. Toronto: Dictionary of Old English Project.

Donald, A. K. (ed.) 1975 [1895]. *Mélusine* (Early English Text Society: Extra Series 68). Millwood, NY: Kraus Reprint.

Dufresne, Monique, Fernande Dupuis & Mireille Tremblay. 2003. Preverbs and particles in Old French. In Geert Booij & Jaap van Marle (eds.), *Yearbook of morphology 2003*, 30–60. Dordrecht, Netherlands: Kluwer.

Durkin, Philip. 2009. *The Oxford guide to etymology.* Oxford: Oxford University Press.

Durkin, Philip. 2014. *Borrowed words: A history of loanwords in English.* Oxford: Oxford University Press.

Dyboski, Roman & Zygfryd Marjan Arend. 1971. *Knyghthode and bataile.* (Early English Text Society: Original Series 201). London: Oxford University Press.

Elenbaas, Marion. 2007. *The synchronic and diachronic syntax of the English verb-particle combination.* PhD thesis, Radboud Universiteit Nijmegen. Utrecht, Netherlands: LOT.

Elenbaas, Marion. 2014. Directional *out of* in the history of English: Grammaticalization and reanalysis. *Journal of Germanic Linguistics* 26(2), 83–126.

Elsness, Johan. 1997. *The perfect and the preterite in contemporary and earlier English.* Berlin: Mouton de Gruyter.

Ernout-Meillet = Ernout, A. & A. Meillet. 1967. *Dictionnaire étymologique de la langue Latine.* Paris: Klincksieck.

Evans, Vyvyan & Melanie Green. 2006. *Cognitive linguistics: An introduction.* Edinburgh: Edinburgh University Press.

Evans, Vyvyan & Paul Chilton (eds.). 2010. *Language, cognition and space: The state of the art and new directions.* London: Equinox.

Fanego, Teresa. 2012. Motion events in English: The emergence and diachrony of manner salience from Old English to Late Modern English. *Folia Linguistica Historica* 33, 29–85.

Fell, Christine E. 1984. Some questions of language. In Niels Lund, Ole Crumlin-Pedersen, & Christine E. Fell (eds.), *Two voyagers at the court of King Alfred: The ventures of Ohthere and Wulfstan together with the description of Northern Europe from the Old English Orosius*, 56–63. York, U.K.: Sessions.

FEW = Walther von Wartburg. 1922–2002. *Französisches etymologisches Wörterbuch*. Available via ATILF: https://apps.atilf.fr/lecteurFEW/.

Field, P. J .C. (ed.). 2001a. *Malory: Text and sources* (Arthurian Studies XL). Cambridge, U.K.: Brewer.

Field, P. J. C. 2001b. Caxton's Roman War. In P. J. C. Field (ed.), *Malory: Text and sources* (Arthurian Studies XL), 126–61. Cambridge, U.K.: Brewer.

Filipović, Luna. 2007a. *Talking about motion: A crosslinguistic investigation of lexicalization patterns*. Amsterdam: John Benjamins.

Filipović, Luna. 2007b. On the nature of lexicalization patterns: A cross-linguistic inquiry. In Nicole Delbecque & Bert Cornillie (eds.), *On interpreting construction schemas: From action and motion to transitivity and causality* (Trends in Linguistics. Studies and Monographs 198), 307–29. Berlin: Mouton de Gruyter.

Filipović, Luna & Kasia M. Jaszczolt (eds.). 2012. *Space and time in languages and cultures. Vol. II: Language, culture, and cognition* (Human Cognitive Processing 37). Amsterdam: John Benjamins.

Fillmore, Charles. 1985. Frames and the semantics of understanding. *Quaderni di Semantica* VI(2), 222–54.

Fischer, Bonifatius. 1989. *Die lateinischen Evangelien vor dem 10. Jahrhundert. Vol. II: Varianten zu Markus*. Freiburg: Herder.

Fischer, Bonifatius & Robert Weber (eds.). 1997. *Biblia sacra: Iuxta vulgatam versionem: Ed. quartam emendatam*. Stuttgart: Deutsche Bibelgesellschaft.

Fischer, Olga. 1992. Syntax. In Norman Blake (ed.), *The Cambridge history of the English language, Vol. 2: 1066–1476*, 207–408. Cambridge, U.K.: Cambridge University Press.

Fischer, Olga & Wim van der Wurff. 2006. Syntax. In Richard M. Hogg & David Denison (eds.), *A history of the English language*, 109–98. Cambridge, U.K.: Cambridge University Press.

Fischer, Olga, Ans van Kemenade, Willem Koopman, & Wim van der Wurff. 2000. *The syntax of early English*. Cambridge, U.K.: Cambridge University Press.

Fleischman, Suzanne. 1990. *Tense and narrativity: From medieval performance to modern fiction*. Austin, TX: University of Texas Press.

Forshall, Rev. J. F. R. S. & Frederic K. H. F. R. S. Madden. 1850. *The Holy Bible: Containing the Old and New Testaments with the Apocryphal Books in the earliest English versions made from the Latin Vulgate by John Wycliffe and his followers*. Oxford: Oxford University Press.

Fotitch, Tatiana. 1950. *The narrative tenses in Chrétien de Troyes: A study in syntax and stylistics*. Doctoral dissertation, Catholic University of America. Washington, DC: The Catholic University of America Press.

Foulet, Lucien. 1946. L'Effacement des adverbes de lieu. *Romania* 75, 1–79.

Foust, Kirstin V. 2002. Forcible entry. In Ronald H. Fritze & William B. Robison (eds.), *Historical dictionary of late medieval England, 1272–1485*, 208. Westport, CT: Greenwood Press.

Fried, Mirjam & Jan-Ola Östman. 2004. Construction grammar: A thumbnail sketch. In Mirjam Fried & Jan-Ola Östman (eds.), *Construction grammar in a cross-language perspective* (Constructional Approaches to Language 2), 11–86. Amsterdam: John Benjamins.

Galbraith, Vivian H. 1970. *The Anonimalle Chronicle, 1333 to 1381*. Manchester, U.K.: Manchester University Press.

Gamillschegg, Ernst. 1997. *Etymologisches Wörterbuch der französischen Sprache* (Sammlung romanischer Elementar- und Handbücher 3). Heidelberg: Winter.

Geeraerts, Dirk. 1993. Vagueness's puzzles, polysemy's vagaries. *Cognitive Linguistics* 4(3), 223–72.

Geeraerts, Dirk. 2010. *Theories of lexical semantics*. Oxford: Oxford University Press.

Gennari, Silvia P., Steven A. Sloman, Barbara C. Malt, & Tecumseh Fitch. 2002. Motion events in language and cognition. *Cognition* 83, 49–79.

Gneuss, Helmut. 1955. *Lehnbildungen und Lehnbedeutungen im Altenglischen*. Berlin: Erich Schmidt.

Gneuss, Helmut. 1991. The Old English language. In Malcolm Godden and Michael Lapidge (eds.), *The Cambridge companion to Old English literature*, 23–45. Cambridge, U.K.: Cambridge University Press.

Godden, Malcolm R. 1992. Literary language. In Richard M. Hogg (ed.), *The Cambridge history of the English language. Vol. I: The beginnings to 1066*, 490–535. Cambridge, U.K.: Cambridge University Press.

Godefroy, Frédéric. 1937–8. *Dictionnaire de l'ancienne langue française et de tous ses dialectes du IXe au XVe siècle*. Paris: Librairie des Sciences et des Arts.

Goldberg, Adele. 1996. Making one's way through the data. In Masayoshi Shibatani & Sandra A. Thompson (eds.), *Grammatical constructions: Their form and meaning*, 29–53. Oxford: Clarendon Press.

Goldberg, Adele E. 1995. *Constructions*. Chicago: University of Chicago Press.

Goldberg, Adele E. 2006. *Constructions at work: The nature of generalization in language*. Oxford: Oxford University Press.

Goldberg, Adele E. 2009. The nature of generalization in language. *Cognitive Linguistics* 20(1), 93–127.

Goldberg, Adele E. & Ray Jackendoff. 2004. The English resultative as a family of constructions. *Language* 80(3), 532–68.

Goldberg, Adele E., Devin M. Casenhiser, & Nitya Sethuraman. 2004. Learning argument structure generalizations. *Cognitive Linguistics* 15(3), 289–316.

Goschler, Juliana & Anatol Stefanowitsch (eds.). 2013. *Variation and change in the encoding of motion events* (Human Cognitive Processing 41). Amsterdam: John Benjamins.

Görlach, Manfred. 1974. *The textual tradition of the South English Legendary* (Leeds Texts and Monographs. New Series 6). Leeds, U.K.: University of Leeds.

Grant, Anthony P. 2009. Loanwords in British English. In Martin Haspelmath & Uri Tadmor (eds.), *Loanwords in the world's languages: A comparative handbook*, 360–83. Berlin: De Gruyter Mouton.

Gsell, Otto. 1982. Las rosas dattan ora—les röses dà fora—le rose danno fuori: Verbalperiphrasen im Ortsadverb im Rätoromanischen und im Italienischen. In Sieglinde Heinz & Ulrich Wandruszka (eds.), *Fakten und Theorien: Festschrift für Helmut Stimm zum 65. Geburtstag*, 71–85. Tübingen: Narr.

Hanna, Ralph. 2002. Alliterative poetry. In David Wallace (ed.), *The Cambridge history of medieval English literature*, 488–512. Cambridge, U.K.: Cambridge University Press.

Harley, Birgit. 1989. Transfer in the written compositions of French immersion students. In Hans W. Dechert & Manfred Raupach (eds.), *Transfer in language production*, 3–19. Norwood, NJ: Ablex.

Harley, Birgit & Mary L. King. 1989. Verb lexis in the written compositions of young L2 learners. *Studies in Second Language Acquisition* 11(4), 415–39.

Haselow, Alexander. 2012. Lexical typology and typological changes in the English lexicon. In Terttu Nevalainen & Elizabeth Closs Traugott (eds.), *The Oxford handbook of the history of English*, 643–53. Oxford: Oxford University Press.

Haspelmath, Martin & Uri Tadmor (eds.). 2009. *Loanwords in the world's languages: A comparative handbook*. Berlin: De Gruyter Mouton.

HC = *Helsinki corpus of English texts*. 1991. Compiled by Matti Rissanen (project leader), Merja Kytö (project secretary); Leena Kahlas-Tarkka, Matti Kilpiö (Old English); Saara Nevanlinna, Irma Taavitsainen (Middle English); Terttu Nevalainen, Helena Raumolin-Brunberg (Early Modern English). Department of Modern Languages, University of Helsinki. http://www.helsinki.fi/varieng/CoRD/corpora/HelsinkiCorpus/

Hendriks, Jennifer. 2012. English in contact: German and Dutch. In Alexander Bergs & Laurel J. Brinton (eds.), *English historical linguistics: An international handbook*, 1659–70. Berlin: De Gruyter.

Herbst, Thomas & Michael Klotz. 2003. *Lexikografie*. Paderborn: Schöningh.

Hickmann, Maya & Stéphane Robert (eds.). 2006. *Space in languages: Linguistic systems and cognitive categories* (Typological Studies in Language 66). Amsterdam: John Benjamins.

Hijazo-Gascón, Alberto & Iraide Ibarretxe-Antuñano. 2013. Same family, different paths: Intratypological differences in three Romance languages. In Juliana Goschler & Anatol Stefanowitsch (eds.), *Variation and change in the encoding of motion events* (Human Cognitive Processing 41), 39–54. Amsterdam: John Benjamins.

Hiltunen, Risto. 1983a. *The decline of the prefixes and the beginnings of the English phrasal verb*. Turku, Finland: Turun Yliopisto.

Hiltunen, Risto. 1983b. Syntactic variation in the early history of the English phrasal verb. In Sven Jacobson (ed.), *Papers from the Second Scandinavian Symposium on Syntactic Variation*, 95–108. Stockholm: Almqvist & Wiksell.

Hiraga, Masako K. 1994. Diagrams and detaphors: Iconic aspects in language. *Journal of Pragmatics* 22 (Special Issue Metaphor and Iconicity), 5–21.

Hohenstein, Jill, Ann Eisenberg, & Letitia Naigles. 2006. Is he floating across or crossing afloat? Cross-influence of L1 and L2 in Spanish-English bilingual adults. *Bilingualism* 9(3), 249–61.

Holthausen, Ferdinand. 1934. *Altenglisches etymologisches Wörterbuch*. Heidelberg: Winter.

HTE = *Historical thesaurus of the OED*. 2009. Ed. by Christian Kay, Jane Roberts, Michael Samuels, & Irené Wotherspoon. Oxford: Oxford University Press. http://www.oed.com

Huber, Judith. 2012. 'No man entreth in or out': How are typologically unsuitable loanverbs integrated into English? In Irén Hegedűs & Alexandra Fodor (eds.), *English historical linguistics 2010: Selected papers from the Sixteenth International Conference on English Historical Linguistics (ICEHL 16), Pécs, 23–27 August 2010* (Current Issues in Linguistic Theory 325), 327–46. Amsterdam: John Benjamins.

Huber, Judith. 2013. Caused-motion verbs in the Middle English intransitive motion construction. In Juliana Goschler & Anatol Stefanowitsch (eds.), *Variation and change in the encoding of motion events* (Human Cognitive Processing 41), 203–22. Amsterdam: John Benjamins.

Iacobini, Claudio. 2009. The role of dialects in the emergence of Italian phrasal verbs. *Morphology* 19, 15–44.

Iacobini, Claudio. 2012. Grammaticalization and innovation in the encoding of motion events. *Folia Linguistica Historica* 46(2), 359–85.

Iacobini, Claudio & Benjamin Fagard. 2011. A diachronic approach to variation and change in the typology of motion event expression: A case study: From Latin to Romance. *Faits de langues: Les cahiers* 3, 152–71.

Iacobini, Claudio & Francesca Masini. 2006. The emergence of verb-particle constructions in Italian: Locative and actional meanings. *Morphology* 16(2), 155–88.

Ibarretxe-Antuñano, Iraide. 2006. *Sound symbolism and motion in Basque*. Munich: LINCOM Europa.

Ihrig, Roscoe M. 1916. *The semantic development of words for 'walk, run' in the Germanic languages*. Chicago: University of Chicago Press.

Ikegami, Yoshihiko. 1970. *The semological structure of the English verbs of motion: A stratificational approach*. Tokyo: Sanseido.

Ingham, Richard (ed.). 2010. *The Anglo-Norman language and its contexts*. Woodbridge, U.K.: York Medieval Press.

Ingham, Richard. 2010a. Later Anglo-Norman as a contact variety of French? In Richard Ingham (ed.), *The Anglo-Norman language and its contexts*, 8–25. Woodbridge, U.K.: York Medieval Press.

Ingham, Richard. 2010b. The transmission of later Anglo-Norman: Some syntactic evidence. In Richard Ingham (ed.), *The Anglo-Norman language and its contexts*, 164–82. Woodbridge, U.K.: York Medieval Press.

Ingham, Richard. 2012. *The transmission of Anglo-Norman: Language history and language acquisition*. Amsterdam: John Benjamins.

Israel, Michael. 1996. The *way* constructions grow. In Adele Goldberg (ed.), *Conceptual structure, discourse, and language*, 217–30. Stanford, CA: CSLI.

Ito, Eiko. 1998. Reflexive verbs in Chaucer. In Jacek Fisiak & Akio Oizumi (eds.), *English historical linguistics and philology in Japan* (Trends in Linguistics. Studies and Monographs 109), 55–77. Berlin: Mouton de Gruyter.

Iwata, Seizi. 2002. Does MANNER count or not? Manner-of-motion verbs revisited. *Linguistics* 40(1), 61–110.

Iwata, Seizi. 2005. Locative alternation and two levels of verb meaning. *Cognitive Linguistics* 16(2), 355–407.

Jaberg, Karl. 1972 [1939]. Considérations sur quelques caractères généraux du romanche. In *Mélanges de linguistique offerts à Charles Bally*, 283–92. Geneva: Slatkine Reprints.

Jean d'Arras. 1972 [1845]. *Melusine: Nouvelle édition, conforme à celle de 1478 revue et corrigée avec une préface par M. Ch. Brunet*. Nendeln, Liechtenstein: Kraus Reprint.

Jorgensen, Alice. 2010. Introduction: Reading the Anglo-Saxon Chronicle. In Alice Jorgensen (ed.), *Reading the Anglo-Saxon Chronicle: Language, literature, history* (Studies in the Early Middle Ages 23), 1–29. Turnhout, Belgium: Brepols.

Kaiser, Rolf. 1937. *Zur Geographie des mittelenglischen Wortschatzes*. Leipzig: Mayer & Müller.

Käsmann, Hans. 1961. *Studien zum kirchlichen Wortschatz des Mittelenglischen, 1100–1350: Ein Beitrag zum Problem der Sprachmischung* (Buchreihe der Anglia 9). Tübingen: Niemeyer.

Kastovsky, Dieter. 1992. Semantics and vocabulary. In Richard M. Hogg (ed.), *The Cambridge history of the English language: Vol. I: The beginnings to 1066*, 290–408. Cambridge, U.K.: Cambridge University Press.

Kay, Christian, Jane Roberts, Michael Samuels, & Irené Wotherspoon. 2009. *Historical Thesaurus of the Oxford English Dictionary: Vol. I*. Oxford: Oxford University Press.

Keenan, Edward L. 2002. Explaining the creation of reflexive pronouns in English. In Donka Minkova & Robert Stockwell (eds.), *Studies in the history of the English language: A millennial perspective* (Topics in English Linguistics 39), 325–54. Berlin: Mouton de Gruyter.

Kemmer, Suzanne. 1993. *The middle voice* (Typological Studies in Language 23). Amsterdam: John Benjamins.

Kisbye, Torben. 1971–2. *An historical outline of English syntax*. 2 vols. Aarhus, Denmark: Akademisk Boghandel.

Kjellmer, Göran. 2002. 'Must down': On non-occurring verbs of motion in Modern English. *Nordic Journal of English Studies* 1(2), 339–53.

Koch, Peter. 2000. Indirizzi cognitivi per una tipologia lessicale dell'italiano. *Italienische Studien* 21, 99–117.

Koch, Peter & Wulf Oesterreicher. 2011. *Gesprochene Sprache in der Romania: Französisch, Italienisch, Spanisch*, 2nd edn. (Romanistische Arbeitshefte 31). Berlin: Mouton de Gruyter.

König, Ekkehard & Peter Siemund. 2000. The development of complex reflexives and intensifiers in English. *Diachronica* 17(1), 39–48.

Kooper, Erik. 2006. *Sentimental and humorous romances* (Teams Middle English Texts Series). Kalamazoo, MI: Medieval Institute Publications.

Kopecka, Anetta. 2004. *Etude typologique de l'expression de l'espace: Localisation et déplacement en Français et en Polonais*. PhD dissertation. Lyon, France: Université Lumière Lyon II.

Kopecka, Anetta. 2006. The semantic structure of motion verbs in French: Typological perspectives. In Maya Hickmann & Stéphane Robert (eds.), *Space in languages: Linguistic systems and cognitive categories* (Typological Studies in Language 66), 8–101. Amsterdam: John Benjamins.

Kopecka, Anetta. 2009. Continuity and change in the representation of motion events in French. In Jiansheng Guo, Elena Lieven, Nancy Budwig, Susan Ervin-Tripp, Keiko Nakamura, & Seyda Özçalişkan (eds.), *Crosslinguistic approaches to the psychology of language: Research in the tradition of Dan Isaac Slobin*, 415–26. New York: Psychology Press.

Kopecka, Anetta. 2013. Describing motion events in Old and Modern French: Discourse effects of a typological change. In Juliana Goschler & Anatol Stefanowitsch (eds.), *Variation and change in the encoding of motion events* (Human Cognitive Processing 41), 163–83. Amsterdam: John Benjamins.

Kornexl, Lucia. 2001. 'Unnatural words?' Loan-formations in Old English glosses. In Dieter Kastovsky & Arthur Mettinger (eds.), *Language contact in the history of English*, 195–216. Frankfurt: Peter Lang.

Koziol, Herbert. 1972. *Handbuch der englischen Wortbildungslehre*, 2nd ed. Heidelberg: Winter.

Kramer, Johannes. 1981. Die Übernahme der deutschen und der niederländischen Konstruktion Verb + Verbzusatz durch die Nachbarsprachen. In Wolfgang Meid & Karin Heller (eds.), *Sprachkontakt als Ursache von Veränderungen der Sprach- und Bewusstseinsstruktur*, 129–40. Innsbruck, Austria: Institut für Sprachwissenschaft.

Laing, Margaret & Roger Lass. 2006. Early Middle English dialectology: Problems and prospects. In Ans van Kemenade & Bettelou Los (eds.), *The handbook of the history of English*, 417–51. Malden, MA.: Blackwell.

Lakusta, Laura & Barbara Landau. 2005. Starting at the end: The importance of goals in spatial language. *Cognition* 96, 1–33.

Langacker, Ronald W. 1990. *Concept, image, and symbol: The cognitive basis of grammar*. Berlin: Mouton de Gruyter.

Langacker, Ronald W. 2000. *Grammar and conceptualization*. Berlin: Mouton de Gruyter.

Langacker, Ronald W. 2009. Constructions and constructional meaning. In Vyvyan Evans & Stéphanie Pourcel (eds.), *New directions in cognitive linguistics*, 225–67. Amsterdam: John Benjamins.

Lange, Claudia. 2007. *Reflexivity and intensification in English: A study of texts and contexts*. Frankfurt: Peter Lang.

Larrañaga, Pilar, Jeanine Treffers-Daller, Françoise Tidball, & Mari-carmen Gil Ortega. 2012. L1 transfer in the acquisition of manner and path in Spanish by native speakers of English. *International Journal of Bilingualism* 16(1), 117–38.

Lenker, Ursula. 1997. *Die westsächsische Evangelienversion und die Perikopenordnungen im angelsächsischen England* (Texte und Untersuchungen zur Englischen Philologie 20). Munich: Fink.

Levin, Beth. 1993. *English verb classes and alternations*. Chicago: University of Chicago Press.

Levinson, Stephen C. 2003. *Space in language and cognition: Explorations in cognitive diversity*. Cambridge, U.K.: Cambridge University Press.

Levinson, Stephen C. & David P. Wilkins (eds.). 2006. *Grammars of space: Explorations in cognitive diversity*. Cambridge, U.K.: Cambridge University Press.

Lewis, Robert E., Mary J. Williams, & Marilyn S. Miller. 2007. *Middle English dictionary: Plan and bibliography*. Ann Arbor, MI: University of Michigan Press.

Liebermann, Felix. 1903. *Die Gesetze der Angelsachsen. Bd. 1*. Halle: Niemeyer.

Lindberg, Conrad. 1984. The language of the Wycliff Bible. In Wolf-Dietrich Bald & Horst Weinstock (eds.), *Medieval Studies Conference Aachen 1983*, 103–10. Frankfurt: Peter Lang.

Lindberg, Conrad. 2007. Revising the Wycliff Bible. In Hans Sauer & Renate Bauer (eds.), *Beowulf and beyond* (Studies in English Medieval Language and Literature 18), 197–204. Frankfurt: Peter Lang.

Liuzza, Roy M. (ed.). 1994. *The Old English version of the Gospels* (Early English Text Society: Original Series 304). Oxford: Oxford University Press.

Liuzza, Roy M. 2000. *The Old English version of the Gospels. Volume two: Notes and glossary* (Early English Text Society: Extra Series 314). Oxford: Oxford University Press.

Lund, Niels. 1984. Introduction. In Niels Lund, Ole Crumlin-Pedersen, & Christine E. Fell (eds.), *Two voyagers at the court of King Alfred. The ventures of Ohthere and Wulfstan together with the description of northern Europe from the Old English Orosius*, 5–14. York, U.K.: Sessions.

Lüttgens, Carl. 1888. *Über Bedeutung und Gebrauch der Hilfsverba im frühen Altenglischen:* Sculan *und* Willan. Dissertation Universität Kiel. Wismar: Hinstorffsche Raths-Buchdruckerei (L. Eberhardt).

Lutz, Angelika. 2002a. When did English begin? In Teresa Fanego, Belén Méndez-Naya, & Elena Seoane (eds.), *Sounds, words, texts and change: Selected papers from 11 ICEHL, Santiago de Compostela, 7–11 September 2000* (Current Issues in Linguistic Theory 224), 145–71. Amsterdam: John Benjamins.

Lutz, Angelika. 2002b. The reflexive in Middle English: Loose ends in the pronominal system. In Peter J. Lucas & Angela M. Lucas (eds.), *Middle English from tongue to text* (Studies in English Medieval Language and Literature 4), 45–68. Frankfurt: Peter Lang.

Lutz, Angelika. 2013. Language contact and prestige. *Anglia* 131, 562–90.

Machan, Tim W. 2012. Language contact and linguistic attitudes in the later middle ages. In Terttu Nevalainen & Elizabeth Closs Traugott (eds.), *The Oxford handbook of the history of English*, 518–27. Oxford: Oxford University Press.

Magennis, Hugh. 2009. Ælfric scholarship. In Hugh Magennis & Mary Swan (eds.), *A companion to Ælfric* (Brill's Companions to the Christian Tradition 18), 5–34. Leiden, Netherlands: Brill.

Maguire, Mandy J., Kathy Hirsh-Pasek, Roberta Michnick Golinkoff, Mutsumi Imai, Etsuko Haryu, Sandra Vanegas, Hiroyuki Okada, Rachel Pulverman, & Brenda Sanchez-Davis. 2010. A developmental shift from similar to language-specific strategies in verb acquisition: A comparison of English, Spanish, and Japanese. *Cognition* 114(3), 299–319.

Mailhammer, Robert & Elena Smirnova. 2013. Incipient grammaticalization: Sources of passive constructions in Old High German and Old English. In Gabriele Diewald, Leena Kahlas-Tarkka, & Ilse Wischer (eds.), *Comparative studies in early Germanic languages: With a focus on verbal categories* (Studies in Language Companion Series 138), 41–70. Amsterdam: John Benjamins.

Mair, Walter N. 1984. Transferenz oder autonome Bildung? Bemerkungen zum Problem der Partikelverben im Ladinischen, Friulanischen, Italienischen und Französischen. *Zeitschrift für romanische Philologie* 100, 408–32.

Malblanc, Alfred. 1968. *Stylistique comparée du français et de l'allemand: essai de représentation linguistique comparée et étude de traduction*, 5th edn. Paris: Didier.

Marchand, Hans. 1969. *The categories and types of present-day English word-formation: A synchronic-diachronic approach*. Munich: Beck.

Marchello-Nizia, Christiane. 2002. Prépositions françaises en diachronie: une catégorie en question. *Lingvisticæ Investigationes* 25(2), 205–11.
Matlock, Teenie. 2004. The conceptual motivation of fictive motion. In Klaus-Uwe Panther & Günter Radden (eds.), *Studies in linguistic motivation*, 221–48. Berlin: Mouton de Gruyter.
Matsunami, Tamotsu. 1985. The Middle English verbs of motion. *Poetica (Tokyo)* 21/22, 71–105.
Mayer, Mercer. 1969. *Frog, where are you?* New York: Dial Press.
McCarthy, Terence. 1996. Malory and his sources. In Elizabeth Archibald & A. S. G. Edwards (eds.), *A companion to Malory*, 75–95. Cambridge, U.K.: Brewer.
McCrea, Carol M. 1976. *Ælfric: His sources and style in the lives of Æthelthryth, Oswald, and Edmund.* PhD dissertation, Fordham University, Ann Arbor, MI: University Microfilms.
McFadden, Thomas & Artemis Alexiadou. 2010. Perfects, resultatives, and auxiliaries in Earlier English. *Linguistic Inquiry* 41(3), 389–425.
McLaren, Mary-Rose. 2002. The London chronicles of the fifteenth century: A revolution in English writing. Woodbridge, U.K.: Brewer.
MED = *Middle English dictionary*. 1952–2001. Ed. by Hans Kurath, Sherman M. Kuhn, John Reidy, & Robert E. Lewis. Ann Arbor, MI: University of Michigan. http://quod.lib.umich.edu/m/med/
Meichtry-Gruber, Annemarie. 2008. *Die Sprache der Wyclif-Bibel: Die Verwendung von Lehnwörtern in den Büchern Baruch, Richter und Hiob*. Frankfurt: Peter Lang.
Mengden, Ferdinand von. 2012. Periods: Old English. In Alexander Bergs & Laurel J. Brinton (eds.), *English historical linguistics: An international handbook*, 19–32. Berlin: De Gruyter.
Meyer, Charles F. 2002. *English corpus linguistics: An introduction.* Cambridge, U.K.: Cambridge University Press.
Mills, C.A. 1911. *Ye Solace of Pilgrimes: A description of Rome, circa A.D. 1450, by John Capgrave, an Austin friar of King's Lynn.* London: Oxford University Press.
Minkoff, Harvey. 1976. Some stylistic consequences of Ælfric's theory of translation. *Studies in Philology* 73(1), 29–41.
Minkova, Donka. 2008. Prefixation and stress in Old English: In memoriam Richard Hogg (1944–2007). *Word Structure* 1(1), 21–52.
Minkova, Donka. 2012. Functional stress-shifting in Old English. Paper read at the *17th International Conference on English Historical Linguistics*, Zürich, 22 August 2012.
Mitchell, Bruce. 1978. Prepositions, adverbs, prepositional adverbs, postpositions, separable prefixes, or inseparable prefixes, in Old English. *Neuphilologische Mitteilungen* LXXIX, 240–57.
Mitchell, Bruce. 1985. *Old English syntax.* Oxford: Clarendon Press.
Mitchell, Bruce. 1995. *An invitation to Old English and Anglo-Saxon England.* Oxford: Blackwell.
Mitchell, Bruce & Fred C. Robinson. 2006. *A guide to Old English*, 7th edn. Malden, MA: Blackwell.
Molencki, Rafał. 2012. *Causal conjunctions in Mediaeval English: A corpus-based study of grammaticalization.* Katowice, Poland: Uniwersytet Śląski/Oficyna Wydawnicza.

Mortelmans, Tanja, Kasper Boye, & Johan van der Auwera. 2009. Modals in the Germanic languages. In Björn Hansen & Ferdinand de Haan (eds.), *Modals in the languages of Europe: A reference work*, 11–69. Berlin: Mouton de Gruyter.

Mossé, Fernand. 1973. *Mittelenglische Kurzgrammatik*. Ismaning: Hueber.

Mustanoja, Taunc. 1960. *A Middle English syntax part I*. Helsinki: Société Néophilologique.

Naigles, Letitia & Paula Terrazas. 1998. Motion verb generalizations in English and Spanish: The influence of language and syntax. *Psychological Science* 9, 363–9.

Naigles, Letitia, Ann Eisenberg, Edward T. Kako, Melissa Highter, & Nancy McGraw. 1998. Speaking of motion: Verb use in English and Spanish. *Language and Cognitive Processes* 13(5), 521–49.

Narasimhan, Bhuvana. 2003. Motion events and the lexicon: A case study of Hindi. *Lingua* 113(2), 123–60.

Navarro, Samuel & Elena Nicoladis. 2005. Describing motion events in adult L2 Spanish narratives. In David Eddington (ed.), *Proceedings of the 6th Conference on Acquisition of Spanish and Portuguese as First and Second Languages*, 102–7. Somerville, MA: Cascadilla Press.

NCA = *Nouveau Corpus d'Amsterdam. Corpus informatique de textes littéraires d'ancien français (ca 1150–1350), établi par Anthonij Dees (Amsterdam 1987), remanié par Achim Stein, Pierre Kunstmann & Martin-D. Gleßgen*. 2011. Ed. by Achim Stein et al. Stuttgart: Institut für Linguistik/Romanistik, version 3. http://www.uni-stuttgart.de/lingrom/stein/corpus/

Nerlich, Brigitte & David D. Clarke. 2003. Polysemy and flexibility: Introduction and overview. In Brigitte Nerlich, Todd Zazie, Herman Vimala, & David D. Clarke (eds.), *Polysemy: Flexible patterns of meaning in mind and language* (Trends in Linguistics. Studies and Monographs 142), 3–30. Berlin: Mouton de Gruyter.

Nevanlinna, Saara, Päivi Pahta, Kirsti Peitsara, & Irma Taavitsainen. 1993. Middle English. In Matti Rissanen, Merja Kytö, & Minna Palander-Collin (eds.), *Early English in the computer age: Explorations through the Helsinki Corpus* (Topics in English Linguistics 11), 33–51. Berlin: Mouton de Gruyter.

Nichols, Ann E. 1971. Ælfric and the brief style. *Journal of English and Germanic Philology* 1(1), 1–12.

Niles, John D. 1981. Compound diction and the style of *Beowulf*. *English Studies* 62(6), 489–503.

Norris, Ralph. 2008. *Malory's library: The sources of the Morte Darthur* (Arthurian Studies lxxi). Cambridge, U.K.: Brewer.

OED = *The Oxford English dictionary*. 2000–. 3rd ed. online. Oxford: Oxford University Press. http://www.oed.com/

Oesterley, Hermann. 1872. *Gesta Romanorum*. Berlin: Weidmannsche Buchhandlung.

Ogura, Michiko. 2000. '*Gewat* + infinitive' and '*uton* + infinitive'. *Neuphilologische Mitteilungen* 101, 69–78.

Ogura, Michiko. 2002. *Verbs of motion in medieval English*. Cambridge, U.K.: Brewer.

Öhlschläger, Günther. 1989. *Zur Syntax und Semantik der Modalverben des Deutschen*. Tübingen: Niemeyer.

Özçalışkan, Seyda. 2004. Typological variation in encoding the manner, path, and ground components of a metaphorical motion event. *Annual Review of Cognitive Linguistics* 2, 73–102.

Özçalişkan, Seyda. 2005. Metaphor meets typology: Ways of moving metaphorically in English and Turkish. *Cognitive Linguistics* 16(1), 207–46.

Papafragou, Anna, Christine Massey & Lila Gleitman. 2002. Shake, rattle, 'n' roll: The representation of motion in language and cognition. *Cognition* 84(2), 189–219.

Papafragou, Anna, Lila R. Gleitman, & Christine Massey. 2006. When English proposes what Greek presupposes: The cross-linguistic encoding of motion events. *Cognition* 98(3), B75–B87.

Paul, Hermann. 1920 [1886]. *Prinzipien der Sprachgeschichte*, 5th edn. Halle: Niemeyer.

Pedersen, Johan. 2009. The construction of macro-events: A typological perspective. In Christopher S. Butler & Javier Martín Arista (eds.), *Deconstructing constructions* (Studies in Language Companion Series 107), 25–62. Amsterdam: John Benjamins.

Peitsara, Kirsti. 1997. The development of reflexive strategies in English. In Matti Rissanen, Merja Kytö, & Kirsi Heikkonen (eds.), *Grammaticalization at work: Studies of long-term developments in English* (Topics in English Linguistics 24), 277–370. Berlin: Mouton de Gruyter.

Petré, Peter & Hubert Cuyckens. 2008. Bedusted, yet not beheaded: The role of *be*-'s constructional properties in its conservation. In Alexander Bergs & Gabriele Diewald (eds.), *Constructions and language change*, 135–69. Berlin: Mouton de Gruyter.

Petré, Peter & Hubert Cuyckens. 2009. Constructional change in Old and Middle English copular constructions and its impact on the lexicon. *Folia Linguistica Historica* 30, 311–66.

PLD = *Patrologia Latina database*. 2006. Electronic version of *Patrologiae cursus completus*, ed. Jacques-Paul Migne. 1844–65. Cambridge, U.K.: Chadwyck-Healey.

Plummer, Charles. 1899. *Two of the Saxon Chronicles parallel (787–1001 A.D.). Vol. II: Introduction, notes, and index*. Oxford: Clarendon Press.

Pope, Mildred. 1956. *From Latin to Modern French with especial consideration of Anglo-Norman*. Manchester, U.K.: Manchester University Press.

PPCEME = *Penn–Helsinki parsed corpus of Early Modern English*. 2004. Compiled by Anthony Kroch, Beatrice Santorini, & Ariel Diertani. http://www.ling.upenn.edu/hist-corpora/

PPCME2 = *Penn–Helsinki parsed corpus of Middle English, second edition*. 2000. Compiled by Anthony Kroch & Ann Taylor. http://www.ling.upenn.edu/hist-corpora/

Quirk, Randolph, Sidney Greenbaum, Geoffrey Leech, & Jan Svartvik. 1985. *A comprehensive grammar of the English language*. London: Longman.

Radden, Günter. 1995. Motion metaphorized: The case of *coming* and *going*. In Eugene H. Casad (ed.), *Cognitive linguistics in the Redwoods: The expansion of a new paradigm in linguistics* (Cognitive Linguistics Research 6), 423–58. Berlin: Mouton de Gruyter.

Radden, Günter & René Dirven. 2007. *Cognitive English grammar*, 2nd edn. (Cognitive Linguistics in Practice 2). Amsterdam: John Benjamins.

Rauh, Gisa. 1981. On *coming* and *going* in English and German. *Papers and Studies in Contrastive Linguistics* 13, 53–68.

Regier, Terry & Mingyu Zheng. 2007. Attention to endpoints: A cross-linguistic constraint on spatial meaning. *Cognitive Science* 31(4), 705–19.

Reshöft, Nina. 2011. Converging evidence in the typology of motion events: A corpus-based approach to interlanguage. In Doris Schönefeld (ed.), *Converging evidence. Methodological and theoretical issues for linguistic research* (Human Cognitive Processing 33), 293–318. Amsterdam: John Benjamins.

Rissanen, Matti. 1999. Syntax. In Roger Lass (ed.), *The Cambridge history of the English language. Vol. III: 1476–1776*, 187–331. Cambridge, U.K.: Cambridge University Press.

Roberts, Jane. 2002. Some thoughts on the representation of early Middle English in the historical thesaurus of English. *Dictionaries: Journal of The Dictionary Society of North America* 23, 180–207.

Roberts, Jane & Christian Kay. 2000. *A thesaurus of Old English*. Amsterdam: Rodopi.

Rohde, Ada R. 2001. *Analysing PATH: The interplay of verbs, prepositions and constructional semantics*. PhD dissertation. Houston, TX: Rice University.

Rohlfs, Gerhard. 1983. *Romanische Lehnübersetzungen aus germanischer Grundlage: Materia romana, spirito germanico*. Munich: Verlag der Bayerischen Akademie der Wissenschaften.

Rothwell, William. 1991. The missing link in English etymology: Anglo-French. *Medium Aevum* 60, 173–96.

Rothwell, William. 1993a. The 'faux franceis d'Angleterre': Later Anglo-Norman. In Ian Short (ed.), *Anglo-Norman anniversary essays*, 309–26. London: Anglo-Norman Text Society.

Rothwell, William. 1993b. The legacy of Anglo-French: *Faux amis* in French and English. *Zeitschrift für romanische Philologie* 109, 16–46.

Rothwell, William. 1998. Arrivals and departures: The adoption of French terminology into Middle English. *English Studies* 79(2), 144–65.

Rowley, Anthony. 2007. Woher? Wohin? Richtung und Ort im Bayerischen. *Akademie Aktuell: Zeitschrift der Bayerischen Akademie der Wissenschaften* 2, 34–9.

Rypins, Stanley. 1924. *Three Old English prose texts in MS Cotton Vitellius A XV* (Early English Text Society: Original Series 161). London: Oxford University Press.

Samson, Annie. 1986. The South English Legendary: Constructing a context. In Peter R. Coss & Simon D. Lloyd (eds.), *Thirteenth century England I: Proceedings of the Newcastle-upon-Tyne Conference 1985*, 185–95. Woodbridge, U.K.: Boydell.

Schaefer, Ursula. 2012. Middle English: Standardization. In Alexander Bergs & Laurel J. Brinton (eds.), *English historical linguistics: An international handbook*, 519–33. Berlin: De Gruyter.

Scheler, Manfred. 1977. *Der englische Wortschatz*. Berlin: Schmidt.

Schendl, Herbert & Laura Wright. 2011. Code-switching in Early English: Historical background and methodological and theoretical issues. In Herbert Schendl & Laura Wright (eds.), *Code-switching in Early English* (Topics in English Linguistics 76), 15–46. Berlin: De Gruyter Mouton.

Schlauch, Margaret. 1952. Chaucer's colloquial English: Its structural traits. *PMLA* 67(7), 1103–16.

Schøsler, Lene. 2008. L'expression des traits manière et direction des verbes de mouvement: Perspectives diachroniques et typologiques. In Elisabeth Stark, Roland Schmidt-Riese, & Eva Stoll (eds.), *Romanische Syntax im Wandel. Festschrift für Wulf Oesterreicher*, 113–32. Tübingen: Narr.

Schwarze, Christoph. 1985. 'Uscire' e 'andare fuori': Struttura sintattica e semantica lessicale. In Annalisa Franchi de Bellis & Leonardo M. Savoia (eds.), *Sintassi e*

morfologia della lingua italiana d'uso: Teorie e applicazioni descrittive. Atti del XVII Congr. Internaz. di Studi, Urbino, 11–13 sett. 1983 (Pubblicazioni della Società di Linguistica Italiana 24), 355–71. Rome: Bulzoni.
Scragg, Donald G. 1991. The nature of Old English verse. In Malcolm Godden & Michael Lapidge (eds.), *The Cambridge companion to Old English literature*, 55–70. Cambridge, U.K.: Cambridge University Press.
Severs, Jonathan B., Peter G. Beidler, Albert E. Hartung, & John E. Wells (eds.). 1919–51. *A manual of the writings in Middle English: 1050–1500*. New Haven, CT: Connecticut Academy of Arts and Sciences.
Shepherd, Stephen H. (ed.). 2004. *Le Morte Darthur or The Hoole Book of Kyng Arthur and of his Noble Knyghtes of the Rounde Table: Authoritative text, sources and backgrounds, criticism*. New York: Norton.
Short, Ian (ed.). 1993. *Anglo-Norman anniversary essays* (Anglo-Norman Text Society: Occasional publications series 2). London: Anglo-Norman Text Society.
Simpson, John. 2000. *Preface to the third edition of the* OED. http://public.oed.com/the-oed-today/preface-to-the-third-edition-of-the-oed/ (8 November 2012).
Simpson, John. 2002. The revolution of English lexicography. *Dictionaries: Journal of The Dictionary Society of North America* 23, 1–15.
Skaffari, Janne. 2012. English in contact: French. In Alexander Bergs & Laurel J. Brinton (eds.), *English historical linguistics: An international handbook*, 1671–86. Berlin: De Gruyter.
Skeat, Walter W. 1891. *Ælfric's Lives of Saints. Vol. 1. Chapters I–XXIII* (Early English Text Society: Original Series 76.82). London: Trübner.
Skeat, Walter W. 1966 [1900]. *Ælfric's Lives of Saints. Vol. 2. Chapters XXIII B–XXXVII* (Early English Text Society: Original Series 94.114). London: Oxford University Press.
Slobin, Dan I. 1996a. From 'thought and language' to 'thinking for speaking'. In John J. Gumperz & Stephen C. Levinson (eds.), *Rethinking linguistic relativity*, 70–96. New York: Cambridge University Press.
Slobin, Dan I. 1996b. Two ways to travel: Verbs of motion in English and Spanish. In Masayoshi Shibatani & Sandra A. Thompson (eds.), *Grammatical constructions: Their form and meaning*, 195–220. Oxford: Clarendon Press.
Slobin, Dan I. 1997. Mind, code, and text. In Joan Bybee, John Haiman, & Sandra A. Thompson (eds.), *Essays on language function and language type: Dedicated to T. Givón*, 437–67. Amsterdam: John Benjamins.
Slobin, Dan I. 2004a. The many ways to search for a frog: Linguistic typology and the expression of motion events. In Sven Strömqvist & Ludo Verhoeven (eds.), *Relating events in narrative: Typological and contextual perspectives*. Vol. 2, 219–57. Mahwah, NJ: Lawrence Erlbaum Associates.
Slobin, Dan I. 2004b. How people move: Discourse effects of linguistic typology. In Carol L. Moder & Aida Martinovic-Zic (eds.), *Discourse across languages and cultures*, 195–210. Amsterdam: Benjamins.
Slobin, Dan I. 2005. Relating narrative events in translation. In Dorit D. Ravid & Hava Bat-Zeev Shyldkrot (eds.), *Perspectives on language and language development: Essays in honor of Ruth A. Berman*, 115–29. Dordrecht, Netherlands: Kluwer.
Slobin, Dan I. 2006. What makes manner of motion salient? Explorations in linguistic typology, discourse, and cognition. In Maya Hickmann & Stéphane Robert

(eds.), *Space in languages: Linguistic systems and cognitive categories* (Typological Studies in Language 66), 59–81. Amsterdam: John Benjamins.

Slobin, Dan I. & Nini Hoiting. 1994. Reference to movement in spoken and signed languages: Typological considerations. *Proceedings of the Annual Meeting of the Berkeley Linguistics Society* 20, 487–505.

Smithers, Geoffrey V. 1957. *Kyng Alisaunder. Introduction, commentary and glossary* (Early English Text Society: Original Series 237). London: Oxford University Press.

Snell-Hornby, Mary. 1983. *Verb-descriptivity in German and English: A contrastive study in semantic fields.* Heidelberg: Winter.

Spisak, James W. & William Matthews (eds.). 1983. *Caxton's Malory. A new edition of Sir Thomas Malory's Le Morte Darthur based on the Pierpont Morgan copy of William Caxton's edition of 1485.* Berkeley, CA: University of California Press.

Stanley, Eric. 2000. *OED* and the earlier history of English. In Lynda Mugglestone (ed.), *Lexicography and the* OED: *Pioneers in the untrodden forest*, 126–55. Oxford: Oxford University Press.

Stanley, Eric. 2002. Celebrating the *Middle English Dictionary. Dictionaries: Journal of The Dictionary Society of North America* 23, 23–47.

Staub, Marianne. 1949. *Richtungsbegriff, Richtungsausdruck: Versuch zu einem Vergleich von deutscher und französischer Ausdrucksweise* (Romanica Helvetica 27). Bern: Francke.

Stefanowitsch, Anatol. 2013. Variation and change in English path verbs and constructions: Usage patterns and conceptual structure. In Juliana Goschler & Anatol Stefanowitsch (eds.), *Variation and change in the encoding of motion events* (Human Cognitive Processing 41), 223–44. Amsterdam: John Benjamins.

Stefanowitsch, Anatol. 2008. R-Relationen im Sprachvergleich: Die Direkte-Rede-Konstruktion im Englischen und Deutschen. In Anatol Stefanowitsch & Kerstin Fischer (eds.), *Konstruktionsgrammatik II: Von der Konstruktion zur Grammatik*, 247–61. Tübingen: Stauffenburg.

Stefanowitsch, Anatol & Ada Rohde. 2004. The goal bias in the encoding of motion events. In Klaus-Uwe Panther & Günter Radden (eds.), *Studies in linguistic motivation*, 249–68. Berlin: Mouton de Gruyter.

Stodnick, Jaqueline. 2010. Sentence to story: Reading the Anglo-Saxon Chronicle as formulary. In Alice Jorgensen (ed.), *Reading the Anglo-Saxon Chronicle: Language, literature, history* (Studies in the Early Middle Ages 23), 91–111. Turnhout, Belgium: Brepols.

Stolova, Natalya. 2008. From satellite-framed Latin to verb-framed Romance: Late Latin as an intermediate stage. In Robert Wright (ed.), *Latin vulgaire—latin tardif VIII. Actes du VIIIe colloque international sur le latin vulgaire et tardif, Oxford, 6–9 septembre 2006*, 253–62. Hildesheim: Olms-Weidmann.

Storjohann, Petra. 2003. *A diachronic constrastive* [sic] *lexical field analysis of verbs of human locomotion in German and English.* Frankfurt: Peter Lang.

Stouff, Louis. 1974 [1932]. *Jean d'Arras: Melusine.* Geneva: Slatkine Reprints.

Strömqvist, Sven & Ludo Verhoeven (eds.). 2004. *Relating events in narrative: Typological and contextual perspectives.* Vol. 2. Mahwah, NJ: Lawrence Erlbaum Associates.

Suchier, Hermann. 1889. *Aucassin und Nicolette.* Paderborn: Schöningh.

Svensén, Bo. 2009. *A handbook of lexicography: The theory and practice of dictionary-making.* Cambridge, U.K.: Cambridge University Press.
Swanton, Michael. 1996. *The Anglo-Saxon Chronicle.* New York: Routledge.
Sweet, Henry. 1883. *King Alfred's Orosius: Old-English text and Latin original* (Early English Text Society: Original Series 79). London: Trübner.
Sweet, Henry. 1888. *History of English sounds.* Oxford: Clarendon Press.
Sweet, Henry. 1892. *A new English grammar.* Oxford: Clarendon Press.
Szumlakowski Morodo, Irene. 2006. *Ich will weg, wohin musst du?* Modalverben mit Richtungsbestimmung und ohne Infinitiv. *Muttersprache* 116(4), 322–47.
Talmy, Leonard. 1975. Semantics and the syntax of motion. In John P. Kimball (ed.), *Syntax and semantics vol. 4*, 181–238. New York: Academic Press.
Talmy, Leonard. 1985. Lexicalization patterns: Semantic structure in lexical forms. In Timothy Shopen (ed.), *Language typology and syntactic description. Vol. 3: Grammatical categories and the lexicon*, 57–149. Cambridge, U.K.: Cambridge University Press.
Talmy, Leonard. 1991. Path to realization: A typology of event conflation. *Proceedings of the Annual Meeting of the Berkeley Linguistics Society* 7, 480–510.
Talmy, Leonard. 2000a. *Toward a cognitive semantics, Vol. I: Concept structuring systems.* Cambridge, MA: MIT Press.
Talmy, Leonard. 2000b. *Toward a cognitive semantics, Vol. II: Typology and process in concept structuring.* Cambridge, MA: MIT Press.
Taylor, Pauline. 1952. *Gerbert de Mez. Chanson de geste du xiie siècle.* Louvain, Belgium: Nauwelaerts.
Tesnière, Lucien. 1976 [1959]. *Éléments de syntaxe structurale,* 2nd edn. Paris: Klincksieck.
Thim, Stefan. 2008. The rise of the phrasal verb in English: A case of Scandinavian influence? In Klaus Stierstorfer (ed.), *Anglistentag 2007 Münster* (Proceedings of the Conference of the German Association of University Teachers of English XXIX), 291–304. Trier: WVT.
Thim, Stefan. 2011. Historical dictionaries of English. *Lexicographica* 27, 63–99.
Thim, Stefan. 2012. *Phrasal verbs: The English verb-particle construction and its history* (Topics in English Linguistics 78). Berlin: De Gruyter Mouton.
Tinkler, John D. 1971. *Vocabulary and syntax of the Old English version in the Paris Psalter.* The Hague: Mouton.
TL = Tobler–Lommatzsch: *Altfranzösisches Wörterbuch.* 1925–2008. Ed. Adolf Tobler & Erhard Lommatzsch. Wiesbaden: Steiner.
TLL = *Thesavrvs lingvae latinae. Editus iussu et auctoritate consilii ab academiis societatibusque diversarum nationum electi.* 1994–. Bayerische Akademie der Wissenschaften, Internationale Thesauruskommission. Berlin: De Gruyter. http://www.degruyter.com/view/serial/36167
TLF = *Trésor de la langue française: Dictionnaire de la langue du XIXe et du XXe siècle (1789–1960).* Published under the direction of Paul Imbs. Centre National de la Recherche Scientifique, Institut National de la Langue Française, Nancy. http://atilf.atilf.fr/tlf.htm
TOE = Roberts, Jane & Christian Kay. 2000. *A thesaurus of Old English.* Amsterdam/Atlanta: Rodopi. http://oldenglishthesaurus.arts.gla.ac.uk/.

Townend, Matthew. 2006. Contacts and conflicts: Latin, Norse, and French. In Lynda Mugglestone (ed.), *The Oxford history of the English language*, 61–85. Oxford: Oxford University Press.

Traugott, Elizabeth Closs. 1992. Syntax. In Richard M. Hogg (ed.), *The Cambridge history of the English language. Vol. I: The beginnings to 1066*, 168–289. Cambridge, U.K.: Cambridge University Press.

Traugott, Elizabeth Closs. 2012. Linguistic levels: Semantics and the lexicon. In Alexander Bergs & Laurel J. Brinton (eds.), *English historical linguistics: An international handbook*, 164–77. Berlin: De Gruyter.

Traugott, Elizabeth Closs & Richard B. Dasher. 2002. *Regularity in semantic change.* Cambridge, U.K.: Cambridge University Press.

Traugott, Elizabeth Closs & Graeme Trousdale. 2013. *Constructionalization and constructional changes.* Oxford: Oxford University Press.

Treffers-Daller, Jeanine & Françoise Tidball. 2015. Can L2 learners learn new ways to conceptualize events? A new approach to restructuring in motion event construal. In Pedro Guijarro-Fuentes, Katrin Schmitz, & Natascha Müller (eds.), *The acquisition of French in multilingual contexts*, 145–184. Bristol, U.K.: Multilingual Matters.

Turville-Petre, Thorlac. 1977. *The alliterative revival.* Cambridge, U.K.: Brewer.

Ungerer, Friedrich & Hans-Jörg Schmid. 2006. *An introduction to cognitive linguistics*, 2nd edn. Harlow, U.K.: Pearson.

Urwin, Kenneth (ed.). 1980–81. *The life of Saint John the Almsgiver.* 2 Vols. (Anglo-Norman Texts 38–9). London: Anglo-Norman Text Society.

van Bergen, Linda. 2012. Let's talk about *uton*. Paper read at the 17th *International Conference on English Historical Linguistics.* Zürich, Switzerland, 25 August 2012.

van Kemenade, Ans & Bettelou Los. 2003. Particles and prefixes in Dutch and English. In Geert Booij & Jaap van Marle (eds.), *Yearbook of Morphology 2003*, 79–117. Dordrecht, Netherlands: Kluwer.

van Riemsdijk, Henk. 2002. The unbearable lightness of GOing: The projection parameter as a pure parameter governing the distribution of elliptic motion verbs in Germanic. *Journal of Comparative Germanic Linguistics* 5, 143–96.

Verkerk, Annemarie. 2013. *Scramble, scurry* and *dash*: The correlation between motion event encoding and manner verb lexicon size in Indo-European. *Language Dynamics and Change* 3, 169–217.

Verspoor, Marjolijn & René Dirven. 2004. *Cognitive exploration of language and linguistics*, 2nd, revised edn. (Cognitive Linguistics in Practice 1). Amsterdam: John Benjamins.

Vezzosi, Letizia. 2012. English in contact: Latin. In Alexander Bergs & Laurel J. Brinton (eds.), *English historical linguistics: An international handbook*, 1703–19. Berlin: De Gruyter.

Vinaver, Eugène. 1929. *Malory.* Oxford: Clarendon Press.

Vinaver, Eugène. 1947. *The Works of Sir Thomas Malory.* Oxford: Clarendon Press.

Visser, Fredericus T. 1963–73. *A historical syntax of the English language.* Leiden, Netherlands: Brill.

Vulchanova, Mila & Emile van der Zee, eds. 2012. *Motion encoding in language and space.* Oxford: Oxford University Press.

Wälchli, Bernhard. 2001. A typology of displacement (with special reference to Latvian). *Sprachtypologie und Universalienforschung* 54(3), 298–323.

Walde-Hofmann = Walde, A. & J. B. Hofmann. 1982. *Lateinisches Etymologisches Wörterbuch*. Heidelberg: Winter.

Wallace, David (ed.). 2002. *The Cambridge history of medieval English literature*. Cambridge, U.K.: Cambridge University Press.

Weman, Bertil. 1933. *Old English semantic analysis and theory with special reference to verbs denoting locomotion*. Lund, Sweden: Gleerup.

Whittaker, William Joseph (ed.). 1895. *Mirror of Justices* (Selden Society 7). London: Quaritch.

Wilcox, Jonathan. 2006. The audience of Ælfric's Lives of Saints and the face of Cotton Caligula A. xiv, fols. 93–130. In A. N. Doane & Kirsten Wolf (eds.), *Beatus Vir. Studies in Early English and Norse manuscripts in memory of Phillip Pulsiano*, 229–63. Tempe, AZ: Arizona Center for Medieval and Renaissance Studies.

Wogan-Browne, Jocelyn (ed.). 2009. *Language and culture in medieval Britain: The French of England c.1100–c.1500*. York, U.K.: York Medieval Press.

Wright, Laura. 2000. Bills, accounts, inventories: Everyday trilingual activities in the business world of later medieval England. In David A. Trotter (ed.), *Multilingualism in later medieval Britain*, 149–56. Woodbridge, U.K.: Brewer.

Wright, Laura. 2011. On variation in medieval mixed-language business writing. In Herbert Schendl & Laura Wright (eds.), *Code-switching in Early English* (Topics in English Linguistics 76), 191–218. Berlin: De Gruyter Mouton.

Wright, William A. 1887. *The Metrical Chronicle of Robert of Gloucester*. London.

Wulff, Stefanie. 2006. Go-V vs. go-and-V in English: A case of constructional synonymy? In Stefan T. Gries & Anatol Stefanowitsch (eds.), *Corpora in cognitive linguistics: Corpus-based approaches to syntax and lexis* (Trends in Linguistics. Studies and Monographs 172), 101–25. Berlin: Mouton de Gruyter.

Wunderli, Peter. 1987. Die Funktion von 'en' im Mittelfranzösischen. Syntax, Semantik und Pragmatik. In Wolfgang Dahmen, Günther Holtus, Johannes Kramer, & Michael Metzeltin (eds.), *Latein und Romanisch. Romanistisches Kolloquium I*. (Tübinger Beiträge zur Linguistik 308), 197–244. Tübingen: Narr.

Xu, Dan. 2006. *Typological changes in Chinese syntax*. Oxford: Oxford University Press.

Ziolkowski, Jan M. 1996. Epic. In F. A. C. Mantello & A. G. Rigg (eds.), *Medieval Latin*, 547–55. Washington, DC: The Catholic University of America Press.

Zlatev, Jordan. 2003. Polysemy or generality? Mu. In Hubert Cuyckens, René Dirven, & John R. Taylor (eds.), *Cognitive approaches to lexical semantics*, 447–94. Berlin: Mouton de Gruyter.

Zlatev, Jordan. 2007. Spatial semantics. In Dirk Geeraerts & Hubert Cuyckens (eds.), *The Oxford handbook of cognitive linguistics*, 318–50. Oxford: Oxford University Press.

Zlatev, Jordan & Peerapat Yangklang. 2004. A third way to travel: The place of Thai and serial verb languages in motion event typology. In Sven Strömqvist & Ludo Verhoeven (eds.), *Relating events in narrative: Typological and contextual perspectives*. Vol. 2, 159–90. Mahwah, NJ: Lawrence Erlbaum Associates.

Zupitza, Julius. 1875–6. *The romance of Guy of Warwick. The second or 15th-century version* (Early English Text Society: Extra Series 25/26). London: Trübner.

INDEX

Printed in the USA/Agawam, MA
November 14, 2017

662530.049